Topics in Nonlinear Dynamics
A Tribute to Sir Edward Bullard

(La Jolla Institute)

AIP Conference Proceedings
Series Editor: Hugh C. Wolfe
Number 46

Topics in Nonlinear Dynamics
A Tribute to Sir Edward Bullard

(La Jolla Institute)

Editor
Siebe Jorna

American Institute of Physics
New York 1978

L.C. Catalog Card No. 78-057870
ISBN 0–88318–145–2
DOE CONF-771234

DEDICATION

This volume is dedicated to Sir Edward (Teddy) Bullard, and it is only fitting that it should contain some account of him and his doings. He was born in Norwich, England, in 1907, and is now 70 years of age. His scientific work started in 1929 when he became a graduate student in Rutherford's Cavendish Laboratory. Here, under the direction of P. M. S. Blackett, he studied the scattering of slow electrons in gases. In this enterprise he was soon joined by another graduate student, H. S. W. Massey, now Secretary of the Royal Society. By 1931, they had found unmistakable evidence of diffraction patterns similar to the rings seen around a street lamp in a fog. The results were explained very naturally by wave mechanics, and were clearly inconsistent with classical views.

It might have been expected that this striking early success would have led to a career in the rapidly expanding field of atomic and nuclear physics. This was not to be; 1931 was the worst year of the economic depression. Bullard could see his Ph. D. approaching, but no prospect of a job. Rutherford's advice was to "take any job you can get."

The job that came along was to help teach surveying to Colonial Service Probationers. The course was run by the charming, elegant, elderly Reader in Geodesy, Colonel Sir Gerald Lenox-Conyngham, who had taken the post after a long military career in India. He and Teddy Bullard got along very well. No more was heard of teaching surveyors. Bullard embarked on a very profitable eight years in which he learned the elements of Earth science and carried out a remarkably diverse series of projects. The success of these was in large part due to the steady support of Lenox-Conyngham, who was not at all disturbed when told that the police were looking for the perpetrators of an explosion which had left a hole in a road in Leicestershire.

The success in carrying out substantial projects in instrument development and in field work, with almost no money, was made possible by Bullard's discovery that Leslie Flavill, a boy of 16 whom Lenox-Conyngham had hired as a clerk, was a most able instrument designer and a mechanic of genius. Bullard maintains Harold Jeffreys, Johnny von Neumann, and Leslie Flavill are the three most intelligent people he has ever met.

At first the work was in gravity measurement, largely because the Department possessed some pendulums. This took Bullard to East Africa in 1933-34. A long paper in the Philosophical Transactions, which contains the observations, and what now seems an erroneous interpretation of the origin of the rift valleys, brought him widespread recognition.

Bullard's next interest was in explosion seismology. He, Tom Gaskell, W. B. Harland, and Kerr-Grant set out to map the paleozoic floor beneath east England. In this work he acquired a good knowledge of the techniques and theory of refraction shooting. This was essential to the next phase of his activities, the study of the seafloor.

The move out to sea was a consequence of a meeting with Dick Field at the IUGG meeting in Edinburgh in 1936. Field was a maverick geologist from Princeton, who can now be seen as the most influential Earth scientist of his generation. It was his enthusiasm and tremendous powers of persuasion which launched marine geology on its spectacular development. He invited Bullard to the U. S. in 1937 and exhorted him to take up the study of the seafloor. Like Harry Hess and Maurice Ewing before him, Bullard was easily converted. He went to sea with Ewing, and when he got home, started similar work on his own. He, Tom Gaskell, and Ben Browne found that the continental shelf south of Ireland was similar to the one examined by Ewing on the other side of the Atlantic. The discovery of the vast basins of sediments beneath the continental shelves has had consequences extending far beyond their scientific interest; they are now a major source of politically secure oil.

War was approaching, but there was still time for one more investigation before it came. Bullard took up the measurement of heat flow on land which had been almost abandoned for 50 years. He spent the winter of 1938-39 in South Africa working in collaboration with L. J. Krige. They made measurements in boreholes, which set a standard in what, despite its apparent simplicity, is a difficult field. War interrupted further development and postponed for 10 years the initiation of similar work at sea.

In November, 1939, Bullard became an "Experimental Officer" in the Naval Scientific Service. After a sharp struggle with the naval scientific establishment in Portsmouth, he found himself in charge of the development of methods of protecting ships from magnetic mines. A few months later he became head of the group developing methods for sweeping all kinds of mines. After 18 months spent on these matters, the mining losses had fallen to less than 10% of those caused by submarines, and Bullard started to look for other ways of helping the Navy. He moved to London and joined Patrick Blackett, who had been his thesis advisor 10 years before in Cambridge. Blackett was starting to apply the methods of what is now known as operations analysis to naval operations. Bullard worked on naval problems for a year or more, and then became diverted to the study of the intelligence that R. V. Jones was uncovering concerning the German development of rockets and flying bombs; and later, to a study of the most economical ways of attacking the firing sites in northern France. In the middle of this work he also acquired the job of protecting the Commander-in-Chief of Fighter Command from politicians. This assignment provided a training in the realities of politics that has proved valuable in many subsequent enterprises.

When the war with Germany ended in May, 1945, Bullard returned to Cambridge. He found a shambles; the door of the laboratory had not been opened for years, most of the equipment had been removed, and what was left was covered with rust. There was no money and no help; his first task was to scrub the floor. During the next four months young men started to return as graduate students, and it became possible to take up geophysics where it had been left six years before. The first priority was to carry out seismic refraction work in the deep sea; the methods for this were developed by Maurice Hill. Other projects were observations in Europe of the seismic waves produced by the explosion of a very large ammunition store in a cave under the island of Heligoland and a continuation of prewar work on heat flow. In 1947, when he was 80 years of age, Lenox-Conyngham retired, and Bullard became head of the Department.

Continual difficulties over the financing of research and over the almost total lack of ship time exasperated him to a point where he decided to leave Cambridge. The final straw was the refusal of the University to let him have a typist for five half-days a week instead of for two half-days. He had just been told of this decision when an emissary from the University of Toronto arrived and asked him to become head of their Physics Department. He accepted, but it proved an unfortunate decision. He and his family did not settle happily in Canada, and he became homesick for Cambridge. A bright interlude was a summer in La Jolla, which he remembers as the most prolific period of his scientific career. He and Art Maxwell, then a graduate student, built the long-postponed equipment for measuring heat flow at sea. He shared my office, and I remember the enthusiasm with which he and Art spent long days in the workshop building this quite complicated machine with their own hands. Watertight equipment for use on the floor of the deep sea was at that time a novelty, and, so far as I know, the heat-flow equipment was the first to use the now indispensable O-ring for its watertight joints. During this summer Bullard also wrote a well-known paper on the westward drift of the Earth's magnetic field.

Just before he came to La Jolla in 1949, Bullard was offered, and had accepted, the Directorship of the National Physical Laboratory in England. He took up the job at the beginning of 1950. He was, I think, a well-liked and effective Director, but his heart was not really in it. Like many of his contemporaries, he suffered from an indecision of purpose. If he was in an academic job where he could get on with his own work, he felt that he should be directing some great and prestigious enterprise; if he let himself accept such a position, he at once started to lament the loss of freedom and opportunity to do his own work.

In fact, he did succeed in doing a substantial amount of his own work while at the N.P.L. A new heat-flow apparatus was built, this time with the help of an outstanding design office and workshop, and was used in the Atlantic. He also wrote a long and elaborate paper on dynamo theory. It now seems likely that the particular dynamo proposed in the paper will not work; nevertheless, the paper had an important influence on the general acceptance of a dynamo in the Earth's core as the cause of its magnetic field. Bullard's paper in this volume is an offshoot of this work.

In 1956, Bullard got what he wanted, and returned to Cambridge, to almost the bottom position on the academic totem pole. This time he managed to resolve his indecision about what kind of life he wanted. His primary commitment was to his own work and to helping a large number of very able graduate students. This was the period when the ideas of seafloor spreading and continental movement were developing rapidly. The Cambridge Department played a large part in this development. Bullard had long been interested in these matters, but it was only after his return to Cambridge that he was fully convinced of the reality of the movements. The critical year was 1963 when Harry Hess, Tuzo Wilson, Drummond Matthews, and Fred Vine were all at Cambridge. After this, none of the group had any doubts, and soon embarked on a campaign to convert unbelievers, particularly in the U. S. An odd consequence of their success is that Bullard, almost for the first time, finds himself a defender of orthodoxy.

While at the N.P.L. and in Cambridge, he spent much time on matters far from his usual scientific interests. Among other things, he became a director of the family brewery (now, alas, taken over by a faceless conglomerate), a director of the British subsidiary of I.B.M., a British representative at the 1958 talks with the Russians on the possibility of a test-ban, and joint chairman of the Anglo-American Ballistic Missile Committee.

In 1964, the University of Cambridge made him Professor of Geophysics. While at Cambridge, he paid regular visits for one or two months each year to Scripps.

In 1974, Teddy retired and came to live in La Jolla and to work at Scripps. He gives widely appreciated lectures, directed nominally to first-year nonscientists, but into which many scientists infiltrate. He still writes on his favorite topics of plate tectonics and the origin of the Earth's magnetic field, though his main interest has shifted to the murky politico-scientific-industrial questions of nuclear waste disposal. He also threatens to write an autobiography.

While he has received many honors and awards, I think what he appreciates most is the friendship of old students and fellow workers in the Earth Sciences. The Director of Scripps recently described him as "a happy geophysicist;" it is, I think, an apt comment.

W. H. Munk
La Jolla, California

PREFACE

A workshop on topics in nonlinear dynamics was held in La Jolla from December 27-29, 1977, under auspices of the La Jolla Institute, in celebration of Sir Edward Bullard's seventieth birthday. Its organizers were Professor A. N. Kaufman, Professor E. W. Montroll, and Professor K. M. Watson.

This book contains the lectures presented there, as well as papers which were felt should be included to improve the balance and also to render the field accessible to those new to it. The contributions by Professors Moser, Berry, Ford and Helleman were kindly made available by R. H. G. Helleman and J. Ford who had planned to include these in a book entitled "Introduction to Nonlinear Mechanics," a project that has now been postponed indefinitely. Space was limited, and we were unfortunately unable to include all the contributions that were offered by Sir Edward's friends and colleagues.

A number of articles are in the nature of reviews and there is some overlap among them. But each treats the subject from a different viewpoint, so that a reader with a mathematical background might read the articles by Moser and Treve, while the more physics oriented might prefer the papers by Berry and Ford. Other articles treat various aspects of nonlinear mechanics. These include applications ranging from circular accelerators and storage rings to ocean waves, mathematical properties of certain nonlinear equations such as the Korteweg-deVries equation governing soliton formation, and schemes for utilizing Lie transforms to simplify perturbation expansions for Hamiltonian systems. In an appendix, Helleman has set down brief descriptions of concepts frequently used in nonlinear mechanics, but which may not be familiar to all readers. Only limited editing has been done, as much to expedite publication as to preserve the individual flavor of each piece. It is hoped that the readers will not find the resulting inconsistencies in format (e.g. of references) too grating.

I thank the authors for their fine contributions and for their assistance with the proofreading. We also express our gratitude to P. J. McIntyre and Judy Gregg for their skilful typing.

The Workshop and these Proceedings were financed by independent research funds of the La Jolla Institute.

Siebe Jorna
La Jolla, California

TABLE OF CONTENTS

NEARLY INTEGRABLE AND INTEGRABLE SYSTEMS *

by

Jürgen Moser **
Courant Institute
New York University
251 Mercer Street
New York, NY 10012

I. INTRODUCTION

The main purpose of my talk at this Physics Conference is to give a survey of certain topics in Hamiltonian dynamics which have been the subject of intensive study by mathematicians, astronomers and physicists of the last two centuries. These topics, which involve questions like the long term stability of the three body problem (Sun-Earth-Moon system) proved to be very difficult to analyze, much less to solve completely. It took long, hard and ingenious work to obtain even the most rudimentary knowledge about the simplest possible models that went one step beyond the Kepler two-body problem and the harmonic oscillator lattices. The mutual interaction of the internal degrees of freedom of a given system, governed by complicated coupled nonlinear differential equations, give rise to resonances that render the traditional perturbation methods divergent. Inevitably, therefore, a certain feeling of discouragement set in, and the need for altogether new techniques and innovative approaches was generally realized. In the last 10-15 years, however, there have been some interesting developments in this direction, from both a mathematical as well as a numerical point of view, which have rekindled active interest in the theory of stability of Hamiltonian systems. It is this recent work that I want to discuss in this lecture. I will try to emphasize the physical significance but I'm afraid my discussion will unavoidably reflect my training as a mathematician.

Let me start by writing down the equations of motion of a Hamiltonian system of n degrees of freedom:

$$\dot{q}_k = \frac{\partial H}{\partial p_k} \; ; \; \dot{p}_k = - \frac{\partial H}{\partial q_k} \quad , \tag{1}$$

---

* Based on an invited lecture at the Eastern Theoretical Physics, Conference Rochester, NY, November 1976. Lecture notes taken by Tassos Bountis.

** Supported in part by the Office of Naval Research, Contract No. N00014-76-C-0301.

ISSN: 0094-243X/78/001/$1.50

$q = (q_1,..,q_n)$; $p = (p_1,..,p_n)$ $k = 1, 2, .., n$ where q_k, p_k are the position and momentum variables respectively, corresponding to the kth degree of freedom, and $H = H(q,p)$ is the Hamiltonian function of the system. We will assume throughout this lecture that any dissipation phenomena can be neglected.

What can we say about the long-time behavior of the system? Obviously, before we attempt to answer this question we have to know something about the nature of the system, i.e. the form of $H(q,p)$. If we are dealing with a set of coupled harmonic oscillators, the solutions of (1) are periodic, or quasi-periodic, depending on whether the mass to frequency ratios of the oscillators are rationally dependent or not. A statistical mechanician is, in general, interested in the behavior of a gas of many particles over a certain period of time. He introduces the notion of the 2n-dimensional "Phase (or Γ-) Space" consisting of points (q,p) each of which completely characterizes the system at a given time t (for N particles in 3 dimensions, $n = 3N$). Then our statistical mechanician considers the orbit of the system in phase space over an arbitrarily long time interval. This orbit, according to the popular notion of "ergodicity", would be expected to wander more or less uniformly over the (2n-1)-dimensional manifold of constant energy to which it is confined. The phase space trajectory of the system point $(q(t), p(t))$ [which is completely determined once the values of q and p are known at t=0] might be restricted to lie for all time on hypersurfaces of dimension smaller than 2n-1, depending on whether the system possesses any global (single valued and analytic in q,p) integrals of the motion other than the energy integral. Thus the property of integrability - or non-integrability for that matter - is related to questions of stability which may be posed in terms of the regions of phase space occupied by different sets of orbits of the system under consideration.

At this point, let us recall the famous theorem of Poincaré on the non-existence of such integrals:[1]

Poincaré was able to prove, under quite general conditions, that a given Hamiltonian system possesses *no* integral of the motion (single valued and) *analytic* in the phase space coordinates as well as some parameters, other than the energy integral.

Prompted by the results of Poincaré's widely acclaimed work, many people developed the following attitude which was destined to become a common belief: That practically any integrable system can be made ergodic under small perturbations.

But let us for the moment leave the realm of generalities and look closer at some important physical situations to which our considerations can apply:[2]

(i) The motion of planetary systems

(ii) The motion of charged particles in strong magnetic fields as in a vacuum chamber of a high energy particle accelerator for instance. Although reliable astronomical observations of the last few hundreds or even thousands of years indicate that the motion of the solar system is certainly not ergodic the question of whether it will remain stable for all time is still open!

In the storage rings of a high-energy accelerator, where even after $\sim 10^{11}$ revolutions one would like the particles still to be in a stable configuration, stability for arbitrarily long times has not been established rigorously.

The fact is that by inquiring about stability *for all time*, as opposed to stability *for long times*, we have posed a severe and difficult mathematical question. In attempting to answer this question from a strictly mathematical point of view we end up having to study complicated Cantor sets of orbits that are not particularly intuitive.

On the other hand it is standard assumption, inspired by statistical considerations, that most Hamiltonian systems are ergodic, i.e. that after small perturbation of the Hamiltonian one will limit on ergodic systems. This would, of course, be in sharp contrast to stability.

It is this type of misconception that the recent work of Kolmogorov[3], Arnol'd[4] and Moser[5] attempts to resolve by establishing some fundamental facts about Hamiltonian systems. The basic result of what is now called by some people KAM theory, is that any systems sufficiently close to an integrable one which are not ergodic, possess a set of positive measure of quasi-periodic solutions, i.e. have a strong stability behavior. The situation is complicated by the fact that stable and unstable behavior can take place for one and the same system in different regions of the phase space.

Let us now turn to a more detailed discussion of near-integrable or stable systems. We will introduce yet another word for them: we will call them *normal* systems if there exists a canonical transformation:

$$(p,q) \leftrightarrow (I,\phi) \;;\; I \equiv (I_1,\ldots,I_n) \;,\; \phi \equiv (\phi_1,\ldots,\phi_n) \;,$$

to normal variables I,ϕ where I is otherwise known as the *action* and ϕ is the *angle* variable, of period 2π.

We write the Hamiltonian of the normal system as a series expansion in some parameter μ:

$$F = F_0(I) + \mu F_1(I,\phi) + \ldots \;, \qquad (2)$$

where μ measures the effect of interaction between the internal degrees of freedom. F is assumed to be real and analytic in all variables and the $F_k(I,\phi)$ are periodic functions in ϕ for all $k\neq 0$.

It is common practice in astronomy to discuss systems without interactions in terms of $F_0(I)$, a function of the n action variables only, which actually are our n integrals of the motion.

So when $\mu=0$, we have an integrable system and the available region in phase space is "foliated" by tori on which all the system trajectories must lie. The angles ϕ_k can be thought of as varying on the surfaces of these tori and their equations of motion

$$\dot{\phi}_k = \frac{\partial F}{\partial I_k} = \omega_k$$

can be directly solved to yield

$$\phi_k = \omega_k t + \beta_k,$$

where the ω_k's are the n unperturbed frequencies associated with our system and β_k are the phases. All solutions are quasi-periodic and can be viewed as Fourier series expansions in $e^{i\langle j,\omega\rangle t}$ summed over appropriate amplitudes and all integer vectors $j = (j_1, j_2, \ldots, j_n)$. The motion will be exactly *periodic* (as opposed to quasi-periodic) if the ω_k's satisfy the condition that they are an integer multiple of a fixed number p.

We now ask the question: what happens when $\mu \neq 0$? Does this tori foliation persist? The answer is: "*not completely*". Only part of the structure survives but it is quite a sizeable part - in the sense of measure. This is, in fact, the content of the KAM theorem that we are now ready to formulate[6].

Consider an integrable system whose flow (or totality of trajectories) is restricted to a (2n-1)-dimensional manifold of constant energy foliated by the (n-1)-dimensional tori that we described above. We will further assume that the unperturbed frequencies satisfy the following conditions:

(i) They are rationally independent

(ii) They satisfy the infinitely many inequalities

$$\left|\sum_{k=1}^{n} j_k \omega_k\right| \geq c|j|^{-\gamma} ,$$

for all integers j_k with $|j| \equiv \sum_{k=1}^{n} |j_k| > 0$ and c, γ certain fixed positive constants.

(iii)
$$\det \begin{bmatrix} \dfrac{\partial \omega_j}{\partial I_k} & \omega_j \\ \omega_k & 0 \end{bmatrix} \neq 0 .$$

We can intuitively understand the third condition as follows: it allows us to vary all frequency ratios ω_k/ω_1 (k=2,..,n) as we change the I's on a pre-assigned energy surface so that the first two irrationality conditions can be satisfied by a choice of the $I_1,..,I_n$.

When a system satisfying the above condition is slightly perturbed, then *most* phase space tori do survive, i.e. *most* solutions remain quasi-periodic, for small values of μ. To be more precise, the solution q and p of our Hamiltonian equations (1) can now be written in terms of analytic functions $\Phi=\Phi(\omega,\phi,\mu)$ which are periodic in ϕ and yield quasi-periodic solutions if we let $\phi=\omega t + \beta$ for some constant vector $\beta \equiv (\beta_1,\beta_2,..,\beta_n)$. We emphasize again that these orbits form a set of large measure and their existence depends crucially on the fact that they are *non-resonant*, i.e. that all unperturbed frequencies are rationally independent.

In 1923, a paper was written by Enrico Fermi[7] which dealt precisely with the kind of systems that we described above. Fermi came to the erroneous conclusion that systems of this type are quasi-ergodic in general, i.e. their orbits cover densely the energy surface. He argued as follows:

Let us assume, Fermi said, that such a system is *not quasi-ergodic* and let us consider a small neighborhood of the constant energy surface. It is clear then, by assumption, that the set of all orbits going through that neighborhood would not cover all of the energy surface. Consider now the boundary of that neighborhood which is a (2n-2)-dimensional manifold consisting itself of orbits and which is therefore invariant under the flow. Fermi proceeds to show that such a manifold cannot exist, which he does by making a generalization of Poincaré's theorem of the non-existence of integrals that we mentioned earlier in the lecture. Poincaré's theorem precludes the existence of *families* of such (2n-2) dimensional manifolds but Fermi goes even further to say that *not a single one* of them is actually there!

So what is wrong with Fermi's argument? It turns out that the set of orbits that comprise the boundaries of Fermi's neighborhoods are dense in the space of all orbits which, of course, is hard to visualize. Let us recall that according to the KAM theorem also the set of invariant tori has a boundary dense in the phase space. For $\mu = 0$ they are obtained when we exclude from all possible orbits the ones for which the frequency ratios are within some small neighborhood of the rationals!

We may think of this situation in phase space as represented by an exceedingly intricate sponge whose solid part is composed of stable quasi-periodic solutions. The hollow spaces of the sponge stand for those regions in phase space where the conditions of the KAM theorem are violated and the solutions are allowed to "leak out", spreading uniformly throughout the sponge.

Finally we might want to ask the question: where does one usually find near-integrable systems? Consider a Hamiltonian system near a stable equilibrium point where the quadratic part of the Hamiltonian has imaginary eigenvalues. Then it is a "fact of life" that sufficiently close to that point the system is nearly integrable. In other words, it possesses n formal integrals that are generally divergent but good enough to approximate the behavior of the system near the equilibrium point. The existence of these n formal integrals was shown by G. D. Birkhoff[8] in the 1920's while their generally divergent nature was later established by C. Siegel[9].

One encounters a very similar picture near a periodic orbit of a Hamiltonian system that is also stable in linear terms, i.e. its associated Floquet multipliers lie on the unit circle. Thus, we find that in sufficiently small neighborhoods near a stable equilibrium point or a stable periodic orbit of a Hamiltonian system, *most* solutions are quasi-periodic. This is what is often referred to as "stability in measure".

We can turn again to our favorite application: the stable motion of charged particles in the storage rings of a high energy accelerator. One might argue that this stability would be the result of dissipation effects caused by an imperfect vacuum. Such an explanation, however, is artificial since it would not account for the large number of revolutions ($\sim 10^{10}$) that the particle beam performs before undergoing any essential distortion.

Let us briefly discuss the unstable orbits of near integrable systems which occupy sets of relatively small measure in phase space. These rather exceptional "leakage" sets were recently studied at CERN in Geneva by Keil[10] who used the accelerator essentially to perform the task of an analog computer!

We note here that these unstable orbits are intimately connected with the concepts of resonance and what has come to be known *as Arnol'd diffusion*.[11-15] It was Arnol'd[14] who in 1964 first gave an example of the fact that in these exceptional sets of unstable orbits there are solutions which gradually escape and eventually diffuse far from the torus from which it started. Then, twenty years later, a student of Arnol'ds, Nekhoroshev[15] gave estimates for the time variation of the action variables of all solutions that can escape at a very slow rate. He found that

$$|I(t) - I(0)| < \mu^b , \qquad (3)$$

as long as $t < \exp(1/\mu^a)$ for some positive constants a, b. It is clear that these escape times become astronomically long as μ approaches zero.

In the way of an example, consider a *non linear* pendulum driven by a small periodic force in the absence of dissipation. We assume that we can write the Hamiltonian associated with this pendulum in the form:

$$H(I,\phi,\mu,t) = \frac{1}{2} I^2 + f(I,\phi,\mu)\delta_p(t) , \qquad (4)$$

where I,ϕ are the action-angle variables, μ is a small coupling parameter and

$$\delta_p(t) \equiv \sum_{j=-\infty}^{+\infty} \delta(t-j) .$$

The perturbation "kicks" the pendulum at unit time intervals by a very small amount that is determined by the function $f(I,\phi,\mu)$. The non-linear nature of the pendulum is assured by the fact that its frequency depends on the amplitude of the motion through the familiar relationship

$$\omega = \frac{\partial H}{\partial I} .$$

If we apply the results of the theory that we have outlined to this simple example, we will find [5,36] that the stability of all solutions is guaranteed, i.e.

$$|I(t) - I(0)| \leq \delta \quad \text{for all } t ,$$

provided that the absolute values of the derivatives of $f(I,\phi,\mu)$, up to order five, are small enough. On the other hand, one can construct a counter-example, for which the derivatives of $f(I,\phi,\mu)$ are small only up to first order and the second derivative is bounded and which experiences a slow build-up of resonances with $I(t)$ ultimately becoming arbitrarily large. This is a rather subtle effect and it actually shows the importance of how smooth the perturbation is (see Appendix).

For systems of more than one degree of freedom rational frequency ratios inevitably give rise to resonances. But these frequency ratios, which are also functions of the amplitudes or the action variables I_k, go through *rational* as well as *irrational* values with varying $I_k(k=1,..,n)$. Thus, as the amplitude of the motion tries to increase, it gets "stuck" in those *non-resonant* regions where all ω_k's are rationally independent. We see, therefore, how in this sense, the nonlinearity of the frequency dependence can have a stabilizing effect on the overall behavior of the system.

We can learn a lot about the nature of the solutions of a system of two degrees of freedom by fixing one of the four phase space coordinates and considering the successive intersections of a number of orbits with the plane defined by two of the remaining variables. In fact, there exists an intimate connection between these so called "surfaces of section" of Hamiltonian systems and the iterates of area-preserving mappings of the plane onto itself[16]. This connection has been greatly explored numerically in the last decade[17-18] and for a very good reason: computing machines perform the algebraic operations needed to iterate a mapping much more rapidly and efficiently than they can integrate the coupled differential equations of motion of the Hamiltonian system.

An interesting example of a numerical investigation of a two dimensional mapping of this type is the one by Miss Rannou[19], operating with integers only so as to always control the round-off error of the computer. She was able to demonstrate by means of very illustrative pictures that by varying the initial conditions and the coupling parameter in the algebraic equations one can make the iterates of her mapping fill out a small region of the plane, spread over the full plane leaving out a small region, or cover uniformly all of the available two dimensional space!

3. INTEGRABLE SYSTEMS

Having discussed near-integrable systems, let us turn our attention to the class of the integrable systems. As we have already seen, these are exceptional from a mathematical point of view. Let us recall that a system of n degrees of freedom is called integrable if it possesses n integrals of the motion that are in "involution", i.e. the Poisson bracket of any two of them vanishes identically.

Right from the start we are met with the following surprising observation: despite the expected sparsity of such systems, a great number of them have cropped up in the recent years in a variety of fields: Nonlinear optics, Statistical Mechanics and Quantum Field Theory to mention but a few. Many of these integrable systems at first sight appear to be completely unrelated to each other but is that really the case? Rather than engaging in speculation, let us examine a few representative examples: Consider the following 1-dimensional potential introduced by Calogero:

$$V(x) = ax^2 + \frac{b}{x^2} \tag{5}$$

where a, b are positive constants.[20] Calogero solved the Schrödinger equation for this potential and found explicit formulae for the eignefunctions and their respective eigenvalues. He was thus led to the conjecture that the corresponding classical problem might have a simple solution.

Mark Adler[21] recently studied a one-dimensional chain of n particles which interact with each other via the Calogero potential (5). He was able to show that this system possesses n integrals in involution that are *rational* in the coordinates p_k, e^{aq_k} (k=1, 2,..,n) and hence its solutions must be at least quasi-periodic in nature. Actually, more is true: all solutions are, in fact, periodic with a period which is independent of the individual solution! In other words the system is linear in some appropriate coordinates, a result that is hardly expected at the outset. Let me mention here that a similar result was obtained independently by Perelomov and Olshanetzky in the Soviet Union[22].

I have studied[23,24] the case a=0 and b>0 which is the case of an inverse-cube repelling force. The solutions are clearly unbounded and therefore less interesting, but a new surprise is in store when we try to solve the associated scattering problem: We find that the scattering is the same as for completely elastic collisions! The first particle fully imparts its velocity to the last one, the second one to the one before last, making all the scattering phase shifts exactly zero. Moreover, the coordinates of the particles are algebraic functions of the time variable t.

So much for this example. Let us now briefly discuss the well-known example of the *Korteweg-de Vries equation:*

$$u_t - 6uu_x + u_{xxx} = 0 \ . \tag{6}$$

This equation is often viewed as describing a Hamiltonian system of an infinite number of dimensions. It can be shown that this system is integrable, i.e. that it possesses infinitely many integrals, or Conservation Laws as they are otherwise called[25]. These integrals are obtained in the form of eigenvalues of linear operators, in the case of the KdV equation a "Schrödinger operator"

$$\phi'' + q\phi = \lambda\phi \tag{7}$$

where $q = u(x,t)$[26].

Solving equation (7) under certain boundary conditions we find an infinite set of eigenvalues λ_i which commute with each other and are indeed "constants of the motion" associated with equation (6).

The fact that the KdV equation is integrable was shown by Faddeev[27] and Zhakarov who introduced *normal coordinates*, identifying the λ_i's above as the *action variables* and the reflection coefficients coming from the spectral theory of (7) as the *angle variables*.

There has been some recent work on this subject by Novikov[28] in Moscow (1974) and Lax[29], McKean and Van Moerbeke[30] at N.Y.U. (1975) who studied the solutions of the KdV equation that are *periodic in x*. McKean and Trubowitz[31] (1976) showed that all such solutions are almost periodic functions of the time with an infinite number of frequencies.

An unexpected relationship between the KdV equation and the Calogero potential (5) was also discovered by Airault, McKean and myself[32]. It turns out that the behavior of the solutons of eq. (6), u(x), that are rational in x is characterized by the position of its poles if written in the form $u(x) = 2 \sum_{j=1}^{n} (x-x_j)^{-2}$. These poles can be shown to satisfy a differential equation whose Hamiltonian is actually one of the integrals of the inverse square potential (5)!

As a third example, I wish to bring up a nonlinear hyperbolic equation studied by Zakharov:

$$u_{tt} = \left(\frac{\partial}{\partial x}\right)^2 (u+u^p+u_{xx}) , \tag{8}$$

often referred to as the Boussinesq equation[33]. For p=2, Zakharov found that this equation possesses an infinite number of integrals, or conservation laws, in involution. He constructed a third order linear operator whose eigenvalues are the infinitely many constants of the motion associated with equation (8). However, this is not the end of the story. The complete solution of the Boussinesq equation, which constitutes the ultimate proof of integrability, is still a subject of on-going research.

If we now make the following change of variables

$$W_x \equiv u ,$$

we can rewrite equation (8) in the form:

$$W_{tt} = W_{xx} + (W^p)_{xx} + W_{xxxx} . \tag{9}$$

For the case p=3, Zakharov discretized equation (9) and pointed out the intimate connection that exists between the Boussinesq equation and the chain of anharmonic oscillators with cubic nonlinearity that Fermi, Pasta and Ulam investigated numerically in the early fifties[34].

The well known result that the FPU System exhibits no signs of ergodicity was also supported by the work of Nishida (1971)[35]. He verified all the conditions of the KAM theory to FPU chains consisting of 2^n particles - for some positive integer n - and found that stability was indeed the general case.

The contribution of rigorous mathematics to the theory of Hamiltonian systems has been clearly emphasized in this lecture. We must bear in mind, however, that although mathematicians have shed some light on the qualitative aspects of Classical Dynamics they have done little to enhance our *quantitative* understanding of the behavior of Hamiltonian systems. The KAM theory for instance can make rigorous stability predictions concerning orbits that are within a small neighborhood of a stable equilibrium point or periodic orbit of an integrable system. But how small is that neighborhood? This is an example where one must resort to computer experiments of the kind that Fermi Pasta and Ulam performed. Let us recall that the regions of stability that they found are much larger than KAM would predict. The explanation for this apparent discrepancy most probably lies in the fact that the FPU system is closely related to the integrable Boussinesq equation. And we come now to the question alluded to earlier in the lecture: why do so many physically interesting systems turn out to be integrable or at least near-integrable?

A vague answer to this question at the present time is as follows: We know that Hamiltonian systems that are in some sense close to integrable ones possess solutions exhibiting strong stability behavior. There is good reason to believe that the examples we have discussed in this lecture all belong to such neighborhood of integrable Hamiltonian systems. This raises the intriguing question, of how one can approximate a given Hamiltonian system by an integrable one.

The infinitely many conservation laws or integrals associated with the Korteweg-de Vries and the Boussinesq equations betray the existence of hidden symmetries whose nature still eludes us. If we could make some sense out of them then maybe we could understand what makes an equation integrable. But for the time being their origin remains unknown.

APPENDIX

Since the example described below (4) is not in the literature and as it is sufficiently simple to be explained we will do so. We shall take for $f(I,\phi,\mu)$ a function independent of I and of the form

$$f(\phi,\mu) = \mu^2\, g(\mu^{-1}\phi) ,$$

where

$$g(x) = x(2|x|-1) \text{ for } |x| \le \frac{1}{2} ,$$

$$g(x+1) = g(x).$$

Moreover we set $\mu = 2\pi/N$ with a large integer N so that f has period μ, hence also period 2π and the following form

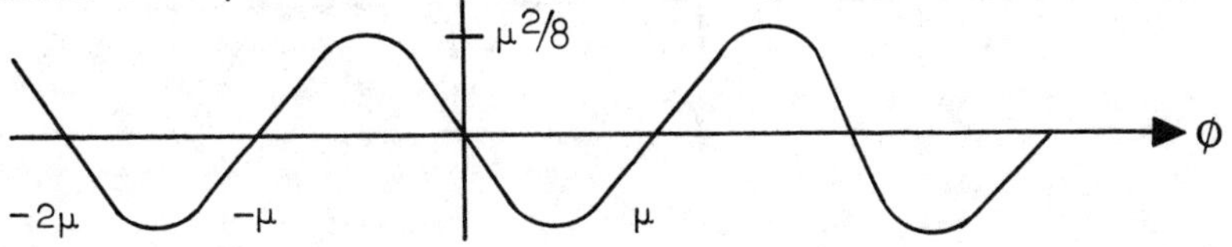

The differential equation for $H = I + f(\phi,\mu)\delta_p$ takes the form

$$\begin{cases} \frac{dI}{dt} = - \frac{\partial H}{\partial \phi} = - \frac{\partial f}{\partial \phi} \delta_p(t) \\ \frac{d\phi}{dt} = I \end{cases}$$

or

$$\ddot{\phi} + \frac{\partial f}{\partial \phi} \delta_p(t) = 0 .$$

Denoting the intial values of $I(t)$, $\phi(t)$ for $t = -0$ by I_0, ϕ_0 and for $t = 1-0$ by I_1, ϕ_1 we find by integration of the equation

$$I_1 = I_0 - \frac{\partial f(\phi_0,\mu)}{\partial \phi} \; ; \; \phi_1 = \phi_0 + I_0 \; ,$$

which we interpret as a mapping of the I,ϕ plane onto itself.

We will establish the instability of the differential equation if we find an orbit for which I under repeated application of this mapping grows over all bounds. Note that the average of $-\partial f/\partial\phi$ is zero and therefore, in general, such escape is not expected.

We denote the image of (I_0,ϕ_0) under the kth iterate of the above mapping by I_k, ϕ_k. Using the fact that

$$\frac{\partial f}{\partial \phi} = \mu \, g' \, (\mu^{-1}\phi) = 4 \, |\phi| - \mu \text{ for } |\phi| < \frac{\mu}{2} \, ,$$

$$\frac{\partial f}{\partial \phi} (\phi+\mu,\mu) = \frac{\partial f}{\partial \phi} (\phi,\mu) \; ,$$

i.e. that $\partial f/\partial\phi$ has the saw-tooth form:

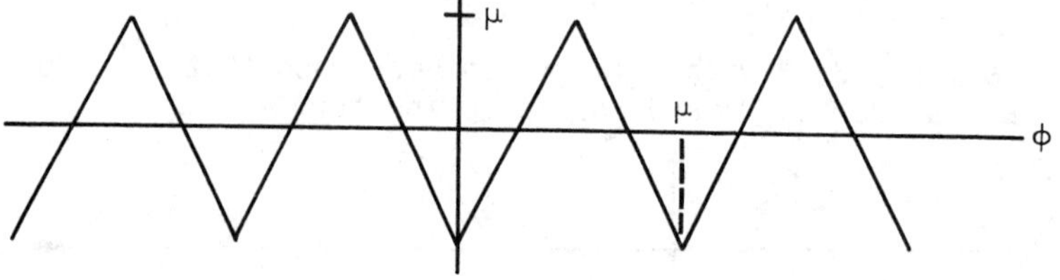

it is easily verified that

$$(*) \qquad I_k = k\mu \ , \ \phi_k = \frac{1}{2} k(k-1)\mu$$

is an orbit. Indeed, since ϕ_k is an integral multiple of μ we have

$$-\frac{\partial f}{\partial \phi} = \mu \text{ for } \phi = \phi_k \ ,$$

and hence from $I_0 = 0$

$$I_{k+1} = I_k + \mu = (k+1)\mu \ .$$

Moreover,

$$\phi_{k+1} - \phi_k = k\mu = I_k \ ,$$

which verifies that (*) defines an unbounded orbit.

we observe that max $|f(\phi,\mu)| \leq 1/8\ \mu^2$, max $|f_\phi(\phi,\mu)| \leq \mu$ and the Lipshitz constant of f_ϕ is 4. It is clear that one can replace f by a smooth function or even a polynomial for which $|f|_{C''} \leq 5$ and which still has the instability phenomenon. One merely has to round off the corners.

The work of Takens[37] shows that one can find functions $f(\phi,\mu)$ for which even $|f|_{C''}$ is small and instability takes place. On the other hand, if $|f|_{C^5}$ is sufficiently small, the solutions are trapped and we have stability!

REFERENCES

1. H. Poincaré, "Les Méthodes Nouvelles de la Mécanique Céleste", Gauthier-Villars, Paris (1892); Dover Press (1957); NASA Translation TT F-450, Washington (1976).
2. J. Moser, "Lectures on Hamiltonian Systems", Memoirs A.M.S. 81, (1968) 1-60. See references listed therein.
3. A. N. Kolmogorov, Intern. Congress Math. Amsterdam, 315. V. I. Arnol'd and A. Avez, "Ergodic Problems of Classical Mechanics" Benjamin Inc., 1968.
4. V. I. Arnol'd, "Proof of a Theorem of A. N. Kolmogorov on the Invariance of Quasi-Periodic Motions Under Small Perturbations of the Hamiltonian", Uspekhi Mat. Nauk 18, No. 5, 13-40 = Russian Math. Surveys 18, No. 5, (1963) 9-36.
5. J. Moser, "On Invariant Curves of Area-Preserving Mappings on an Annulus" Nachr. Akad, Wiss., Göttingen: Math-Phys. Kl. IIa, No. 1, (1962) 1-20.

6. For a proof and a more detailed exposition see C. Siegel and J. Moser, "Lectures on Celestial Mechanics", Springer, New York, 1971.
7. E. Fermi, Physik. Z. 24, (1923) p. 261.
8. G. D. Birkhoff, "Dynamical Systems", American Math. Society, 1927; revised edition 1966.
9. C. Siegel, "On the Integrals of Canonical Systems", Ann. Math. 42, (1941), pp. 806-822.
10. E. Keil, "Incoherent Space Charge Phenomena in the ISR", Proc. of 8th Intern. Conf. on High Energy Accel., CERN, Geneva, 1971, p. 372.
11. B. V. Chiricov "Universal Instability of Many-Dimensional Oscillator Systems", Physics Reports (Phys. Lett. Sect. C),to appear. In this review article the work reported in refs. 12-15 is outlined.
12. B. V. Chiricov, "Research Concerning the Theory of Nonlinear Resonance and Stochasticity", Preprint 267, Institute of Nuclear Physics, Novosibirsk. [English Translation, CERN Trans. 1971.]
13. G. V. Gadiyak, F. M. Izraelev, B. V. Chiricov, "Preliminarv Numerical Experiments on the Arnol'd Diffusion", Institute of Nuclear Physics, Novosibirsk, 1974.
14. V. I. Arnold, Dokl. Adak. Nauk. SSSR, 156 (1964) 9.
15. N. N. Nekhoroshev, Funkt. Analizievo Prilozheniya 5 (1971) 82 (Transl. in Funct. Analysis).
16. J. M. Greene, "Two-Dimensional Measure-Preserving Mappings", J. Math. Phys. 9, No. 3 (1968), p. 760.
17. M. Hénon, Quart. Appl. Math. 27, (1969) pp. 291-312.
18. C. Froeschlé,J. P. Scheidecker, Phys. Rev. 12 A, (1975) pp. 2137-2143.
19. F. Rannou, "Numerical Study of Discrete Plane Area-Preserving Mappings", Astron. and Astrophys. 31, 1974, pp. 289-301.
20. F. Calogero, "Solution of the One-Dim. N-body Problems with Quadratic and/or Inversely Quadratic Pair Potentials", J. Math. Phys. 12, No. 3, (1971) 419.
21. M. Adler, "Some Finite Dimensional Integrable Systems and Their Scattering Behavior", Math. Res. Center Technical Summary Report #1718, University of Wisconsin, Madison, Feb. 1977.
22. M. A. Olshanetzky and A. M. Perelomov, "Explicit Solution of the Calogero Model in the Classical Case and Geodesic Flows on Symmetric Spaces of Zero Curvature", Lettere Al Nuovo Cimento 16,(1976) 33.
23. J. Moser, "Three Integrable Hamiltonian Systems Connected with Isospectral Deformations", Adv. in Math. Vol.16, No. 2, May (1975).
24. J. Moser, "The Scattering Problem for Some Particle Systems on the Line", Conference on Dynamical Systems Rio de Janeiro,Summer 1976 (to appear).
25. R. M. Miura, C. S. Gardner, M. D. Kruskal "KdV Equation and Generalizations II. Existence of Conservation Laws and Con-

stants of the Motion", J. Math. Phys. 9, (1968) 1204-1209.
26. C. S. Gardner, J. M. Greene, M. D. Kruskal, R. M. Miura, Comm. Pure Appl. Math. 27, (1974) 97.
27. L. Faddeev, V. Zakharov, "KdV Equation Completely Integrable Hamiltonian System", Funkt. Analizievo Prilozheniya Vol. 5 (1971), pp. 18-27.
28. S. Novikov, "The Periodic Problem for the KdV Equation" Funkt. Anal. Pril. Vol. 8 (1974) pp. 54-66. Transl. in Funct. Anal. Vol. 8, (1974), pp. 236-246.
29. P. Lax, "Periodic Solutions of the KdV Equation" Comm. Pure Appl. Math. Vol. 28 (1975) pp. 141-188.
30. H. P. McKean, P. Van Moerbeke, "The Spectrum of Hill's Equation", Inventiones Math. Vol. 30 (1975) pp. 217-274.
31. H. P. McKean and E. Trubowitz, Comm. Pure and Appl. Math. Vol. 29, (1976) pp. 143-226.
32. H. Airault, H. P. McKean, J. Moser "Rational and Elliptic Solutions of the KdV Equation and a Related Many-Body Problem", Comm. Pure Appl. Math. Vol. 30, (1977) 95-140.
33. V. Zakharov, "On Stochastization of One Dimensional Chains of Nonlinear Oscillators" Zh. Eksp. Teor. Fiz. 65, 219-255.
34. E. Fermi, J. Pasta, S. Ulam, p. 978 in "Collected Papers of Enrico Fermi", Vol. II, Univ. of Chicago Press, Chicago, 1965. Reprinted in "Nonlinear Wave Motion" ed. by A. C. Newell, Lectures in Appl. Math., Vol. 15, Am. Math Soc. Providence, R.I. (1974).
35. T. Nishida, Memoirs of Faculty of Engineering, Kyoto Univ., (1971).
36. H. Rüssmann, Uber invariante Kurven differenzierbarer Abbildungen eines Kreisringes, Nachr. Akad. Wiss., Göttingen, math. phys. Kl. II, 67-105, 1970.
37. F. Takens, A C^1 - counter example to Moser's twist theorem, Indag. Math. 33, 379-386, 1971.

REGULAR AND IRREGULAR MOTION

M. V. Berry
H. H. Wills Physics Laboratory
Tyndall Avenue
Bristol BS8 1TL, England

1. INTRODUCTION

The aim here is to describe some recent advances in the understanding of classical systems. As is often the case, these advances have been made with the aid of, and have led to, deep theorems in pure mathematics, couched in language unfamiliar to physicists. Concurrently, numerical experiments, extensively carried out by astronomers, have illuminated similar problems from a different point of view. Reviews bringing these approaches together are now appearing along with second-order reviews like this one. The bibliography is intended as a guide; further references may be found in the works cited and in other articles in this volume.

Over the last century attention has shifted from the computation of individual orbits towards the qualitative properties of families of orbits. For example, the question of whether a given orbit is stable can only be answered by studying the development of all orbits whose initial conditions are in some sense "close to" those of the orbit being studied. More generally, one can consider all orbits of a given system - defined by a given Hamiltonian - and inquire whether all, or "almost all" or "most" or "hardly any" are stable. More generally still, one can consider all possible Hamiltonian functions within some class, and seek the "generic" or typical behavior of its family of orbits. Finally, one can ask whether Hamiltonian systems display properties typical of the wider class of "dynamical systems" which may be dissipative and not describable by a Hamiltonian function (they do not).

One motivation for such studies is the feeling that after three hundred years we really ought to know what Newton's equations are telling us about the qualitative behavior of conservative systems with two degrees of freedom. And yet the orbits of a point mass m in the potential (fig. 1)

$$V = \frac{r^2}{2} + ar^3 \sin 3\theta$$

have only begun to be understood in detail in the last decade or so.

Another motivation is the desire to know whether the solar system, and the galaxy, are stable under the mutual perturbations of their constituents, or whether they will eventually collapse or disperse to infinity.

Another motivation is in the foundations of statistical mechanics. In that subject no attempt is made to follow the detailed motion of all the constituents of a complicated system of many inter-

ISSN: 0094-243X/78/016/$1.50

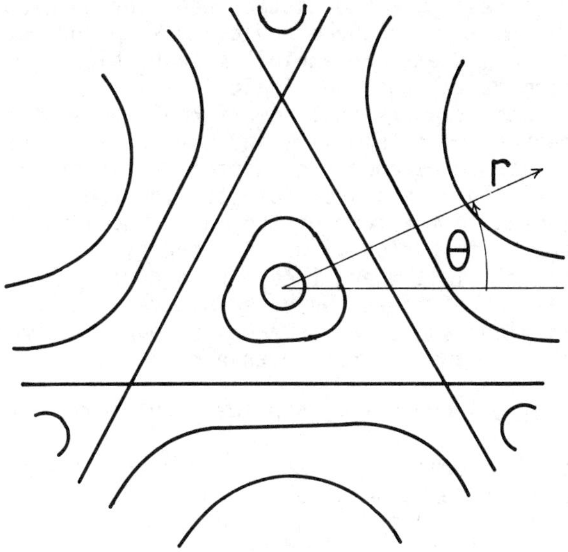

Fig. 1

acting bodies. Instead, we are content with a knowledge of the macroscopic observables, which are assumed to involve long-time averages over the motion of the bodies. Even these are too difficult to calculate, and a further assumption is made - the famous "ergodic hypothesis". This is that over the course of time the system explores the whole of the region of phase space that is energetically available to it (the "energy surface"), and eventually covers this region uniformly. Time averages can thus be replaced by averages over the energy surface in phase space, and statistical mechanics becomes a going concern.

But is the ergodic hypothesis true? No. In general it is false. It is not the case that all systems explore the whole of the available energy surface in phase space. In the standard "integrable" problems of the classical mechanics textbooks - the Kepler problem, n-dimensional harmonic oscillators, etc. - we shall see that systems explore infinitesimal fractions of the energy surfaces. Perhaps ergodic behavior appears the moment an "integrable" system is subjected to a "non-integrable" perturbation? Fermi believed this, but it too is false as we shall see. Then what systems are ergodic? This is still not known, although at last it has been shown that the system consisting of two or more interacting hard spheres is ergodic, so that statistical mechanics is valid in this case.

These results might seem somewhat meagre, justifying the theoretical physicists' traditional skepticism about the value of "ergodic theory". But a by-product of the analysis has been the discovery of a vast realm of "stochastic" behavior between the extremes of the integrable and the ergodic and some understanding of how deterministic systems can exhibit motion which in some respects is as random as the tossing of a coin.

Yet another motivation is that some formulations of classical and quantum mechanics are so similar, any advance in classical mechanics ought to lead to advances in quantum mechanics. In particular, any transition of classical orbits from regular to random ought to be reflected in the form of the wave functions for quantum states, and in the distribution of quantum energy levels, especially in the semi-classical limit where these are densely distributed.

Proper discussion of these subjects involves topology, number theory, smooth mappings and other branches of pure mathematics. My treatment will be intuitive rather than rigorous.

2. INTEGRABLE SYSTEMS AND INVARIANT TORI*

We begin with integrable systems for which the solution of Newton's equations can be reduced to the solution of a set of simultaneous equations, followed by integrations over single variables. We restrict ourselves to (nondissipative) systems describable by a Hamiltonian function

$$H(\underline{q},\underline{p}) ,$$

where there are N degrees of freedom and, $\underline{q} = (q_1 \ldots q_i \ldots q_N)$ is the system's configuration and $\underline{p} = (p_1 \ldots p_N)$ is the canonical momentum vector conjugate to $\underline{q}$. Occasionally, we may allow H to have an explicit time dependence but almost always we shall treat conservative systems. The history $\underline{q}(t)$ of the system is found by solving Hamilton's equations

$$\dot{\underline{q}} = \nabla_{\underline{p}} H \qquad (2.1.a)$$

$$\dot{\underline{p}} = - \nabla_{\underline{q}} H \qquad (2.1.b)$$

with given initial conditions $\{\underline{q}(0), \underline{p}(0)\}$. Of course, the first Hamilton equation (a) can be used to obtain $\underline{p}$ when $\underline{q}$ and $\dot{\underline{q}}$ are known (eq. for a non-relativistic mass point $\underline{p}=m\dot{\underline{q}}$), so it is natural and useful to describe the system's motion in *phase space* $(\underline{q}, \underline{p})$ and from here on that is what we shall do.

A completely integrable system is one where there exist N independent analytic single-valued first integrals, that is N functions

$$F_m(\underline{q}, \underline{p}) \qquad 1 \leq m \leq N$$

* App. 26 of Ref. 2 and Ch. 2 of Ref. 4.

that are constant along each trajectory of the system. For a conservative system one of these can be taken as the energy - the Hamiltonian itself. For a particle acted on by central forces only, three integrals of motion are the components of angular momentum

$$\underline{L} = \underline{q} \wedge \underline{p} , \tag{2.2}$$

where $\underline{q}$ is measured from the centre of force.

Along any trajectory, the F_m's take constant values f_m, and the N equations

$$F_m(\underline{q}, \underline{p}) = f_m \tag{2.3}$$

can be solved for $\underline{p}$ in terms of $\underline{q}$ and the f's. We can consider the F's to be the new momenta

$$\underline{\overline{p}} = \underline{F} \; (=\{F_m\})$$

in a canonical transformation to new variables $(\underline{\overline{q}}, \underline{\overline{p}})$ in phase space. Since we know the $\underline{\overline{p}}$'s to be constant, the new Hamilton equation (2.1.b) shows that the new Hamiltonian cannot depend on the new $\overline{q}$'s, so the first Hamiltonian equation gives

$$\dot{\underline{q}} = \nabla_{\overline{\underline{p}}} \, (H(\underline{\overline{p}})) = \nabla_{\underline{f}} \, H(\underline{f}) = \text{const.} \tag{2.4}$$

or

$$\underline{\overline{q}}(t) = \nabla_{\underline{f}} \, H(\underline{f})t + \underset{\substack{\uparrow \\ \text{constants}}}{\underline{\delta}} \quad . \tag{2.5}$$

The problem is solved if we can express $\underline{\overline{q}}$ in terms of the old coordinates $\underline{q}$. This is achieved by demanding that the transformation

$$(\underline{q}, \underline{p}) \rightarrow (\underline{\overline{q}} \; \underline{\overline{p}})$$

be canonical. This in turn can be accomplished with the generating function

$$S(\underline{q},\underline{\overline{p}}) = \int_{\underline{q}_0}^{\underline{q}} \underline{p}(\underline{q},\underline{\overline{p}})\cdot\underline{dq} = \int_{\underline{q}_0}^{\underline{q}} \underline{p}(\underline{q},\underline{f})\cdot\underline{dq} = S(\underline{q},\underline{f}) , \tag{2.6}$$

where $\underline{p}(\underline{q},\underline{f})$ is obtained from (2.3). Standard theory yields

$$\underline{\overline{q}} = \nabla_{\underline{f}} \, S(\underline{q},\underline{f}) , \tag{2.7}$$

thus the solution $\underline{q}(t;\underline{f},\underline{\delta})$ via (2.5) (the $\underline{f},\underline{\delta}$ are the 2n constants required to define a solution). All exactly-soluble systems in classical mechanics are integrable in this sense.

Of course, it is necessary for the constants of motion $F_m(\underline{q},\underline{p})$ to be independent and in addition the following condition must also hold:

$$\nabla_{\underline{p}} F_m \cdot \nabla_{\underline{q}} F_n - \nabla_{\underline{p}} F_n \nabla_{\underline{q}} F_m$$

$$\equiv \text{Poisson bracket } \{F_m, F_n\} = 0 \text{ for all } m, n \tag{2.8}$$

The F's are then said to be "in involution".

The existence of the N F_m's implies that each trajectory of the system can explore at most an N-dimensional manifold M in the 2N-dimensional phase space. Except for the trivial case of N=1, this is smaller than the energy surface E, which has 2N-1 dimensions. Therefore the ergodic hypothesis is false in general. A table might help here:

Number of degrees of freedom:	1	2	3	N
Dimensionality of phase space	2	4	6	2N
" E	1	3	5	2N-1
" M	1	2	3	N

It will be important to know what sort of manifold M is, and we show now that M is an N-*dimensional torus*: construct the following N vector fields V_m in *phase space:*

$$V_m \equiv (\nabla_{\underline{p}} F_m \, , \, -\nabla_{\underline{q}} F_m) \, , \tag{2.9}$$

i.e. the V_m's have 2N components.

On each M, defined by fixing the f_m's in (2.3), the V_m's are smooth and independent (because the F_m's are), and moreover the V's are *parallel* to M, since by virtue of (2.8) each V_m is perpendicular to all normals to M:

$$V_m \cdot (\nabla_{\underline{q}} F_n \, , \, \nabla_{\underline{p}} F_m)$$
$$= \{F_m, F_n\} = 0 \text{ for all } m,n. \tag{2.10}$$

We restrict ourselves to "bound" motion in which the region of accessible phase space is finite. Then M is a compact manifold. Now we make use of a theorem in topology which states that a compact manifold "parallelizable" with N smooth independent vector fields must be an N-torus. This is intuitively obvious from Fig. 2. Q.E.D.

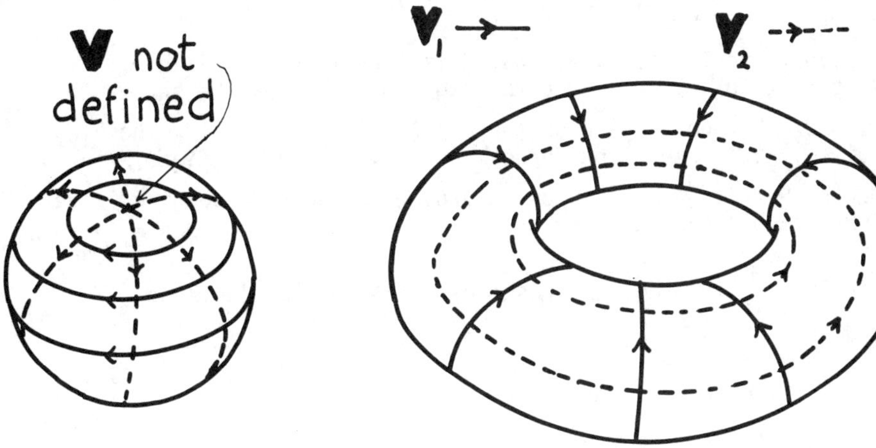

Fig. 2

(Colloquially, if M were hairy, we could comb it without singularity in N ways.)

These tori are called *invariant tori* because an orbit starting out on one remains on one forever. It is natural to coordinatize phase space using the $(\underline{\bar{q}},\underline{\bar{p}})$ defined earlier, since then $\underline{\bar{p}} = \underline{f}$ defines which torus we are on and $\underline{\bar{q}}$ are coordinates on the torus. But there are many ways of doing this, corresponding to taking not the original F_m's but any functions of them (which are still constants of motion). However, there is one standard way, that leads to the introduction of so-called *action-angle variables*. These are topologically natural, and widely employed in analytical mechanics. The variables have the following symbols:

$$\underline{\Theta} = \underline{\bar{q}} = \textit{angles on } \text{torus}$$
$$\underline{I} = \underline{\bar{p}} = \textit{actions of } \text{torus} \qquad (2.11)$$

The definition is based on the observation that the generating function in (2.6), $S(\underline{q},\underline{\bar{p}})$, is generally multivalued because the momenta $\underline{p}$ defined by (2.3) can be multivalued - when a system returns to a given point $\underline{q}$ it need not have the same momentum $\underline{p}$. Therefore S depends on the path $\underline{q}_0 \to \underline{q}$. But the path must certainly lie on M since otherwise (cf. 2.3) $\underline{p}(\underline{q},\underline{f})$ is not defined. Now for different paths $\underline{q}_0\to q$ on M that are deformable into one another S is certainly the same, because it is the solution of a Hamilton-Jacobi equation

$$H(\underline{q},\nabla_{\underline{q}}S(\underline{q},\underline{f})) = H(\underline{f}) \,. \tag{2.12}$$

This means roughly speaking, that S is locally single-valued. It follows that for closed circuits $q_0\to q_0$ on M that can be shrunk to zero, S=0. But on an N-torus there are N independent irreducible circuits γ_i that cannot be shrunk to zero, and this defines N increments ΔS that S can gain on returning to the same point $\underline{q}$. The action variables are defined by

$$I_i \equiv \frac{1}{2\pi}\int_{\gamma_i} \underline{p}\cdot d\underline{q} = \frac{1}{2\pi} \times \text{ sum of areas of N projections}$$

$$\text{of } \gamma_i \text{ on planes } q_1p_1, q_2p_2, \ldots q_Np_N \tag{2.13}$$

$$= \frac{\Delta S}{2\pi} \text{ for ith circuit on } M$$

This defines the N I's in terms of the N f's and vice versa. For a conservative system the energy H is one of the f's and hence can be expressed as a function of the I's:

$$H = H(\underline{I}) \,. \tag{2.14}$$

The I's define the torus M. The coordinates *on* M are the "angles" $\underline{\Theta}$ canonically conjugate to $\tilde{I}$:

$$\underline{\Theta} \equiv \nabla_I \, S(\underline{q},I). \tag{2.15}$$

The reason for calling them angles is that Θ_i changes by 2π, and $\Theta_{i\neq j}$ do not change as we traverse circuit γ_i on M:

$$\begin{aligned}(\Delta\Theta_i)_{\gamma_i} &= \Delta_{\gamma_i}\frac{\partial S}{\partial I_i}(q,I)\\ &= \frac{\partial}{\partial I_i}\Delta_{\gamma_i}S \qquad (2.16)\\ &= \frac{\partial}{\partial I_i}2\pi I_j = 2\pi\delta_{i,j}\,.\end{aligned}$$

This means that in the canonical transformation

$$\begin{Bmatrix} \underline{q} \\ \underline{p} \end{Bmatrix} \mapsto \begin{Bmatrix} \underline{\Theta} \\ \underline{I} \end{Bmatrix} , \qquad (2.17)$$

the $\underline{q}$'s and $\underline{p}$'s are *periodic* functions of $\underline{\Theta}$ with period 2π:

$$\underline{q} = \sum_{\underline{m}} \underline{q}_{\underline{m}}(\underline{I}) e^{i\underline{m}\cdot\underline{\Theta}} , \quad \underline{p} = \sum_{\underline{m}} \underline{p}_{\underline{m}}(\underline{I}) e^{i\underline{m}\cdot\underline{\Theta}} \qquad (2.18)$$

where $\underline{m}$ is a N-dimensional lattice vector (i.e. integer components). Soon we shall see what a pleasant picture of the motion this gives. Meanwhile, we illustrate the formalism with a few *examples*:

Swing (fig. 3)

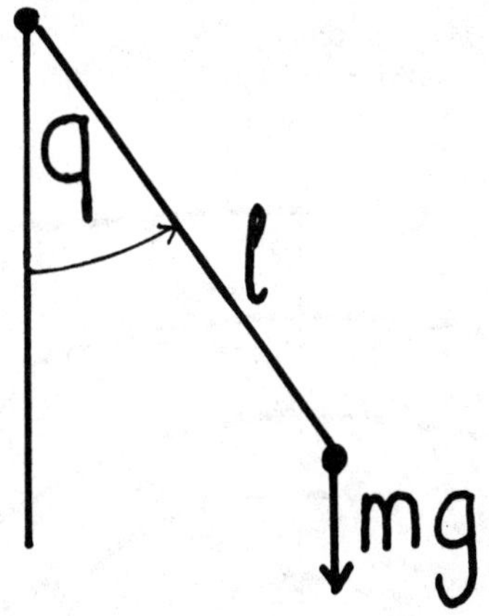

Fig. 3

The motion is one-dimensional:

Lagrangian

$$L = \frac{1}{2} m(\ell\dot{q})^2 - mg\ell(-\cos q) , \qquad (2.19)$$

KE PE

$$P = \frac{\partial L}{\partial \dot{q}} = m\ell^2\dot{q} ;$$

Hamiltonian

$$H = p\dot{q} - L = \frac{p^2}{2m\ell^2} + mg\ell(-\cos q) , \qquad (2.20)$$

which yields the equations of motion

$$\dot{q} = \frac{p}{m\ell^2}\ , \ \dot{p} = -mg\ell\ \sin q\ , \tag{2.21}$$

or

$$\ddot{q} = -\ \frac{g}{\ell}\ \sin q\ , \tag{2.22}$$

as is well known. There is one constant of motion, H itself ($\equiv$f). Therefore, the manifold M is

$$\frac{p^2}{2m\ell^2} - mg\ell\ \cos q = f\ ,$$

$$\text{i.e.}\quad p(q,f) = \sqrt{2m\ell^2(f+mg\ell\ \cos q)}\ . \tag{2.23}$$

All these curves M (1-dimensional) in phase space are closed because $q+2\pi \leftrightarrow q$ so that phase space is really cylindrical (fig. 4).

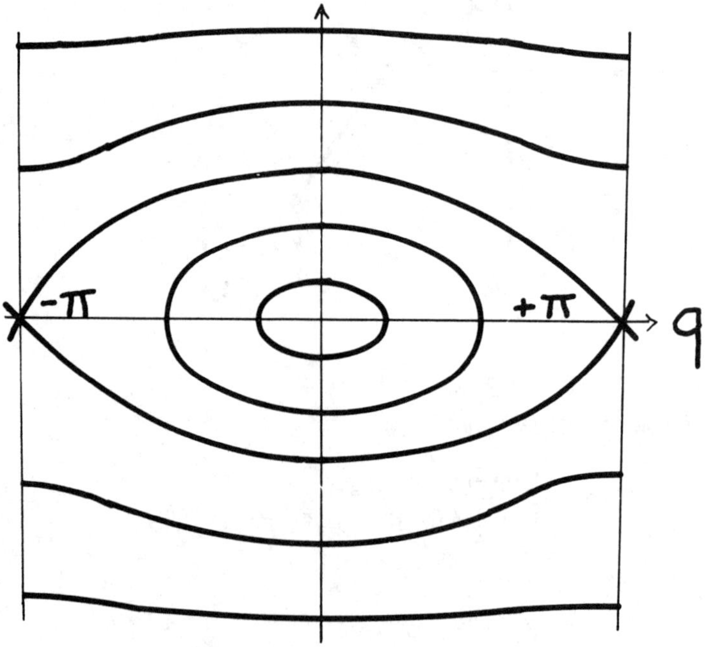

Fig. 4

The closed curves are the tori. These are of two kinds:

"librations" between fixed limits of q, for $f < mg\ell$ (fig. 5)

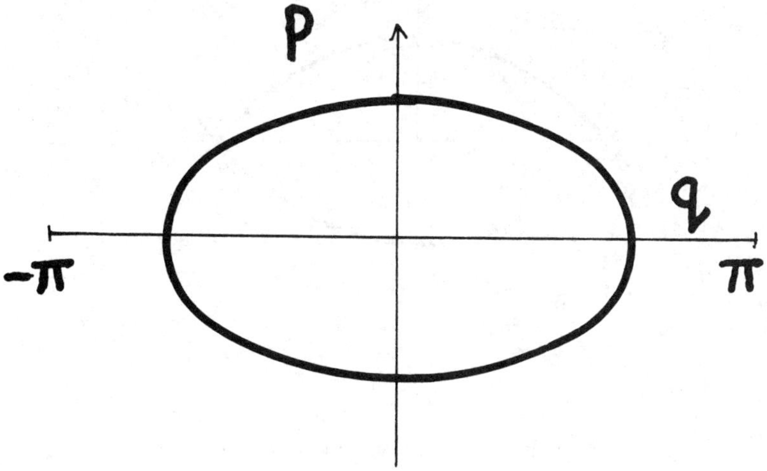

Fig. 5

and "rotations" with q increasing or decreasing forever for $f >$ $mg\ell$ (fig. 6)

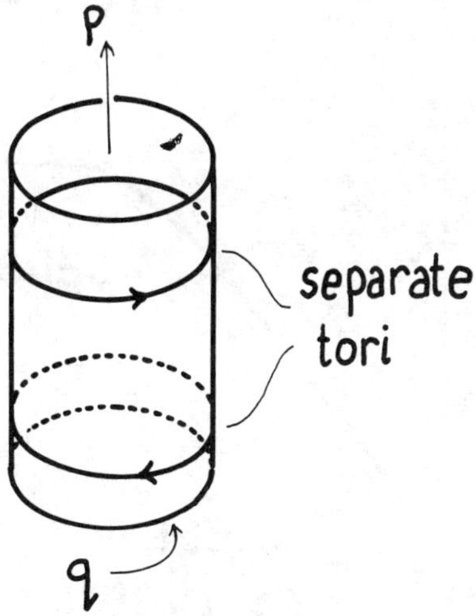

Fig. 6

These are separated by a self crossing curve (fig. 7) corresponding to $f = mg\ell$.

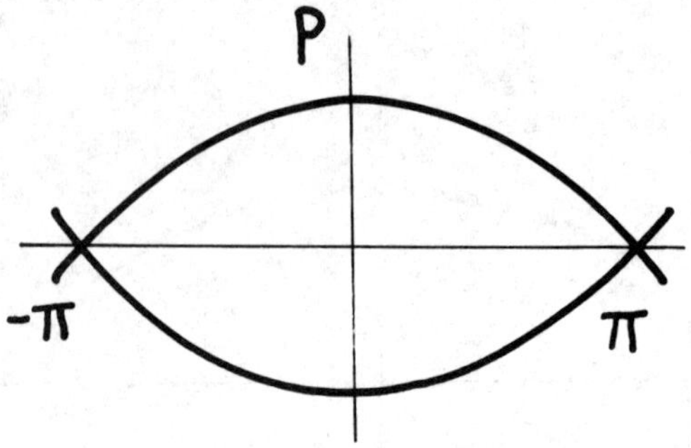

Fig. 7

The action is

$$I \equiv \frac{1}{2\pi} \oint \underline{p} \cdot d\underline{q} = \frac{\text{Re}}{\pi} \int_0 \sqrt{2m\ell^2 (f + mg\ell \cos q)} \quad , \qquad (2.24)$$

which implicitly defines the "action" Hamiltonian (fig. 8).

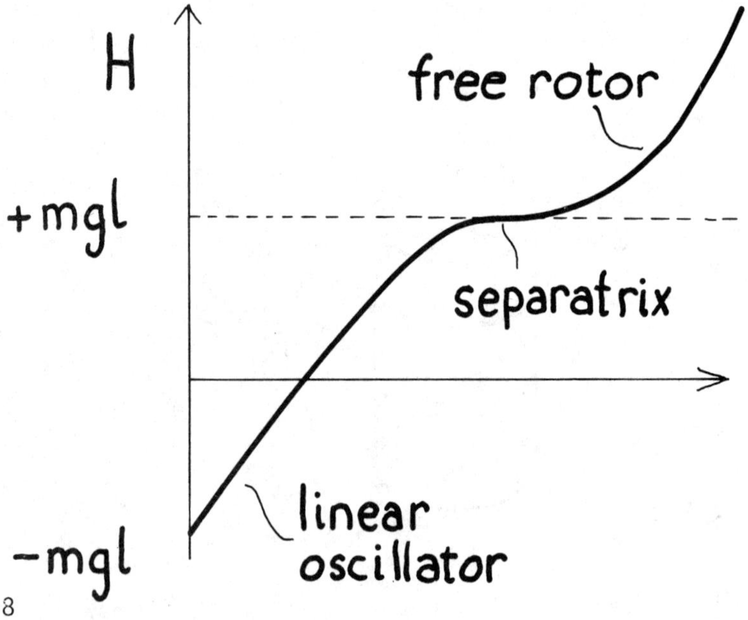

Fig. 8

<u>2-D HARMONIC OSCILLATOR</u>

Here,

$$H(\underline{q},\underline{p}) = \frac{p_1^2}{2} + \frac{p_2^2}{2} + \frac{\omega_1^2 q_1^2}{2} + \frac{\omega_2^2 q_2^2}{2} \ . \tag{2.25}$$

The two constants of motion are (energy in each mode)

$$F_1 = \frac{p_1^2}{2} + \frac{\omega_1^2 q_1^2}{2} \ , \ F_2 = \frac{p_2^2}{2} + \frac{\omega_2^2 q_2^2}{2} \ . \tag{2.26}$$

The q_1 and q_2 motions are uncoupled, hence these give the irreducible circuits γ_1 and γ_2, and (fig. 9):

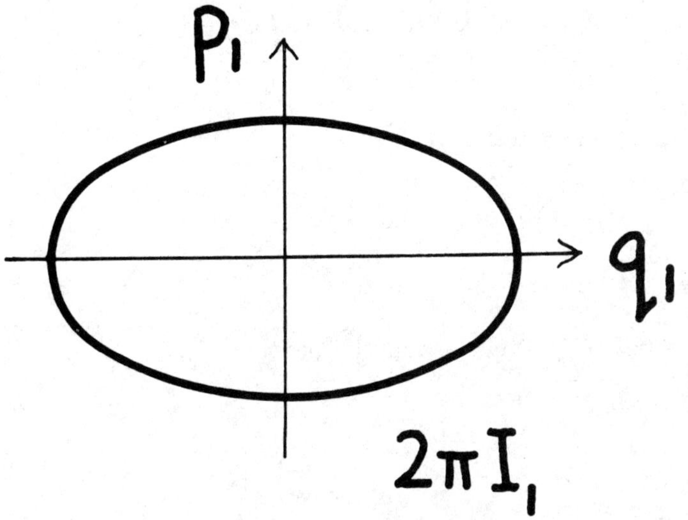

Fig. 9

$$I_1 = \frac{1}{2\pi} \int_{\gamma_1} p_1 dq_1 = \frac{1}{2\pi} \oint \sqrt{2(F_1 - {}_1^2 q_1^2)}\, dq_1$$

$$= \frac{1}{2\pi} \times \text{area of ellipse}$$

$$= \frac{1}{2} \sqrt{2F} \sqrt{\frac{2F}{\omega_1^2}} = \frac{F_1}{\omega_1} \ , \ I_2 = \frac{F_2}{\omega_2} \ . \tag{2.27}$$

Therefore, the Hamiltonian in action variables is

$$H(\underline{I}) = F_1 + F_2 = I_1\omega_1 + I_2\omega_2 \tag{2.28}$$

and I_1 and I_2 are related as in Fig. 10.

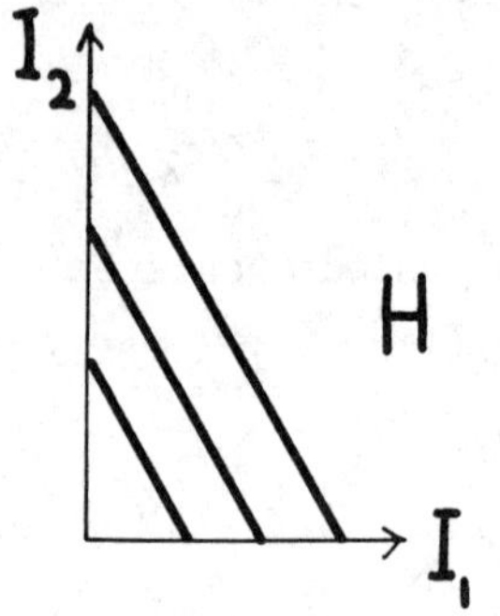

Fig. 10

PLANE MOTION UNDER CENTRAL FORCE

The Hamiltonian is

$$H = \frac{p_r^2}{2m} + \frac{p_\Theta^2}{2mr^2} + V(r) \; , \tag{2.29}$$

for the geometry of Fig. 11 and the potential of Fig. 12.

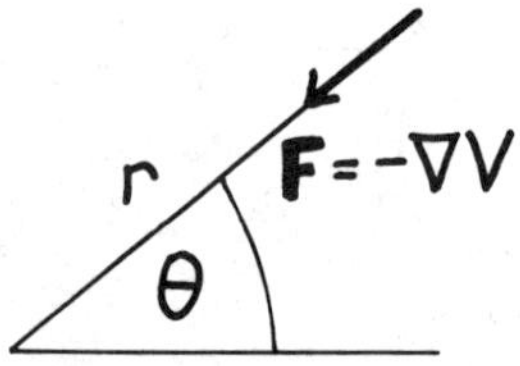

Fig. 11

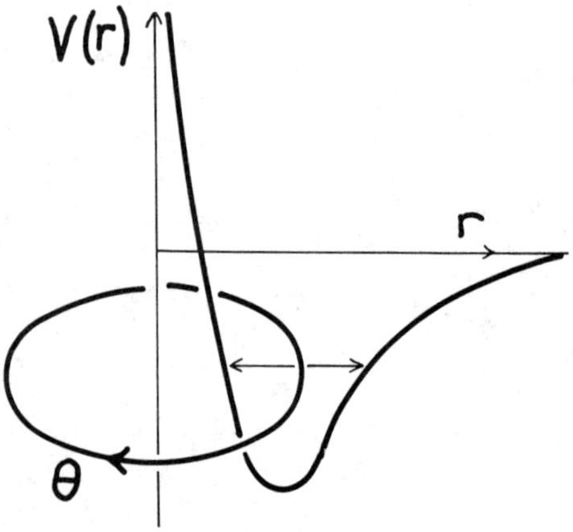

Fig. 12

Constants of motion:

$$F_1 = H, \ F_2 = p_\Theta(=mr^2\dot{\Theta}) \ . \tag{2.30}$$

Irreducible circuits are libration in r and rotation in Θ.

$$I_2 = \frac{1}{2\pi\gamma_2} \int p_\Theta \, d\Theta = \frac{F_2}{2\pi} \int_0^{2\pi} d\Theta = F_2 \ , \tag{2.31}$$

$$I_1 = \frac{1}{2\pi} \oint p_r \, dr$$

$$= \frac{\mathrm{Re}}{\pi} \int_0^\infty [2m(F_1 - \frac{I_2^2}{2mr^2} - V(r))]^{1/2} \, dr \ , \tag{2.32}$$

which defines the Hamiltonian (fig. 13) as

$$H(I_1, I_2) = F_1(I_1, I_2) \ . \tag{2.33}$$

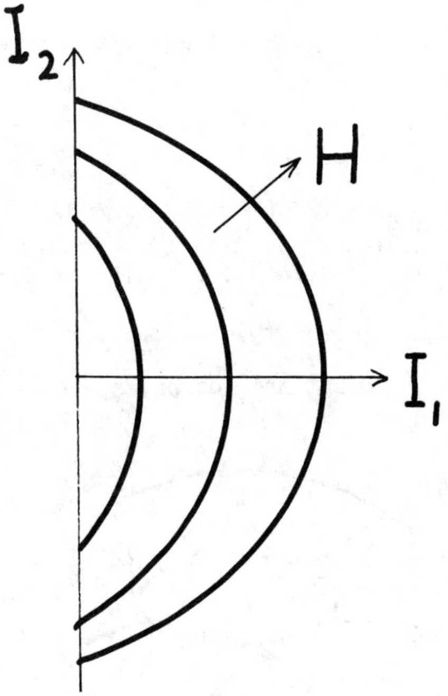

Fig. 13

In terms of general action-angle theory, the Hamiltonian (2.14) yields the equations of motion

$$\underline{I} = \text{const.}, \quad \underline{\dot{\Theta}} = \nabla_{\underline{I}} H(I) = \text{const.}\ , \tag{2.34}$$

or

$$\underline{\Theta}(t) = \underline{\omega}(\underline{I})t + \underline{\delta}\ , \tag{2.35}$$

where $\underline{\delta}$ represents N constants and

$$\underline{\omega}(\underline{I}) \equiv \nabla_{\underline{I}} H(\underline{I}) \text{ (the frequency vector on the torus } \underline{I}) \ . \tag{2.36}$$

Thus (2.18) gives the motion as

$$\underline{q}(t) = \sum_{\underline{m}} \underline{q}_{\underline{m}}(\underline{I})e^{i(\underline{m}\cdot\underline{\omega}t+\underline{m}\cdot\underline{\delta})}$$
$$\underline{p}(t) = \sum_{\underline{m}} \underline{p}_{\underline{m}}(\underline{I})e^{i(\underline{m}\cdot\underline{\omega}t+\underline{m}\cdot\underline{\delta})} \tag{2.37}$$

This shows that the system's orbit on M is *multiply periodic* with the N periods

$$T_i = \frac{2\pi}{\omega_i}\ . \tag{2.38}$$

for a circuit round Θ_i.

If the orbit on M is *closed*, i.e. if for some τ

$$\underline{\Theta}(\tau) = \underline{\Theta}(0) + 2\pi N\ , \tag{2.39}$$

then it does not fill M but only occupies a one-dimensional region on M (fig. 14).

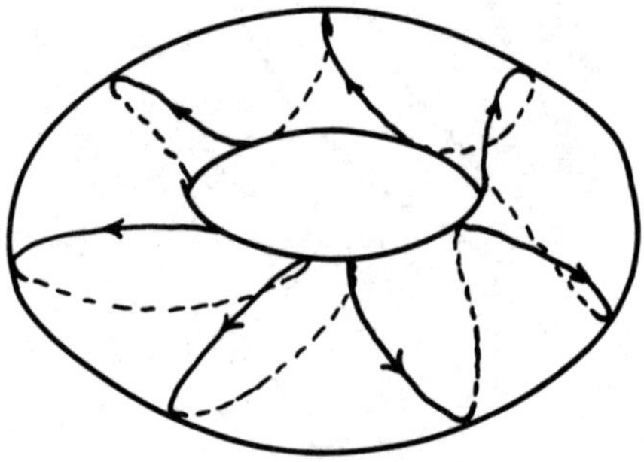

Fig. 14

If the orbit *never closes*, it traverses a helix on M which covers it densely after infinite time (fig. 15). This is "ergodicity on M" though not on the energy surface, and not of any stochastic character.

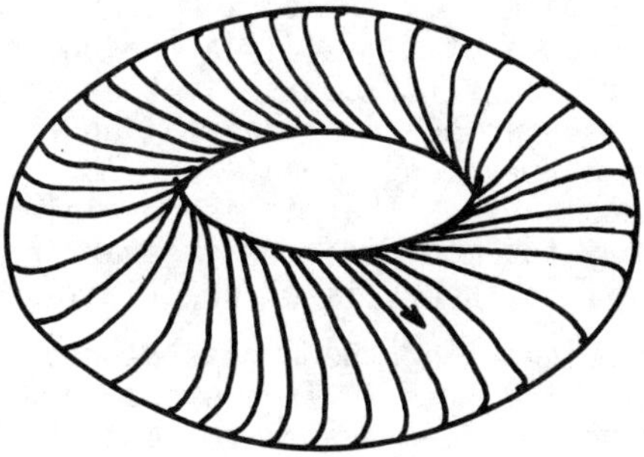

Fig. 15

Now which of these situations is typical? To get closure, the frequencies $\underline{\omega}$ must be commensurable; this condition can be written in either of the following ways:

$$\left.\begin{array}{l}\underline{\omega} = \underline{N}\,\omega_b \quad (\underline{N}\text{ is a vector with integer components}) \\ \text{or N-1 relations } \underline{\omega}\cdot\underline{m} = 0 \text{ hold } (\underline{m}\text{ a finite non-zero vector} \\ \text{with integer components).}\end{array}\right\} \qquad (2.40)$$

In general, closure occurs after N_1 circuits of Θ_1, N_2 circuits of $\Theta_2 \ldots N_N$ circuits of Θ_N, with period

$$\tau = \frac{2\pi}{\omega_b} = \frac{2\pi N_i}{\omega_i} = \frac{N_i}{T_i} \;. \qquad (2.41)$$

From number theory (as we shall see later) commensurability is the exception rather than the rule and (2.40) holds only for a set of $\underline{\omega}$'s, i.e. tori $\underline{I}$, of zero measure. (If s relations $\underline{\omega}\cdot\underline{m} = 0$ hold, where s<N-1, the orbit is not closed but inhabits a submanifold of dimensionality N-s on M. The exceptional tori form a (dense) set of measure zero on which the tori are closed or partially closed.)

3. CANONICAL PERTURBATION THEORY FOR NONINTEGRABLE SYSTEMS*

Is integrability the rule or the exception? If all systems were integrable, the constants of motion $F_m(\underline{q},\underline{p})$ would always exist, and our inability to determine them for all but the simplest

*Some of this material is covered in: Ch. 9 of Ref. 5, Ch. 4 of Ref. 4, Ch. 1 of Ref. 1, Sec. 2-3d of Ref. 17, and Ref. 9.

problems would merely reflect our lack of analytical ingenuity. But it might be that integrable Hamiltonians form a very small set (possibly of zero measure), and the slightest perturbation of such a Hamiltonian would render the F_m non-existent (except the energy for conservative cases) and destroy the tori M so that each system trajectory would in the course of time fill a region in phase space of dimensionality greater than M (of course if the perturbation were small, the system, if started on or near an unperturbed torus M, would stay near M for a very long time).

Astronomers realized long ago that it is possible to devise a formal perturbation theory for modified tori M starting from an "unperturbed" torus M_0. This theory is of practical usefulness in celestial mechanics for calculating the orbits of heavenly bodies over *long but finite periods of time.* In the case of a planetary orbit, for example, the unperturbed system consists of the two-body problem of that planet moving in the Sun's field. This is the easily integrable Kepler problem. The perturbations come from the attractions of the other planets, principally Jupiter. If we "switch on" the mass M_J of Jupiter, and ignore the effect on Jupiter's orbit of the planet (say the earth ⊕) being considered (i.e. regard ⊕ as a "test body"), then we have the simplest case of the "plane restricted three body problem", with M_J as the perturbation away from integrability. We shall study this in more detail later.

But these methods only ensure the existence of tori M if the perturbation series converge *for infinite time,* and we shall see that the convergence is a very delicate matter indeed. Most physicists, when they thought about the matter, have tended to feel, with Fermi, that the slightest perturbation of an integrable system would destroy the tori - i.e. the series for M would diverge. Confirmation of this opinion would go a long way towards validating the ergodic hypothesis and hence statistical mechanics. Landau, however, thought that all systems are in principle integrable - i.e. phase space is filled with tori M. In fact, what happens is that "most" tori persist under perturbation, albeit in distorted form. Some are destroyed, however, and these form not a "set of measure zero" but a finite set which grows with the perturbation. The destroyed tori are distributed among those which are preserved, in a pathological manner. These assertions were rigorously proved by Moser and Arnol'd in 1962, on the basis of suggestions by Kolmogorov in 1954. They form what has become known as the "KAM theorem" which is one of the few certain statements in this subject. The proof is long, intricate and subtle, but the ideas on which it is based are most instructive and we shall concentrate on them.

Suppose that we have an integrable system. Then its phase space can be coordinatized by action-angle variables $\underline{I}$, $\underline{\Theta}$ and the Hamiltonian is a function $H_0(\underline{I})$ of $\underline{I}$ only. We illustrate this with another example to recall the theory: a particle in a two-dimensional box with sides a, b (fig. 16).

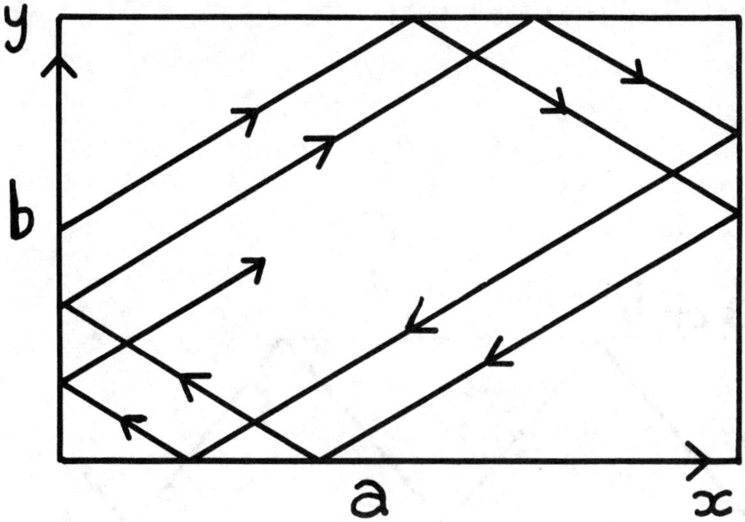

Fig. 16

The constants of motion are the x and y *speeds* $|v_x|$, $|v_y|$ which are unaffected by collisions. The actions are

$$I_1 = \frac{1}{2\pi} \oint p_x dx = \frac{1}{2\pi} \times m \; |v_x| \times 2\,a \;,$$

$$|v_x| = (\pi/ma) I_1 , \; |v_y| = (\pi/mb) I_2 \;,$$

and

$$H_0 = \frac{m}{2} \left(|v_x|^2 + |v_y|^2 \right) = \frac{\pi^2}{2m} \left(\frac{I_1^2}{a^2} + \frac{I_2^2}{b^2} \right) . \qquad (3.1)$$

The frequencies $\underline{\omega}$ are

$$\omega_1 = \frac{\partial H}{\partial I_1} = (\pi^2/ma^2) I_1 , \; \omega_2 = (\pi^2/mb^2) I_2 \; . \qquad (3.2)$$

The coordinates are given in terms of $\underline{\Theta}$ by

$$\begin{pmatrix} \underline{q} \\ \underline{p} \end{pmatrix} = \sum_{\underline{m}} \begin{pmatrix} \underline{q}_{\underline{m}}(\underline{I}) \\ \underline{p}_{\underline{m}}(\underline{I}) \end{pmatrix} e^{\, i \underline{m} \cdot \underline{\Theta}} \;, \qquad (3.3)$$

where q_m and p_m follow readily from a Fourier analysis of the motion, which in this case is as shown in Fig. 17.

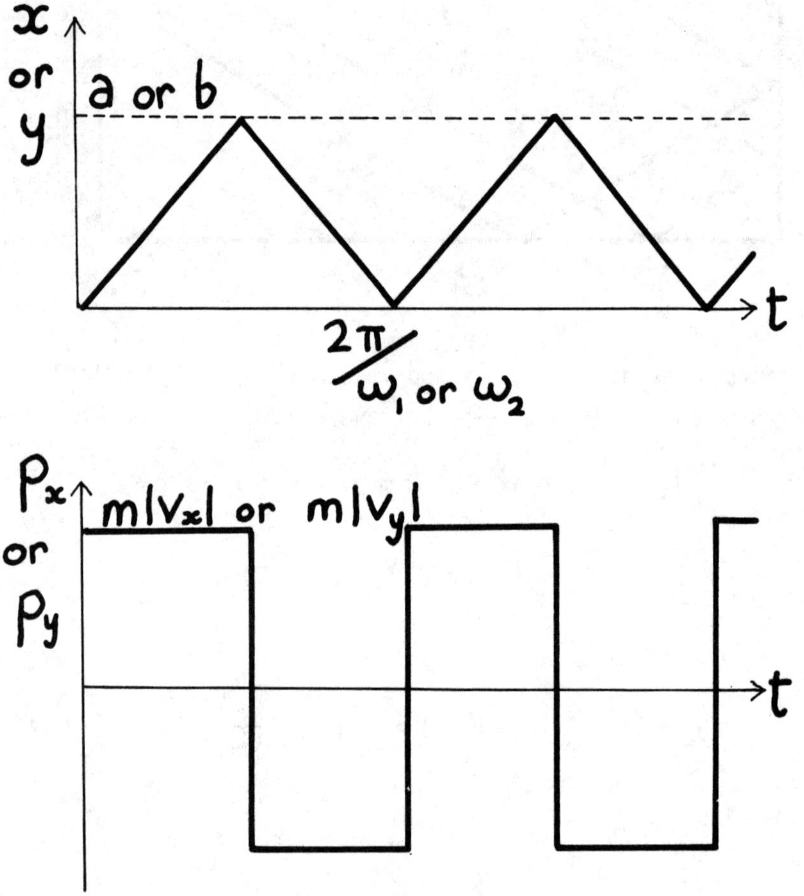

Fig. 17

Perturb the system H_0 with a nonintegrable perturbation $\varepsilon H_1(\underline{I},\underline{\Theta})$. For example, we could give the box a curved floor by making H_1 an extra force

$$H_1 = x^2y^2 = \left(\sum_{\underline{m}} x_{\underline{y}}(\underline{I})e^{i\underline{m}\cdot\underline{\Theta}}\right)^2 \quad \left(\sum_{\underline{m}} y_{\underline{m}}(\underline{I})e^{i\underline{m}\cdot\underline{\Theta}}\right)^2 . \tag{3.4}$$

In the new system,

$$H(\underline{I},\underline{\Theta}) = H_0(\underline{I}) + \varepsilon H_1(\underline{I},\underline{\Theta}) , \tag{3.5}$$

$\underline{I}$ and Θ are perfectly good canonical coordinates but they are no longer action-angle variables because $\underline{\Theta}$ appears in H, so the $\underline{I}$ are not constants of the motion.

If tori exist in this system, there must be new action-angle variables $\underline{I}'$, $\underline{\Theta}'$ such that

$$H(\underline{I},\underline{\Theta}) = H'(\underline{I}') , \tag{3.6}$$

and the new variables must be related to the old by a canonical transformation generated by a function $S(\underline{\Theta},\underline{I}')$,

i.e.

$$\begin{pmatrix}\underline{I}\\ \underline{\Theta}\end{pmatrix} \leftarrow \begin{pmatrix}\underline{I} = \nabla_{\underline{\Theta}} S\\ \nabla_{\underline{I}'} S = \underline{\Theta}'\end{pmatrix} \rightarrow \begin{pmatrix}\underline{I}'\\ \underline{\Theta}'\end{pmatrix} . \tag{3.7}$$

Substitution in (3.6) for $\underline{I}$ gives

$$H(\nabla_{\underline{\Theta}} S(\underline{\Theta},\underline{I}'), \underline{\Theta}) = H'(\underline{I}') \tag{3.8}$$

as the condition S must satisfy. Thus the question of the continuing existence of tori reduces to the question of whether (3.8) can be solved.

It is natural to seek a solution S in powers of the perturbation parameter ε, and the "zeroth-order" term must be $\underline{\Theta}\cdot\underline{I}'$. This generates the identity ($\underline{\Theta}' = \underline{\Theta}$; $\underline{I}' = I$). Thus we write

$$S = \underline{\Theta}\cdot\underline{I}' + \varepsilon S_1(\underline{\Theta},\underline{I}') + \ldots \quad . \tag{3.9}$$

We must substitute this into (3.8), where H is given by (3.5):

$$H_0(\underline{I}' + \varepsilon\nabla_{\Theta} S_1 + \ldots) + \varepsilon H_1(\underline{I}' + \ldots, \underline{\Theta}) = H'(E') , \tag{3.10}$$

or (to first order in ε)

$$H_0(\underline{I}') + \varepsilon(\nabla_{I'} H_0(\underline{I}')\cdot\nabla_{\underline{\Theta}} S_1 + H_1(\underline{I}',\underline{\Theta})) = H'(\underline{I}') \, . \qquad (3.11)$$

Now we note that

$$\nabla_{I'} H_0(\underline{I}') = \underline{\omega}_0(\underline{I}') = \text{frequency vector of unperturbed motions,} \qquad (3.12)$$

and

$$H_1(\underline{I},\underline{\Theta}) = \sum_{\underline{m}} H_{1\underline{m}}(\underline{I}) e^{i\underline{m}\cdot\underline{\Theta}} \, , \qquad (3.13)$$

since H_1 is a function of $\underline{p}$ and $\underline{q}$ and these are periodic in $\underline{\Theta}$, and also S_1 is periodic in $\underline{\theta}$ with an arbitrary constant term that we set equal to zero. Hence,

$$S_1(\underline{\Theta},\underline{I}') = \sum_{m\neq 0} S_{1\underline{m}}(\underline{I}') e^{i\underline{m}\cdot\underline{\Theta}} \, . \qquad (3.14)$$

Thus, we equate Fourier coefficients and obtain

$$(\underline{m}=0) \qquad H'(\underline{I}') = H_0(\underline{I}') + \varepsilon H_{1\underline{0}}(I') + \ldots \text{ (the new H)} \qquad (3.15)$$

$$(m\neq 0) \qquad S_{1\underline{m}}(\underline{I}') = + \frac{iH_{1\underline{m}}(\underline{I}')}{\underline{m}\cdot\omega_0(\underline{I}')} + \ldots \qquad . \qquad (3.16)$$

The generator of the new tori is therefore

$$S(\underline{\Theta},\underline{I}') = \underline{\Theta}\cdot\underline{I}' + \varepsilon i \sum_{\underline{m}\neq 0} \frac{H_{1\underline{m}}(\underline{I}')}{\underline{m}\cdot\omega_0(\underline{I}')} e^{i\underline{m}\cdot\underline{\Theta}} + \ldots \quad . \qquad (3.17)$$

It looks as if we are in business - just continue this process to infinite order in ε, and we are left with S and hence the tori. But this is hopelessly naive! The flies in the ointment are the quantities $\underline{m}\cdot\underline{\omega}_0$. If the frequencies $\underline{\omega}_0$ of motion on the unperturbed torus are commensurable - i.e. if the orbits are *closed* and the fundamental frequencies are in *resonance* - then there are always terms $\underline{m}$ for which (cf 2.40)

$$\underline{\omega}_0\cdot\underline{m} = 0 \, . \qquad (3.18)$$

For these $\underline{m}$, the terms in (3.17) are infinite and the series diverges. Worse still, even for incommensurable $\underline{\omega}_0$ it is always possible to find $\underline{m}$'s for which $\underline{\omega}_0 \cdot \underline{m}$ is as small as we like (fig. 18). For ever larger $\underline{m}$'s in the sums in the series (3.17), we must occasionally encounter even smaller $\underline{m}\cdot\underline{\omega}_0$ terms. This makes us doubt whether (3.17) *ever* converges. There are really two doubts: the convergence of the sums $\sum_{\underline{m}}$ and of the series in powers of ε. This is the notorious "problem of small divisors" that plagued celestial mechanics. It does not arise for systems with one degree of freedom.

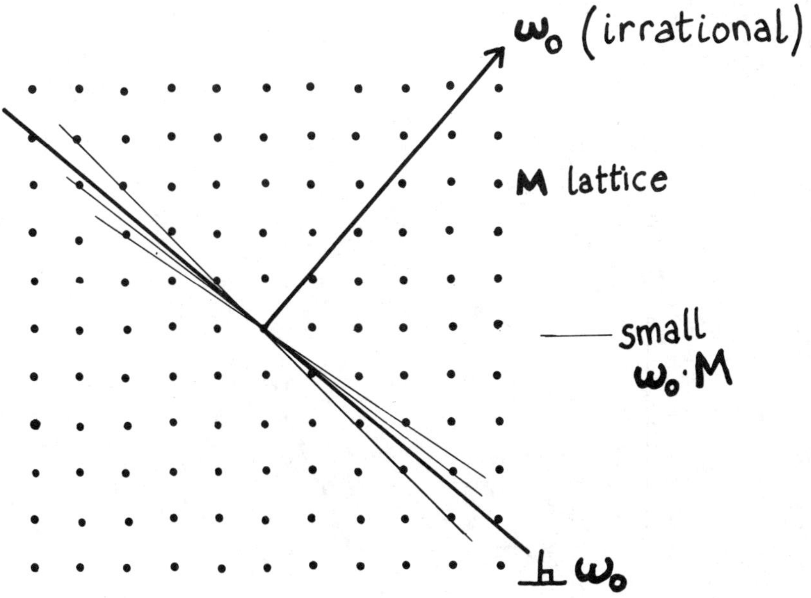

Fig. 18

In astronomical practice the unperturbed motions of interest often lie on tori that are not close to low-order commensurable ones (although there are also astonishing commensurabilities as we shall see later). - i.e. $\underline{\omega}_0 \cdot \underline{m}$ is only small for very large $\underline{m}$ (this means that when considering the Earth's motion, say, we use the fact that its frequency is 11.86 times that of its principal perturber, Jupiter). For these large $\underline{m}$ the Fourier coefficients $H_{1\underline{m}}$ of the perturbation are very small, so that if we cut off the sums before these $\underline{m}$'s are reached, and work to only a few terms in ε, the motion can in practice be predicted for a long time. But the success of these predictions in no way bears on the question of whether tori exist, since this concerns motion over *infinite* times. (For the molecular motions of interest in statistical mechanics, one second corresponds to 10^{13} collisions ("cycles") which for a planetary system would take 1000 times the age of the universe.)

Although perturbation series like (3.17) are familiar to most physicists they are a very crude tool for studying the delicate problems arising from the small denominators. The central feature of KAM's technique is the replacement of (3.17) by a series of successive approximations to the suspected new torus that has a vastly improved convergence. We shall not delve deeply into these technicalities, but only illustrate them with the following example:

Finding the zero of a function

Suppose we want the position x of the zero of a function f(x) (fig. 19).

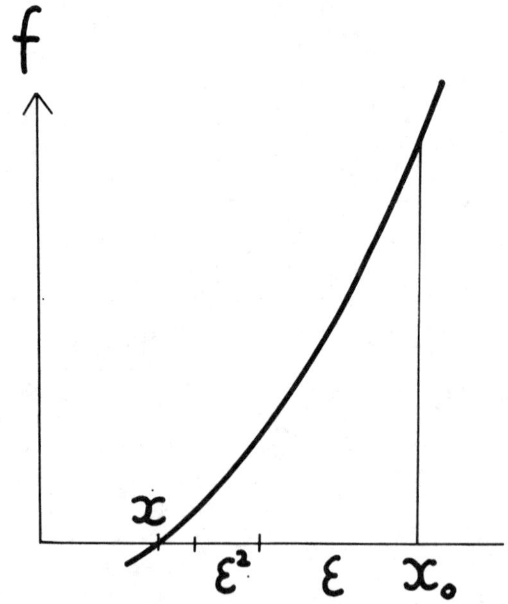

Fig. 19

We start with a guess: the zero is at x_0 - the "unperturbed" value- and proceed to refine the guess. First we use perturbation theory, analogous to (3.17). Write

$$f(x) = 0, \text{ i.e. } f(x_0 + (x-x_0)) = \sum_{n=0}^{\infty} f_n \frac{(x-x_0)^n}{n!} = 0, \quad (3.19)$$

where f_n is the n-th derivative of f at x_0. Rearranging, we get

$$(x-x_0) + \frac{(x-x_0)^2}{2}\frac{f_2}{f_1} + \frac{(x-x_0)^3}{6} + \ldots = \frac{-f_0}{f_1} \equiv \varepsilon \,. \quad (3.20)$$

By standard "reversion of series" we can get the deviation $x-x_0$ from the zero in terms of the "perturbation" ε:

$$x-x_0 = \varepsilon + \varepsilon^2 \left(\frac{-f_2}{2f_1}\right) + \varepsilon^3 \left(2\left(\frac{f_2}{2f_1}\right)^2 - \frac{f_3}{6f_1}\right) + \ldots \,. \quad (3.21)$$

This is the analogue of (3.17) for the mechanical problem.

It is also a pretty silly way to find a zero. Much better is *Newton's method* based on iteration (fig. 20).

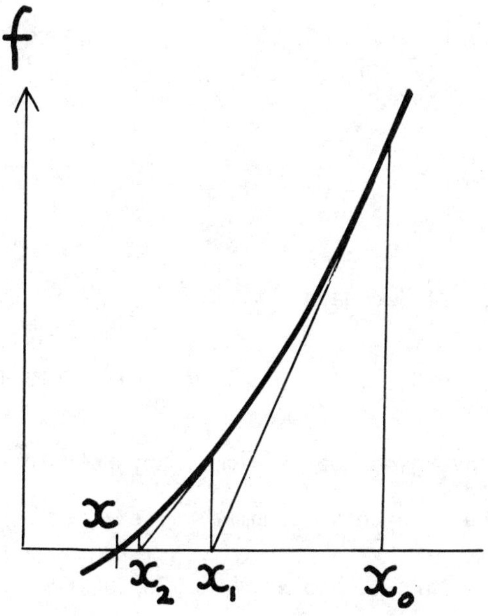

Fig. 20

Starting at x_0, we obtain the next approximation x_1 from

$$f(x) = f(x_0+x-x_0) \sim f(x_0)+(x_1-x_0)f'(x_0) = 0$$

i.e. (3.22

$$\varepsilon_1 \equiv (x_1-x_0) = \frac{-f(x_0)}{f'(x_0)} = \varepsilon \text{ of } (3.20),$$

and the next as

$$\begin{aligned}\varepsilon_2 &\equiv (x_2-x_1) = \frac{-f(x_1)}{f'(x_1)} ,\\ &\vdots \\ \varepsilon_n &\equiv x_n-x_{n-1} = -\frac{f(x_{n-1})}{f'(x_{n-1})} .\end{aligned} \tag{3.23}$$

etc.

How quickly does this converge? We can estimate ε_{n+1} in terms of ε_n by

$$\begin{aligned}\varepsilon_{n+1} &= \frac{-f(x_n)}{f'(x_n)} \simeq \frac{-[f(x_{n-1})+\varepsilon_n f'(x_{n-1})+\frac{1}{2}\varepsilon_n^2 f''(x_{n-1})]}{f'(x_{n-1})+\varepsilon_n f''(x_{n-1})} \\ &\approx -\frac{1}{2}\varepsilon_n^2\left[\frac{f''(x_{n-1})}{f'(x_{n-1})}\right] = 0(\varepsilon_n^2) .\end{aligned} \tag{3.24}$$

$$\therefore \quad \varepsilon_1 = \varepsilon,\ \varepsilon_2 = 0(\varepsilon^2), \varepsilon_3 = 0(\varepsilon^4), \varepsilon_4 = 0(\varepsilon^8), \ldots\ \varepsilon_n = 0(\varepsilon^{2^{n-1}}) , \tag{3.25}$$

so that instead of (3.21) we have

$$x-x_0 = \sum_1^\infty \varepsilon_n = \varepsilon + 0(\varepsilon^2) + 0(\varepsilon^4) + 0(\varepsilon^8) + 0(\varepsilon^{16}) + \ldots \tag{3.26}$$

whose astonishing convergence beats almost any pathology of the approximated function f.

It is amusing to show this by numerical example. Let

$$f(x) = \tan x-1, \quad \text{so } x = \frac{\pi}{4} = .785398164 . \tag{3.27}$$

We take $x_0 = 1$. Then the two methods give

	Perturbation			Newton		
n	x_n	$x_n - x_\infty$	ε^{n+1}	x_n	$x_n - x_\infty$	ε^{n^2}
0	1	.214602	.214602	1	.214602	.214602
1	.837277868	.051880	.0461	.837277868	.051880	.0461
2	.796040059	.010642	.0099	.788180293	.002782	.0021
3	.787025592	.001627	.0021	.785405918	.000008	.000004
∞	.785398164	0	0	.785398164	0	0

The reason for this astonishing "quadratic convergence" is that at each stage f is evaluated at the current approximation x_n rather than at the zero-order approximation x_0 as in the perturbation series (3.21).

Precisely the same device is employed by KAM in the mechanical problem. Each new torus generated by the previous approximation is itself made the basis of the next approximation, rather than (as in 3.17) expressing all approximations in terms of the unperturbed torus $(\underline{I},\underline{\Theta})$ with Hamiltonian $H_0(\underline{I})$. The accelerated convergence thus obtained was subjected to a searching analysis by KAM.

Their central result is that the process of generating "perturbed" tori does in fact converge, for small but finite ε, almost always. Therefore, most trajections continue *for all time* on the tori M of dimensionality N, and do not explore the whole 2N-1 dimensional energy surface. How then can ergodicity come about? What of statistical mechanics? The answer lies in the qualifications "almost always" and "most". For it turns out that unperturbed tori in the neighborhood of those on which orbits are *closed* (or partially closed) are almost all destroyed. These orbits lie on tori with commensurable frequencies, i.e. those for which

$$\underline{\omega}_0 \cdot \underline{m} = 0 \text{ for some } \underline{m}, \text{ or } \underline{\omega}_0 = \underline{N}\omega_b \ . \qquad (3.28)$$

These "destroyed" tori are those giving rise to the small denominators in (3.17). But we have already seen that for *any* $\underline{\omega}_0$ there are, infinitely close by, "rational" tori satisfying (3.28). So, are we any better off? Surely after destroying all rational tori (3.28) and those near them, there are none left at all? Not so. To understand this, and KAM's specification of the "width" of the destroyed regions, we must learn a little of the mathematics of rational and irrational numbers. Then we shall give some striking astronomical illustrations of the KAM theorem, and describe numerical experiments that show what happens in the "gaps" where tori have been destroyed. Finally we attempt to explain how *stochastic* features enter into the motion in these gaps, which therefore provide the gateway to statistical mechnaics, growing with ε and eventually filling the whole phase space.

4. THE ARITHMETIC OF TORUS DESTRUCTION[14]

Consider (3.28) for two degrees of freedom (the simplest non-trivial case). The second of the equations gives the frequency ratio σ of the tori bearing closed orbits as

$$\frac{\omega_{01}}{\omega_{02}} \equiv \sigma = \frac{N_1}{N_2} = \frac{r}{s} \qquad \text{(r and s are integers)} . \qquad (4.1)$$

Thus σ is a rational number (r,s is simply a more convenient notation for N_1, N_2). A torus with incommensurable frequencies has irrational σ and cannot be written in the form (4.1). But it can be approximated *arbitrarily closely* by rational σ's. Take $\sigma=\pi$, for instance. Then a series of approximations can be generated by successive truncations of the decimal expansion:

$$\sigma=\pi = 3.141592654 \ldots \approx r/s = \frac{3}{1}, \frac{31}{10}, \frac{314}{100}, \frac{3142}{1000}, \frac{31416}{10000}, \ldots . \qquad (4.2)$$

The better approximations have larger values of r and s. In fact for these decimal expansions

$$|\sigma - \frac{r}{s}| < \frac{1}{s} . \qquad (4.3)$$

Actually it is possible to do better; to approximate irrational tori much more closely by rationals. The point is that decimals, while useful for computation, are very "impure" representations of numbers σ, in that the arithmetic nature of σ is contaminated with the special properties of the base 10, and the same holds for any other base. The best representation for arithmetic purposes is the *continued fraction* for σ, written as

$$\sigma = a_0 + \cfrac{1}{a_1 + \cfrac{1}{a_2 + \cfrac{1}{a_3 + \ldots}}} \qquad \begin{array}{l}(a_0 \text{ integer } (-\infty\ldots 1,0,+1\ldots)\\ a_1,a_2\ldots \text{natural numbers}\\ \quad (1,2,3,\ldots)\end{array} \qquad (4.5)$$

This is unique, and can be derived by subtracting the integral part a_0 of σ, reciprocating, subtracting the integral part a_1 of the number thus obtained, reciprocating, ...a_2 .... etc. For π we obtain

$$\pi = 3 + \cfrac{1}{7 + \cfrac{1}{15 + \cfrac{1}{293 + \cdots}}} \tag{4.6}$$

The successive approximants of the continued fraction, namely,

$$\sigma_n \equiv \frac{r_n}{s_n} = a_0 + \cfrac{1}{a_1 + \cfrac{1}{a_2 + \cfrac{\ddots}{\cfrac{1}{a_n}}}} \quad , \tag{4.7}$$

define a sequence r_n/s_n of rational approximations to σ. It can be shown that these are *best* approximations in the sense that no rational r/s with $s \leq s_n$ is closer to σ than r_n/s_n. Simple algebra based on (4.7) shows that the sequence always converges to σ, and that the successive r_n/s_n are alternately greater and less than σ (fig. 21).

Moreover,

$$\left|\sigma - \frac{r_n}{s_n}\right| < \frac{1}{s_n s_{n-1}} \quad , \tag{4.8}$$

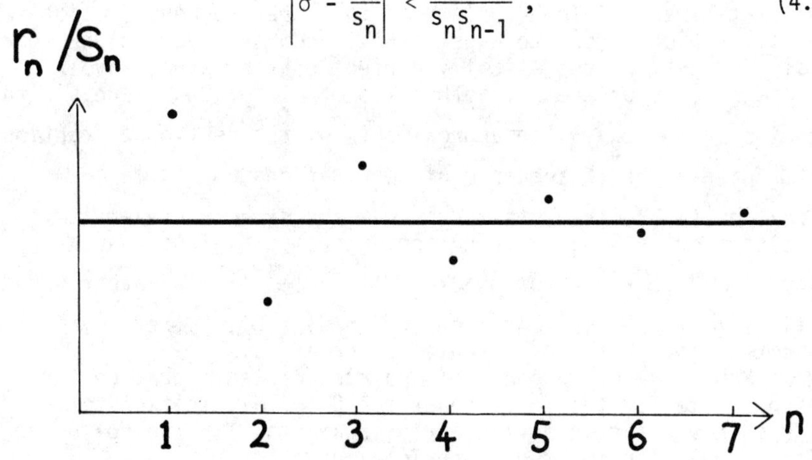

which is much better than (4.3). To illustrate this, we get for π

$$\frac{r_0}{s_0} = 3, \quad \frac{r_1}{s_1} = \frac{22}{7} = 3.1429 \ldots \text{ (2 places)} \quad \frac{r_2}{s_2} = \frac{333}{106} = 3.14151 \text{ (3 places)} \tag{4.9}$$

$$\frac{r_3}{s_3} = \frac{355}{113} = 3.1415929 \text{ (6 places)}$$

The last result was known to Lao-Tze (604-531BC).

Two digressions: (1) obviously the continued fraction for σ converges faster if the sequence $a_0 \; a_1 \ldots$ diverges faster, and vice versa. Thus, the sequence for π, which soon contains the large integer $a_4 = 293$, converges very fast indeed (see 4.9) and the *slowest* convergence, corresponding to the irrational σ worst approximated by rationals, is given by the number

$$\sigma = \cfrac{1}{1 + \cfrac{1}{1 + \cfrac{1}{\ddots}}} = 0.618033989 \ldots = \frac{\sqrt{5}-1}{2} = \text{golden mean} \tag{4.10}$$

(2) Much of the theory of continued fractions was developed in the 17th and 18th centuries in connection with orrery technology. An orrery is a mechanical model of the solar system: turn a handle and a system of gearwheels moves the planets around at the correct proportionate speeds. The problem was that the frequency ratios are not all rational (or, to experimental accuracy, very high-order rational), so that a theoretically perfect gearing would involve inordinate numbers of teeth (tooth ratio ω_1/ω_2 for two planets with frequencies ω_1 and ω_2). The approximants of the continued fraction for ω_1/ω_2 give a "best" sequence of gear rations.

Equation (4.8), then, tells us that for *any* σ whatever it is possible to find rationals r/s differing from σ by less than a quantity of orders s^{-2}. For *particular classes* of σ, sharper bounds exist (i.e. $|\sigma - r/s| < 0(1/s^3)$ or $(\sigma - r/s) < 0(e^{-s})$, etc.), but (4.8) is the best that can be achieved for *all* σ.

Now, KAM prove convergence of the accelerated iteration-perturbation scheme for the torus generator S for all initial tori whose frequency ratio is sufficiently irrational for the following relation to hold (in the two-dimensional case):

$$\left|\frac{\omega_{01}}{\omega_{02}} - \frac{r}{s}\right| > \frac{K(\varepsilon)}{s^{2.5}}\ , \text{ for all integers } r \text{ and } s\ , \tag{4.11}$$

where K is a number, independent of r and s, that tends to zero with the perturbation ε. The tori excluded by (4.11), which are mostly destroyed, are those satisfying

$$\left|\frac{\omega_{01}}{\omega_{02}} - \frac{r}{s}\right| < \frac{K(\varepsilon)}{s^{2.5}}\ , \text{ for some } r \text{ and } s\ . \tag{4.12}$$

This is a more restrictive condition than (4.8) (which applies for all $\sigma = \omega_{01}/\omega_{02}$) and we can expect therefore that after these tori are destroyed there will still be some left.

We can quite easily show that this is in fact the case. Without loss of generality, we can consider all initial tori whose frequency ratios lie in a range of size unity, and moreover take this to be the range $0 \leq \omega_{10}/\omega_{20} < 1$. We delete from this line all segments satisfying (4.12):

0	1/5	1/4	1/3	2/5	1/2	3/5	2/3	3/4	4/5	1 $\rightarrow \dfrac{\omega_{01}}{\omega_{02}}$.
	$\frac{K}{55.9}$	$\frac{K}{32}$	$\frac{K}{15.6}$	$\frac{K}{55.9}$	$\frac{K}{5.7}$	$\frac{K}{55.9}$	$\frac{K}{15.6}$	$\frac{K}{32}$	$\frac{K}{55.9}$	K

Thus we delete $K/s^{2.5}$ about each rational r/s on the range 0 to 1. The total length deleted is thus

$$\sum_{s=1}^{\infty} \frac{K}{s^{2.5}}\, s = K \sum_{s=1}^{\infty} \frac{1}{s^{1.5}} \simeq K \tag{4.13}$$

(s is the number of r-values with r/s on 0 to 1)

which tends to zero with K and hence with ε. This is actually an over-estimate of the "measure" of the destroyed tori, because we have included separately rationals r/s whose deleted neighborhoods overlap. The destroyed tori would still have finite measure if (4.11) were less sharp. More precisely, $K/s^{2.5}$ could be replaced by K/s^{μ} where $\mu>2$. Analogous results were proved by KAM for all degrees of freedom N. The precise value of K is not determined by KAM, nor is it proved what happens to motion in the gaps.

We repeat the main result: in a perturbed system, most orbits lie on tori in phase space. Those that do not, form a small but finite set pathologically distributed in phase space near each unperturbed torus that supported closed or partially closed orbits. From a physical point of view the motion in high-order "gaps" (large s) is hard to study, because these gaps are very narrow

$(O(s^{-2.5}))$, and random *further* perturbations (which must always be present in view of the fact that no real system can be isolated) will probably push the system out of the gap and onto a nearby torus. But the low-order gaps, resulting from low-order resonances among the unperturbed frequencies, are relatively wide and give rise to observable and computable effects as we shall now see.

5. ASTRONOMY OF THE GAPS BETWEEN TORI[11,15,16]

Let us apply the KAM theory to the simplest case of the "plane, circular, restricted three-body problem". Consider three bodies moving under their mutual gravitation: the "attractor" A with mass M, the "perturber" P with mass m and the "test body" T with mass μ (fig. 22).

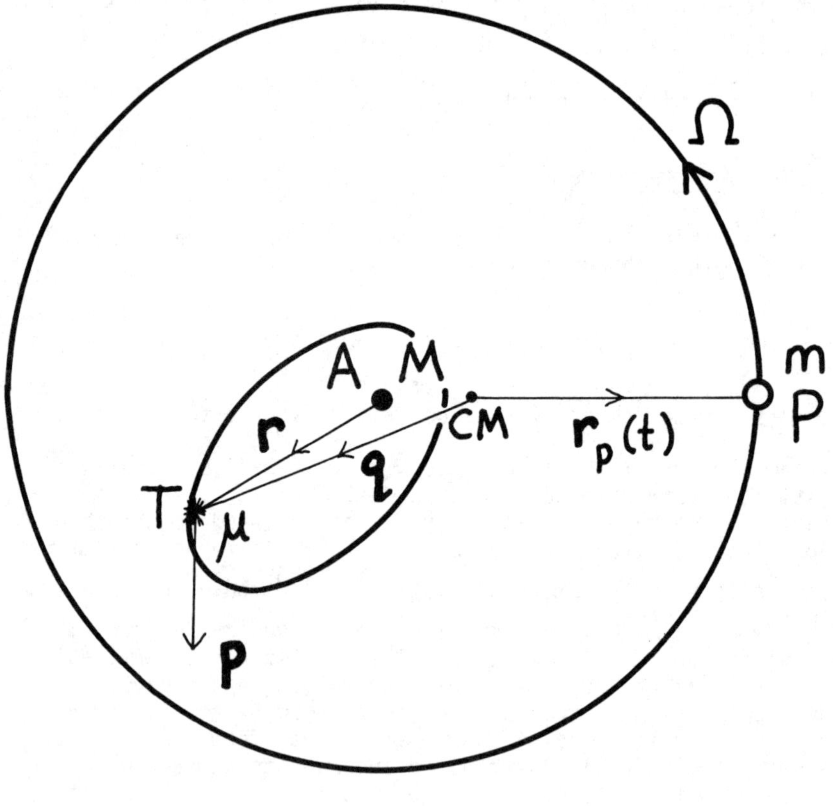

Fig. 22

The "restricted" problem has $\mu \ll m$ and $\mu \ll M$, so that the effect of T on the motion of A and P can be neglected. The two-body motion of A and P is easily solved, and we have to find the motion of T in the known field of A and P. We consider only the case of where A, T, P move in a fixed *plane*, and we let P move about its center of mass with A in a *circle* rather than the more general ellipse. Finally we consider P to be a perturbation on T, i.e. $m \ll M$, so that T's motion is dominated by A. When m=0, T moves in a Kepler ellipse about A. What happens to T when we switch on P? This is the simplest nonintegrable problem in celestial mechanics, and Poincaré realized it to be of crucial importance for theoretical dynamics.

The Hamiltonian for T is

$$H(\underline{q},\underline{p},t) = \frac{|\underline{p}|^2}{2\mu} - \frac{GM\mu}{r} - \frac{Gm\mu}{|\underline{q}-\underline{r}_p(t)|} \quad . \tag{5.1}$$

The coordinate $\underline{q}$ (measured from the center of mass of A and P) has two components because T moves in a plane. $\underline{r}_p(t)$ is the known moving position of the perturbator P (also measured from the center of mass), and $\underline{r}$ is the vector from A to T. It is very awkward to work with this time dependent Hamiltonian, and we can make the system conservative by viewing T's motion from a frame rotating with the angular velocity, $\underline{\Omega}$, of P; in this frame P is at rest (fig. 23). The new Hamiltonian can be shown by elementary methods to be

$$H(\underline{q},\underline{p}) = \frac{|\underline{p}|^2}{2\mu} - \underline{p}\cdot(\underline{\Omega}\times\underline{q}) - \frac{GM\mu}{\underline{r}} - \frac{Gm\mu}{|\underline{q}-\underline{r}_p|} \quad . \tag{5.2}$$

The last term is the non-integrable perturbation H_1, with P's mass m playing the role of the small parameter ε.

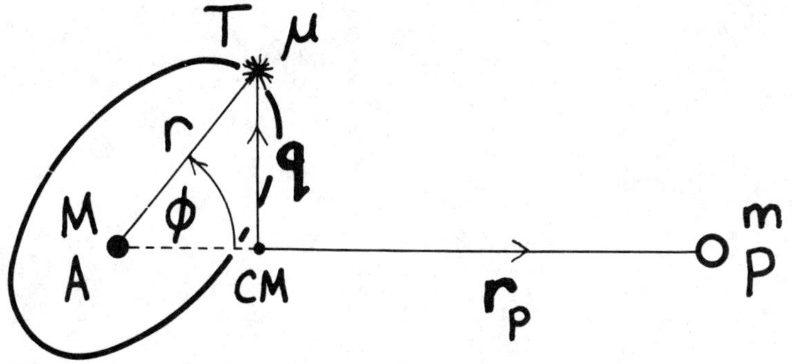

Fig. 23

In polar coordinates r, ϕ, p_r, p_ϕ, the unperturbed Hamiltonian is

$$H_0(\underline{q},\underline{p}) = \frac{p_r^2 + \frac{p_\phi^2}{r^2}}{2\mu} - \Omega P_\phi - \frac{GM\mu}{r} \quad . \tag{5.3}$$

(We have used the fact that $q \to r$ as $m \to 0$.) Neither t nor ϕ appears in this, so the two constants of motion are p_ϕ and H_0 itself. Specifying both of these defines a torus *M*. The actions I_ϕ and I_r of *M* expressed in terms of p_ϕ and H_0 are as follows

$$I_\phi = \frac{1}{2\pi}\int_0^{2\pi} p_\phi \, d\phi = P_\phi$$

$$I_r = \frac{1}{2\pi}\oint p_r dr = \frac{R_e}{\pi}\int_0^{\infty} dr \sqrt{2m(H_0 + I\Omega_\phi + \frac{\mu GM}{r}) - \frac{I_\phi^2}{r^2}} \tag{5.4}$$

$$= -I_\phi + \frac{GM\mu^2}{\sqrt{-2\,(H_0 + \Omega I_\phi)}} \; . \tag{5.6}$$

The Hamiltonian H_0 in action variables for this rotating frame is therefore

$$H_0\,(\underline{I}) = -\frac{\mu^3 M^2 G^2}{2(I_r + I_\phi)^2} - \Omega I_\phi \; . \tag{5.7}$$

The unperturbed frequencies $\nabla_{\underline{I}} H_0$ are

$$\omega_{0\phi} = -\Omega + \frac{\mu^3 G^2 M^2}{(I_r + I_\phi)^3} \; , \; \omega_{0r} = \frac{\mu^3 G^2 M^2}{(I_r + I_\phi)^3} \; . \tag{5.8}$$

This is easily interpreted, since

$$\frac{\mu^3 G^2 M^2}{(I_r + I_\phi)^3} \equiv \omega_T \tag{5.9}$$

is the unperturbed frequency of T's Kepler motion in the *non*-rotating frame, in which both r and ϕ motions have the same frequency because of the well-known degeneracy of motion in the inverse-square law force in which all orbits are closed. Therefore (5.8) can be written in the obvious form

$$\omega_{0\phi} = -\Omega + \omega_T, \quad \omega_{0r} = \omega_T , \tag{5.10}$$

and the frequency ratio is

$$\frac{\omega_{0\phi}}{\omega_{0r}} = 1 - \frac{\Omega}{\omega_T} . \tag{5.11}$$

The KAM theorem can be applied to Hamiltonian (5.2) with m as perturbation. It shows that the motion of T continues to lie on an invariant torus in almost all cases if $m \ll M$. But are destroyed tori near motions of T with rational $\omega_{0\phi}/\omega_{0r}$ or, from (5.11), with rational Ω/ω_T, corresponding to *resonance* between the periods of P and T. In the solar system there are two near-continuous distributions of test bodies T where the gaps between tori are observed.

The first case is the *asteroid belt* between Mars and Jupiter. The attractor is the Sun, the perturber is Jupiter and the test mass is any asteroid. According to (5.11) and KAM, we should expect gaps in the asteroid belt at distances from the Sun where T's unperturbed frequency ω is commensurable with Jupiter's frequency ω_J. These gaps were indeed observed by Kirkwood in 1866. Modern data show them very clearly (fig. 24).

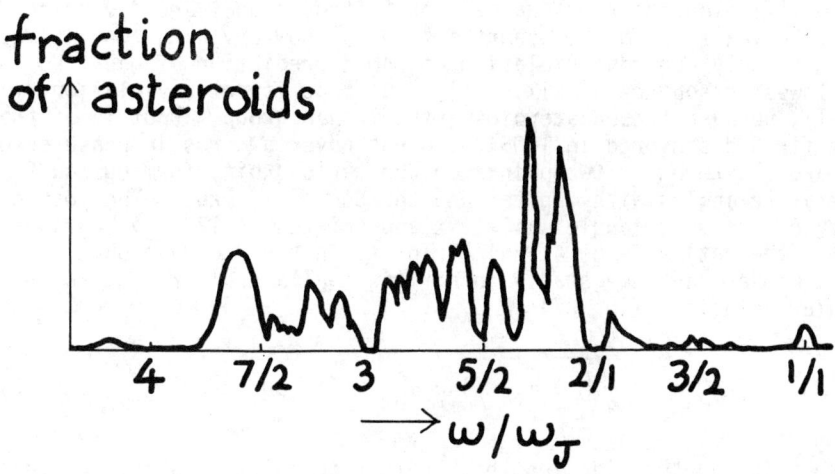

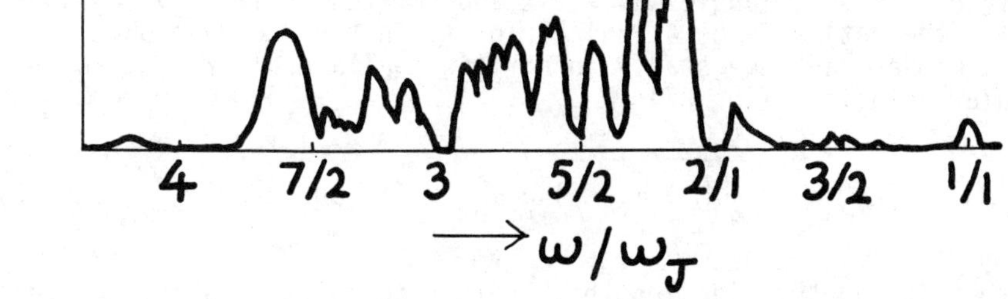

Fig. 24

The early stages of orbit instability beginning as unperturbed circles near the resonance $\omega/\omega_J = 2$ are shown clearly in computations by Franklin in 1973: the orbit turns into an ellipse whose semimajor axis a oscillates in length (fig. 25).

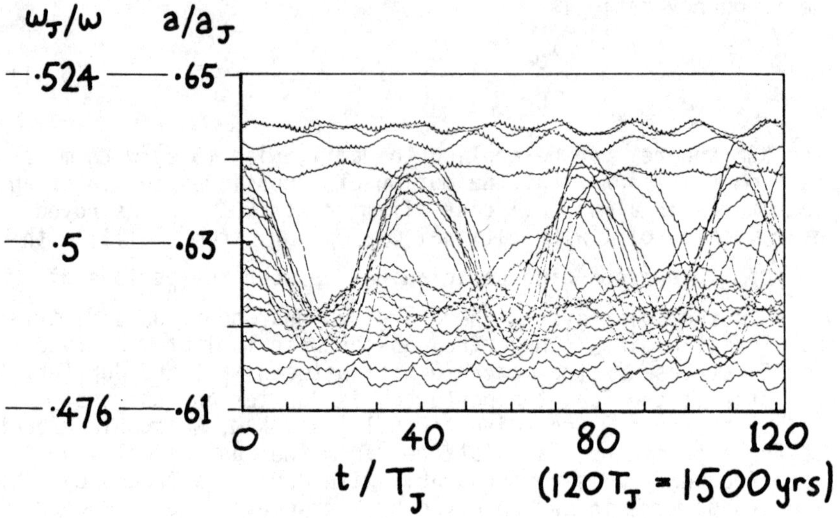

Fig. 25

The oscillations near resonance look as though they are diverging, but 1500 years is far too short a time to show anything conclusive.

The only apparent violation of KAM's prediction of gaps is the lowest resonance of all: 1:1. But this is not a violation at all, because these asteroids - the Trojan group, about 15 of them (the first discovered in 1906) - do not cover a torus in phase space but are clustered at two points on Jupiter's orbit, forming equilateral triangles with Jupiter and the Sun (fig. 26). The possibility of such a triangle was first appreciated in 1772 by Lagrange as (he thought) a mathematical curiosity in the 3 body problem: *any* 3 masses can move stably in a rigid equilateral triangle with angular velocity (fig. 27)

$$\Omega = \sqrt{\frac{G}{r^3}(M + m + \mu)} \approx \omega_J \qquad \text{(in our case)}\,, \qquad (5.12)$$

where r is now the side length (Ω refers to rotation about the center of mass).

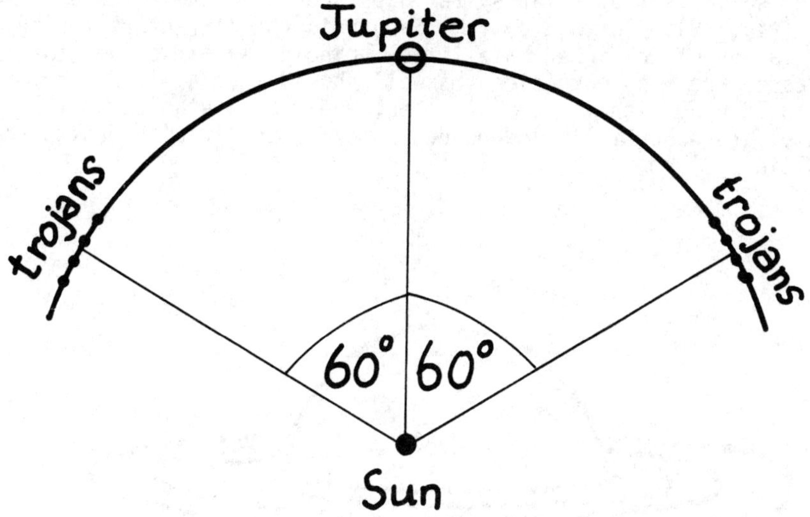
Jupiter
trojans
trojans
60°
60°
Sun

Fig. 26

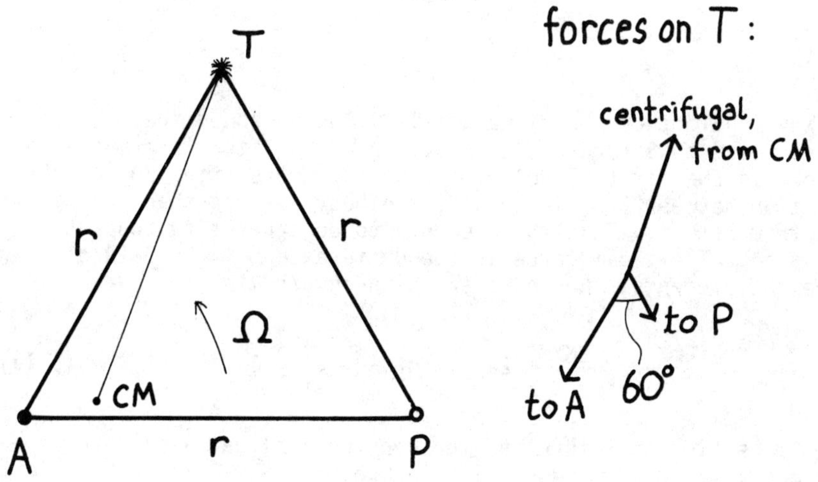
T
r
r
Ω
CM
A
r
P
forces on T:
centrifugal,
from CM
to P
to A
60°

Fig. 27

These "Lagrangian points" on Jupiter's orbit correspond not to a torus'-worth of orbits but to two *isolated* closed orbits.

The second set of solar-system gaps occurs in the *rings of Saturn* (fig. 28). In this system, Saturn is the attractor, the perturber is any of the inner satellites, principally Mimas, and the test masses are ring particles (Maxwell showed that a rigid ring, or a liquid ring with viscous stresses, would be unstable, so the particles are essentially independent, moving in circular orbits and not colliding.)

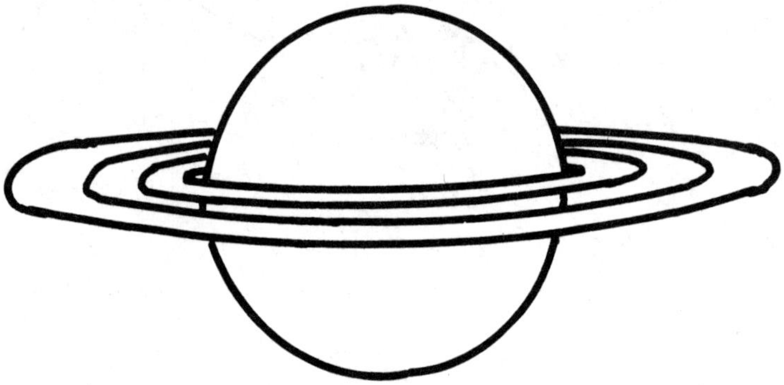

Fig. 28

Why should there be rings at all? Roche showed that any large body so close to Saturn would be disrupted by Saturn's gravity (or not form in the first place). Disruption occurs (fig. 29) if the attraction between two elements of the body is less than the tidal force from Saturn. Take the elements to be spheres in contact (radius a). The tidal force is the difference between Saturn's force on the two elements. Hence, disruption occurs if

$$\frac{Gm^2}{(2a)^2} < \frac{2GMm}{r^3}(2a) , \quad \text{or} \quad \frac{16M}{mr^3} a^3 > 1 . \qquad (5.13)$$

If the densities of Saturn and the hypothetical satellite are assumed equal, this gives ($m \propto a^3$, $M \propto R^3$)

$$r < 16^{1/3} R = 2.52R, \qquad (5.14)$$

$$= 152{,}300 \text{ km} .$$

Janus, the tiny innermost satellite of Saturn (discovered in 1967) just makes it; its distance is r = 156,800 km!

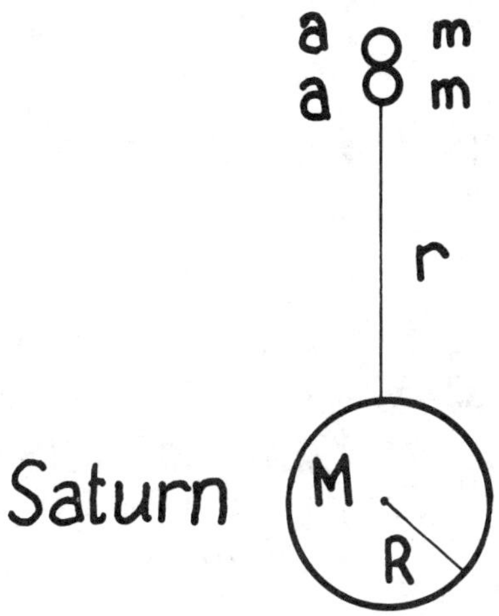

Fig. 29

The ring system lies wholly within the Roche limit, and has the structure shown in Figure 30.

The "3ω mimas" resonance is very close to the gap between ring B and the crepe ring, but "$2\omega_{\text{mimas}}$" and "$3\omega_{\text{enceladus}}$" lie just inside the "Cassini division" between the main A and B rings. However, Frankin has shown (1973) that the effect of the mass in ring B moves both these resonances right into the gap (roughly, it is as if Saturn were a bit heavier).

A present-day theoretical physicist might naturally think of making a computer model of the rings. This leads to difficulties, however, well expressed by Franklin:

"I began with the naive hope that all one had to do was to take a planet, put a large number of massless ring particles around it, introduce the inner satellites.... turn on a machine and after a while discover that the satellite perturbations, particularly ones near resonance, had sculptured the ring. I was set to lease the movie rights....to watch a nice uniform ring quickly being sculptured into Saturn's ring. Well, the movie has yet to be made... The problem is simply that things happen on an enormously slow time scale, not slow cosmologically, but very slow when it comes to the allowances made by administrators of computing facilities".

I want to digress a bit now, to discuss something that confused me while I was checking these Saturnian ring gaps. If one considers Tethys, the next satellite beyond Enceladus, one obtains almost *precisely* the same ring gaps as those from Mimas. The reason is

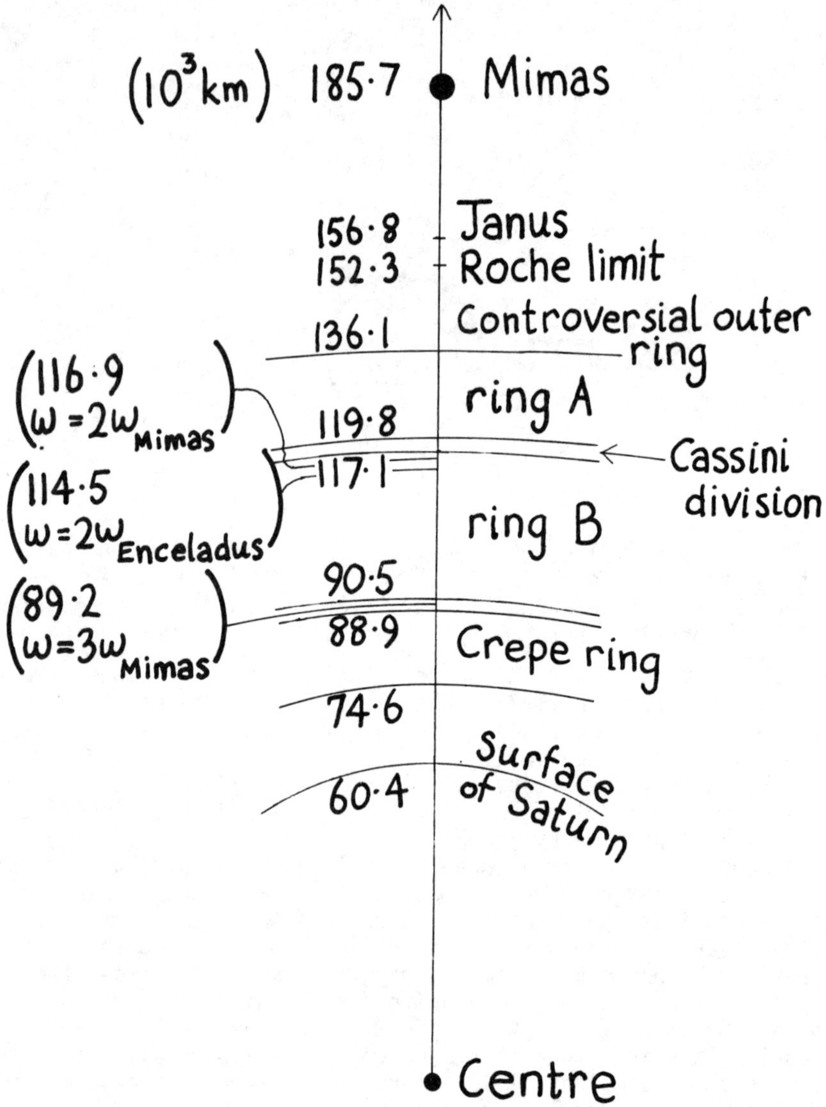
(10^3 km)
185·7
Mimas
156·8
Janus
152·3
Roche limit
136·1
Controversial outer ring
116·9
ω = 2ω Mimas
119·8
ring A
117·1
Cassini division
114·5
ω = 2ω Enceladus
ring B
90·5
89·2
ω = 3ω Mimas
88·9
Crepe ring
74·6
Surface of Saturn
60·4
Centre

Fig. 30

that, to one part in 10^3, $\omega_{Mimas} = 2\omega_{Tethys}$. This seems precisely to contradict the KAM theorem, which as we have seen eliminates tori made up of closed orbits. In reality however, such commensurabilities do not contradict the theorem, for at least two reasons. Firstly, the Mimas-Tethys system is *interactive* rather than perturbative - each body affects the other - whereas in the Mimas-ring system and the Tethys-ring system the ring particles are passive test masses, responding to the satellites' field but not affecting it. Secondly, the KAM theorem does *not* show that closed orbits are destroyed, but only that the tori composed of them are destroyed - *isolated* closed orbits can and do exist (cf. the Trojan asteroids).

Nevertheless, it does seem surprising that of all possible orbits, most of which are unclosed and lie on undestroyed tori, Mimas and Tethys should choose a closed orbit. One's surprise is increased by the recent discovery that commensurability is apparently the rule rather than the exception in the solar system! Careful analysis by Roy and Ovenden in the 1950's and Molchanov in 1966 is claimed to show that this cannot be due to chance. It seems that a *complete* set of commensurabilities $\underline{\omega}\cdot M=0$ exists for each of the following four systems: Saturns' satellites, Jupiter's satellites, Uranus's satellites, and the planets themselves.

Let us examine this for the nine planets: I(Mercury)-IX(Pluto). Then, to a close approximation

$$M_{rn}\omega_n = 0 \qquad (r = 1 \text{ to } 8,\ n = 1 \text{ to } 9)\ ,$$

where

$$M_{rn} = \begin{matrix} 1 & -1 & -2 & -1 & 0 & 0 & 0 & 0 & 0 \\ 0 & 1 & 0 & -3 & 0 & -1 & 0 & 0 & 0 \\ 0 & 0 & 1 & -2 & 1 & 1 & 1 & 0 & 0 \\ 0 & 0 & 0 & 1 & -6 & 0 & -2 & 0 & 0 \\ 0 & 0 & 0 & 0 & 2 & -5 & 0 & 0 & 0 \\ 0 & 0 & 0 & 0 & 1 & 0 & -7 & 0 & 0 \\ 0 & 0 & 0 & 0 & 0 & 0 & 1 & -2 & 0 \\ 0 & 0 & 0 & 0 & 0 & 0 & 1 & 0 & -3 \end{matrix}$$

The error is measured by $\Delta\omega_n/\omega_n$, where ωn is the planet's measured frequency and $\Delta\omega_n$ the deviation when (5.15) is used to calculate the frequency, taking (say) Jupiter's frequency (ω_V) as the standard. Then:

Planet n	$\Delta\omega_n/\omega_n$
I	.0004
II	.0015
III	.0031
IV	.0031
V	0
VI	.0068
VII	-.0118
VIII	.0075
IX	-.0025

It all seems pretty accurate - the commensurability integers in (5.15) are all small (≤ 7) and so are the errors. Perhaps the whole solar system is in resonance! From a historical viewpoint, this discovery discredits the Bode-Titius law, namely:

distance of planet from Sun is

$$\alpha\ 3\mathrm{x}2^n + 4,\ (n = -\infty,1,2,\ldots) \qquad (5.16)$$

and supports Pythagoras' theory of planetary harmony, namely volume of "crystal sphere"

$$\alpha(\text{distance})^3\ \alpha(\text{period})^2 \text{ with } \alpha \text{ an integer .} \qquad (5.17)$$

Suppose this is correct. Then we must ask why is the solar system in a state which seems to have zero a priori probability (closed orbit for the whole motion)? The beautiful theory has been elaborated by Molchanov and Goldreich that the *non-Hamiltonian effects* of viscous dissipation and tidal friction would slowly *pull into resonance* a system started out with random multiply periodic initial conditions (i.e. on some surviving incommensurable KAM torus). The "short" term motion is accurately Hamiltonian, but over cosmological times the system would drift across tori (the "constants" $\underline{I}$ would vary) and into a periodic state as a result of the dissipation. Most of the dissipation is thought to have occured in the early stages of the solar system's evolution as a result of friction from the then relatively dense interplanetary gas. This has now largely condensed, so that the system is almost entirely Hamiltonian but with a motion that was selected by non-Hamiltonian processes. (Dissipation violates Liouville's theorem, so that an ensemble of plane solar systems distributed uniformly over an energy surface in the 9x2x2 = 36 dimensional phase space can eventually all become concentrated on the small set of periodic orbits.)

But all of this is controversial. It may be that as an indication of resonance the equations (5.15) are deceptive. For one thing, only nine frequencies ω_i are involved, even though mutual perturbations break the Kepler degeneracy and lead to two frequencies per planet instead of one. More seriously, a little arithmetic shows that (5.15) implies that all ω_n are integer multiples of a basic frequency

$$\omega_b = \frac{\omega_V}{210} = \frac{\omega_{Jupiter}}{210} , \qquad (5.18)$$

so that the resonant period of the solar system is $210T_{Jupiter}$ ~

2500 years. After this time the planets should have returned to their original configuration. But the errors $\Delta\omega/\omega$ will spoil this prediction, to the extent that the Earth (say) will be about 7 revolutions out of phase after one of these so-called resonant periods. In other words, the commensurability that seemed so close is in reality so poor that the resonance loses phase coherence before one "revolution"! Another criticism of the "goodness of resonance" is based on the claim that almost *any* nine randomly-chosen frequencies can be made to satisfy a low-order resonance equation like (5.5) to comparable accuracy, but this claim is controversial.

6. SURFACES OF SECTION AND AREA PRESERVING MAPPINGS[2,10,13]

Now we describe an important technique, originally suggested by Poincaré, for studying the breakdown of integrability and the motion in gaps between destroyed tori. The method best suits two dimensional problems and we consider only three dimensions here. Then the phase space $\underline{q},\underline{p}$ is four-dimensional and the energy "surface" E (i.e. the "surface" $H=E$) is three-dimensional. (It is a mistake to think of this as being like ordinary three-dimensional position space, because it is non-Euclidean, closed, and may be multiply-connected (cf., two-dimensional surfaces in three-dimensional space).)

The x surface of section S_x is the intersection of E with $y=0$, and has (p_x,x) as coordinates. Then specifying the position of a system on S_x completely specifies its state, apart from a sign, because p_x, x and y $(=0)$ are specified, and for the usual Hamiltonians quadratic in p_i the value of p_y is determined by the condition $H(x,y=0,\ p_x,p_y)=E$ up to a sign. We define this sign to be positive. (The y surface of section S_y is defined in a precisely analogous manner).

A system started out on S_x will subsequently cross it repeatedly, because in a bound system y will repeatedly oscillate through zero and half these zeros will have positive p_y. Thus an initial point $X_0 \equiv (x_0 p_{x0})$ on S_x will subsequently cross S_x at $X_1 \equiv (x_1 p_{x1})$, $X_2, X_3 \ldots$ (fig. 31).

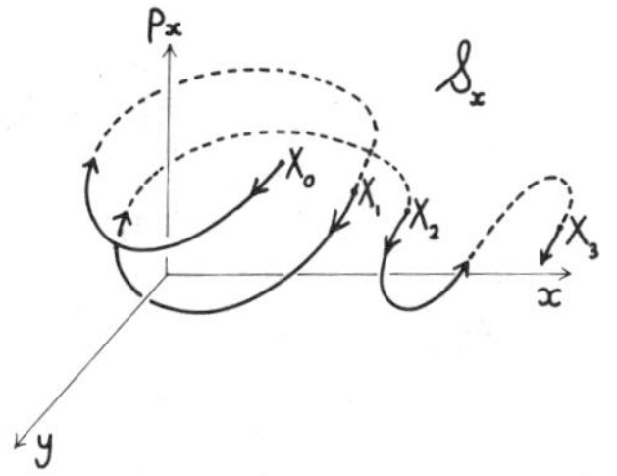

Fig. 31

At each new crossing the whole plane S_x is mapped onto itself, because every point X_0 maps onto some new point X_1. (The time taken for each iteration is different for each point X, but this is irrelevant for our purposes.) Call the mapping T: $X_1 \equiv T(X_0)$ (fig. 32).

An important property of the mapping of S_x is that it is *area-preserving*:

$$\frac{|d\underline{X}_1|}{|d\underline{X}_0|} \quad \frac{|dp_{x1}dx_1|}{|dp_{x0}dx_0|} = 1 . \tag{6.1}$$

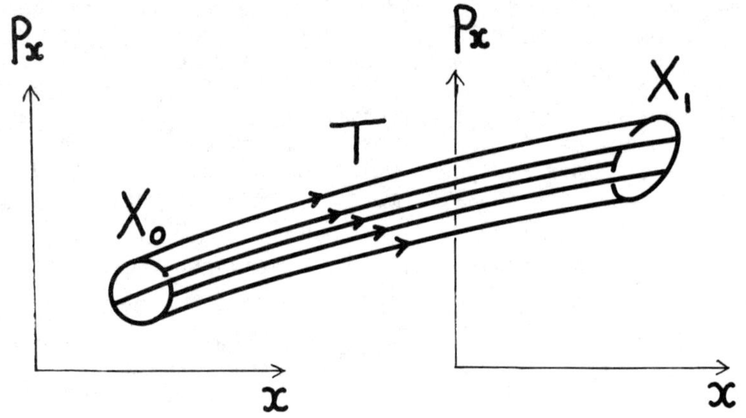

Fig. 32

This follows from the Hamiltonian character of the dynamics between "collisions" with S_x: during the motion, $\underline{q}$ and $\underline{p}$ remain canonical. Therefore, the system's state $(\underline{q}_1, \underline{p}_1)$ at time t, can be obtained from its state $(\underline{q}_0 \underline{p}_0)$ at t_0 by a canonical transformation - the motion itself can be thought of as the unfolding of a (time-dependent) canonical transformation with generator S:

$$\left.\begin{aligned} &\overset{t_0}{(\underline{q}_0 \underline{p}_0)} \leftarrow S(\underline{q}_0, \underline{p}_1 \; , t_o, t_1) \rightarrow \overset{t_1}{(\underline{q}_1{}^1, \underline{p}_1)}, \\ &\underline{p}_0 = \nabla_{\underline{q}_0} S \; , \\ &\underline{q}_1 = \nabla_{\underline{p}_1} S \; . \end{aligned}\right\} \tag{6.2}$$

In particular,

$$p_{x0} = \frac{\partial S}{\partial x_0}, \quad x_1 = \frac{\partial S}{\partial p_{x1}}, \quad \frac{\partial p_{x0}}{\partial p_{x1}} - \frac{\partial x_1}{\partial x_0} = \frac{\partial^2 S}{\partial p_{x1} \partial x_0}, \qquad (6.3)$$

from which (6.1) follows immediately.

The usefulness of these surfaces of section lies in the fact that the iterates $X_1, X_2 \ldots$ of an initial point X_0 reveal whether or not the motion is *integrable*. If it is the system does not explore all of E but only a 2-torus M. This torus intersects S_x in a smooth closed curve C (fig. 33). All iterates X_j must lie on C, and after sufficiently many iterations the form of C usually becomes apparent.

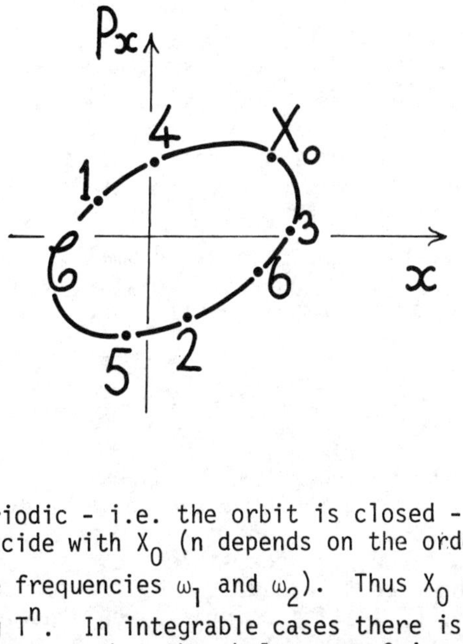

Fig. 33

If the motion is periodic - i.e. the orbit is closed - then some iterate X_n will coincide with X_0 (n depends on the order of commensurability of the frequencies ω_1 and ω_2). Thus X_0 is a *fixed point* of the mapping T^n. In integrable cases there is a torus-full of these closed orbits, so that the whole curve C is made up of these fixed points. Most curves C are, however, sections of irrational tori, and are generated by all the iterates of *any* point X_0. The curves C are *invariant curves* of the mapping, because T maps them onto themselves:

$$T(C) = C \qquad (6.4)$$

For *non-integrable* motion, tori do not exist, so that the system explores a three-dimensional region of E. Therefore its crossings of S_x cover not a curve but a *two-dimensional region of* S_x (Fig. 34).

Then,

$$H = \frac{1}{2}(p_1^2 + p_2^2 + p_3^2) + e^{-(Q_1-Q_3)} + e^{-(Q_2-Q_1)} + e^{-(Q_3-Q_2)} - 3. \tag{6.5}$$

This is really a two-dimensional problem, since $p_1+p_2+p_3$ is a constant of the motion, that reflects our freedom to add an extra rigid translation to any motion. It is possible to make a (non-intuitive) change of variables (canonical transformation) to make the problem mathematically identical to a particle moving in a two-dimensional potential:

$$H(\underline{q},\underline{p}) = \frac{p_x^2 + p_y^2}{2} + \frac{1}{24}[e^{2y + 2\sqrt{3}x} + e^{2y-2\sqrt{3}x} + e^{-4y}] - \frac{1}{8}. \tag{6.6}$$

The equations of motion of trajectories for fixed E were studied by Ford and his collaborators, who presented their results as "x=0" surfaces of section S_y. The perturbation parameter can be thought of as E itself in this case, since for small E the particle can explore only the xy region near the origin, where H is an integrable system - the isotropic harmonic oscillator:

$$H(\underline{q},\underline{p}) \overset{\underline{q}\to 0}{\to} \frac{p_x^2 + p_y^2 + x^2 + y^2}{2} \; . \tag{6.7}$$

Figure 36 shows the equipotential contours.

For the oscillator, the Hamiltonian is integrable. It was thought that as E increased, and more and more of the nonquadratic regions of the well become accessible, more and more "rational" tori would be observed to break up - on the surface of section the "orbits" $X_0, X_1 \ldots$ of initial points X_0 would no longer be on smooth curves C. But this behavior was not found. Instead the iterates stayed firmly on smooth curves for E=1, E=256 (fig. 37), and all E up to the computer's limit of 56000!

It looks as though the system is behaving like an integrable system! The above pictures do not show the individual intersections of the curves C by the orbit $X_0 \ldots$, but the curves themselves. These curves are identical with computer-calculated "analytic" curves obtained by a perturbation theory analogous to that employed in Chapter 2. Analogous, but not the same, because the "unperturbed" Hamiltonian in this case is an equal-frequency oscillator, for which *all* orbits are closed. But a special perturbation theory can be devised to deal with this single massive degeneracy (it is enormously elaborate) and there is a corresponding variant of the KAM

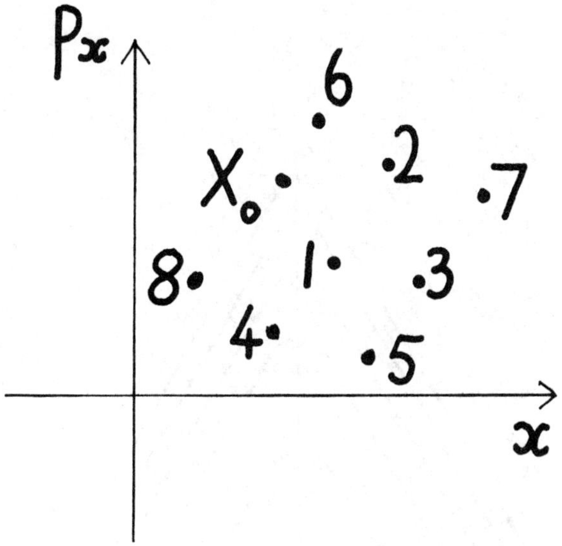

Fig. 34

This will become more and more apparent as the number of iterations increases.

With computers, many iterations of T can be carried out for a range of starting-points X_0 on S_x. Such experiments can never yield proofs of the existence or nonexistence of invariant curves C, because it is always conceivable that high order iterates X_n might diverge from a curve suggested by early iterates, or that iterates apparently randomly filling a part of S_x might really all lie on some definite but complicated curve C. However, when interpreted in the light of the KAM theorem and some further analysis that we shall discuss later, the experiments are extremely instructive.

The first example, that led to a surprise, is the *three-particle Toda lattice*. This is perhaps the simplest non trivial "solid". Three particles move on a "ring with exponential forces" (fig. 35).

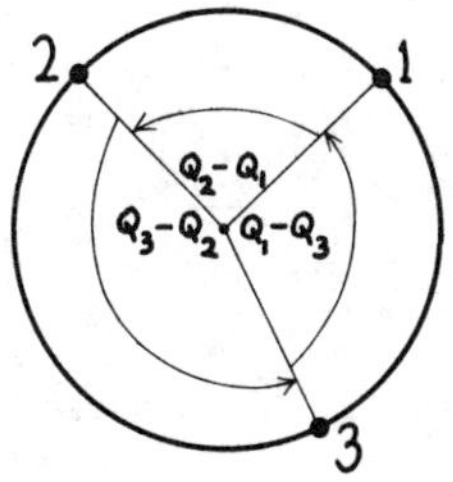

Fig. 35

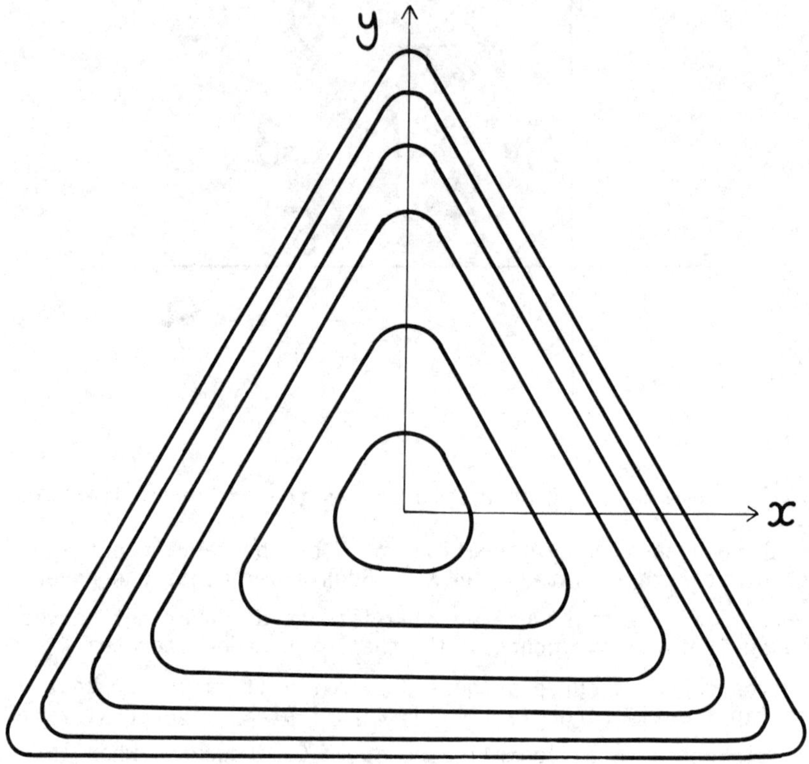

Fig. 36

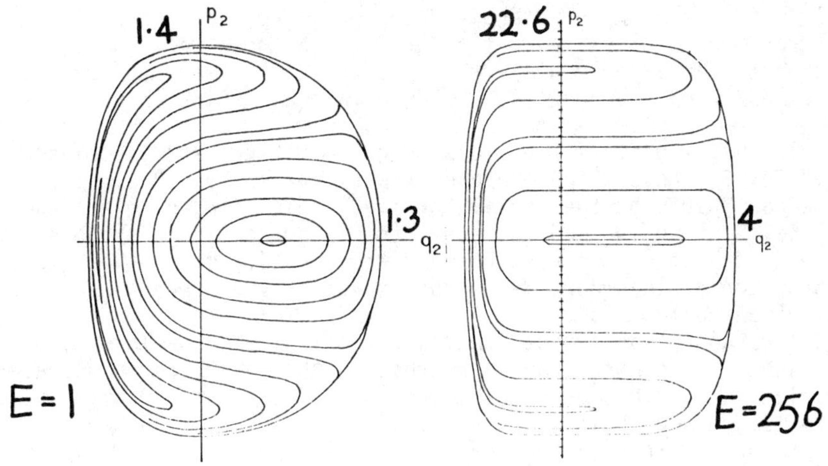

Fig. 37

theory: "most" perturbed orbits will still lie on tori. The perturbation theory gives a series for an extra constant of the motion $F(\underline{p},\underline{q})$, additional to H. It was calculated through eighth-order terms in the p's and q's. These trajectory computations suggest that the series converges - i.e. that F exists.

And indeed it does; Hénon found the following analytic form for it:

$$F(\underline{p},\underline{q}) = 8p_x(p_x^2 - 3p_y^2) + (p_x + \sqrt{3}p_y)\, e^{2y-2\sqrt{3}x} + (p_x - \sqrt{3}p_y)\, e^{2y+2\sqrt{3}x} - 2p_x\, e^{-4y}\,, \qquad (6.8)$$

$$\xrightarrow[\underline{p},\underline{q}\to 0]{} \quad 12(yp_x - xp_y) \quad .$$

Therefore this constant "evolves" out of the angular momentum which is conserved for small E. The invariant curves could have been computed much more easily simply by using (6.6) and (6.8) with $x=0$, eliminating p_x in terms of E and finding $p_y(y)$ in terms of E and F.

Now comes the surprise. Instead of Toda's Hamiltonian (6.6), look at the *Hénon-Heiles potential*, which is its truncation after third-order terms:

$$H(\underline{q},\underline{p}) = \frac{p_x^2 + p_y^2 + x^2 + y^2}{2} + x^2y - \frac{y^3}{3} . \qquad (6.9)$$

This is the system whose potential contours were sketched on fig. 1. It has been employed as a simulation of a three-atom solid, a vibrating triatomic molecule - which is really the same thing - and the "Hartree" averaged field seen by a star moving in the galaxy.

This system differs from Toda's in that it has a "dissociation energy" E=1/6, above which the energy surface is unbounded. Therefore we can "perturb" the oscillator (E=0) only with energies from 0 to 1/6. Bearing in mind, however, the integrability discovered for the Toda lattice for vast E's and the identity of the Hamiltonian's through cubic order, we do not expect very different behavior in the Hénon-Heiles case.

But we do get different behavior. On fig. 38, the left hand column shows the eighth-order-perturbation-theory-generated surfaces of section for various energies computed by Gustavson. Orbits would seem to lie on smooth invariant curves as a result of the new constant of the motion given by the perturbation theory. The right-hand column shows the "exact" trajectories through the surfaces of section. For E=1/24 and E=1/12 the mapping plane is covered with invariant curves identical with those given by perturbation theory.

Above E $\approx$ 1/9, however, Hénon and Heiles found that there are some orbits that do not seem to lie on invariant curves - the tori have been destroyed! The difference in behavior is dramatic: *all* the random looking dots on each of the E=1/8 and E=1/6 pictures were generated by one trajectory as it crossed S_y while wandering ergodically through E. At the same time, some tori persist, even up to (and actually above) the dissociation energy.

The kind of behavior is what the KAM theorem led us to expect. To begin to understand the destruction of tori in more detail, however, we must learn a bit more about mappings.

E = 1/24

E = 1/12

E = 1/8

E = 1/6

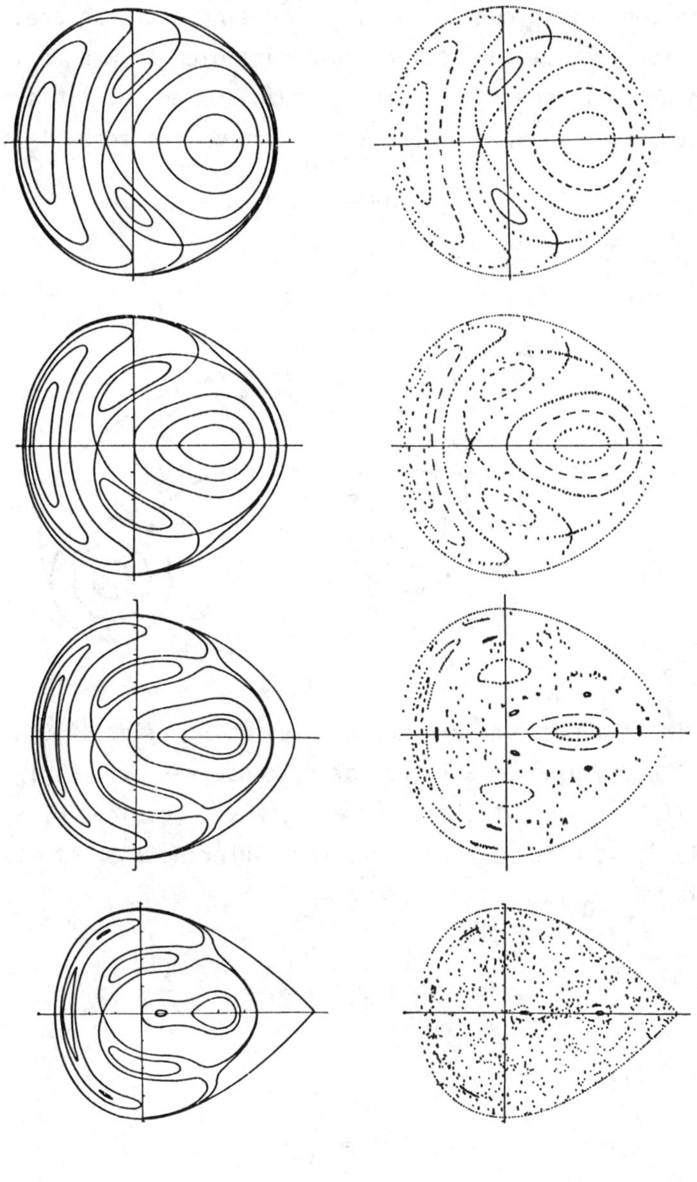

Fig. 38

7. TWIST MAPPINGS, FIXED POINTS, IRREGULARITY*

The tori of integrable systems intersect the surface of section S_x in concentric curves which in the simplest case are closed. It is natural to employ action-angle variables I_x, Θ on S_x, (fig. 39). Then $\rho \equiv \sqrt{2} I_x$ and Θ are polar coordinates on S_x. This is consistent, since the area of the S_x section of the torus I_x is (cf. 2.13).

$$\oint \underline{p}\, d\underline{q} = \oint p_x d_x = \pi \rho^2 = 2\pi I_x . \tag{7.1}$$

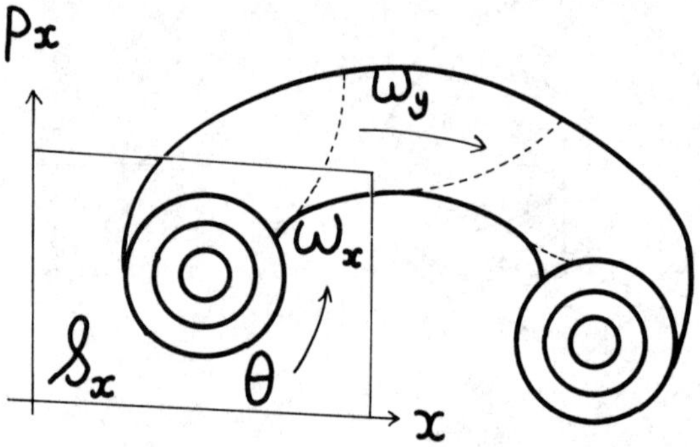

Fig. 39

The invariant curves are now circles on S_x (fig. 40).

The mapping of any trajectory conserves ρ (i.e. I_x) but changes Θ. If $t = 2\pi/\omega_y$ is the interval between crossings of S_x, then the angle Θ_1 at which a trajectory next intersects S_x if its initial angle was Θ_0 is

$$\Theta_0 = \Theta_1 + \omega_x t = \Theta_1 + 2\pi \frac{\omega_x}{\omega_y} . \tag{7.2}$$

Now

$$\frac{\omega_x}{\omega_y} \equiv \alpha(\rho) \tag{7.3}$$

* For reference, see Sec. 20 and App. 27 of Ref. 2, Secs. II.4 and III.6 of Ref. 17, Sec. 5 of Ref. 10.

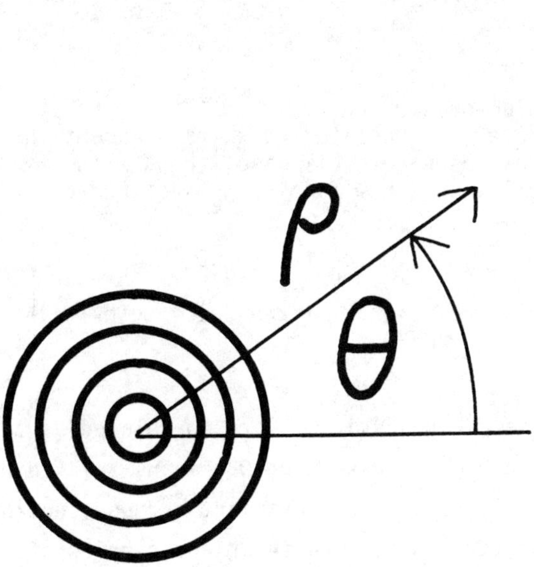

Fig. 40

is the frequency ratio on the torus considered. We have written α as a function of ρ only, because we are working at fixed E, so that specifying ρ (i.e. I_x) also determines the other action I_y. Thus, we have reduced an integrable system to the following "twist mapping" T

$$\left.\begin{aligned} \rho_1 &= \rho_0 \\ \Theta_1 &= \Theta_0 + 2\pi\alpha(\rho_0) \end{aligned}\right\} \begin{pmatrix}\rho_1\\ \Theta_1\end{pmatrix} = T\begin{pmatrix}\rho\\ \Theta\end{pmatrix} . \tag{7.4}$$

$\alpha(\rho)$ is known as the "rotation number". We assume it to be a smooth function of ρ. Circles map to circles, radii to curved arcs through the origin 0 (fig. 41).

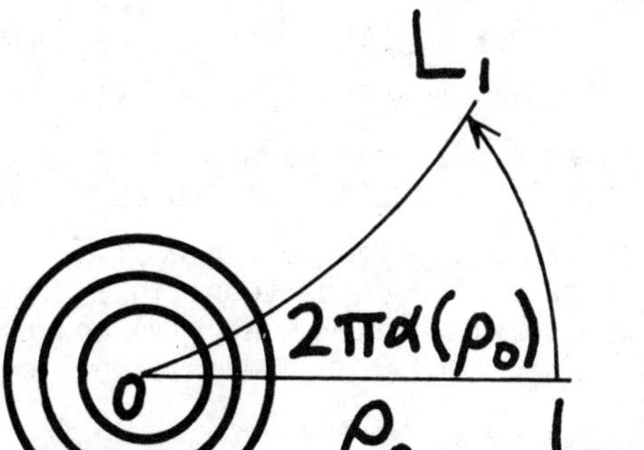

Fig. 41

Obviously T is area-preserving.

So far we have just restated known facts about integrable systems, but in a new language. Likewise, the KAM theorem can be restated: Perturb T into a new mapping T_ε, defined as

$$\left.\begin{aligned} \rho_1 &= \rho_0 + \varepsilon f(\rho_0,\Theta_0) \\ \theta_1 &= \Theta_0 + 2\pi\alpha(\rho_0) + \varepsilon g(\rho_0,\Theta_0) \end{aligned}\right\} \begin{pmatrix}\rho_1\\ \Theta_1\end{pmatrix} = T_\varepsilon\begin{pmatrix}\rho\\ \Theta\end{pmatrix}, \tag{7.5}$$

where f and g have period 2π in Θ_0, are so related as to preserve area, and have f=g=0 at $\rho_0=0$ so that 0 remains a fixed point. Then most points in S_x lie on smooth invariant curves (sections of tori) of T, that are distortions of the invariant circles of T. The only possible exceptions are near "commensurable" tori on which $\alpha(\rho_0)$ was rational. This is more restricted than our earlier statement of the KAM theorem (it applies only to 2 dimensions) and also more general in that T and T_ε can be *any* area-preserving mappings that need not originate in a Hamiltonian, and α, f and g need not be analytic but only smooth in the first 333 derivatives (this number has been reduced since the first proof by Moser).

The advantage of stating the problem in the geometric language of mappings is that it enables us to understand a little more about what happens in the "gaps" where commensurable tori (closed orbits) existed. Consider a "rational" unperturbed circle C,

$$\alpha(\rho_0) = \frac{r}{s} \qquad \text{(r, s integers).} \tag{7.6}$$

Every point on C is a fixed point of T^s, since

$$T^s\begin{pmatrix}\rho_0\\ \Theta_0\end{pmatrix} = \begin{pmatrix}\rho_0\\ \Theta_0 + 2\pi s\,\alpha(\rho_0)\end{pmatrix} = \begin{pmatrix}\rho_0\\ \Theta_0 + 2\pi r\end{pmatrix} = \begin{pmatrix}\rho_0\\ \Theta_0\end{pmatrix}. \tag{7.7}$$

KAM tells us nothing about what happens to this circle of fixed points under T_ε. We might expect them all to be destroyed. This is not the case; in general, an even multiple of s, i.e. 2ks (k=1, 2,3 ...), fixed points remain under perturbation. This is the *Poincaré-Birkhoff fixed point theorem*, which we now prove.

Consider two circles C_+ and C_- between which lies the circle C on which α=r/s. On C_+, α > r/s, and on C_-, α < r/s. Therefore, T^s maps C_+ anti-clockwise, C_- clockwise, and C not at all (fig. 42).

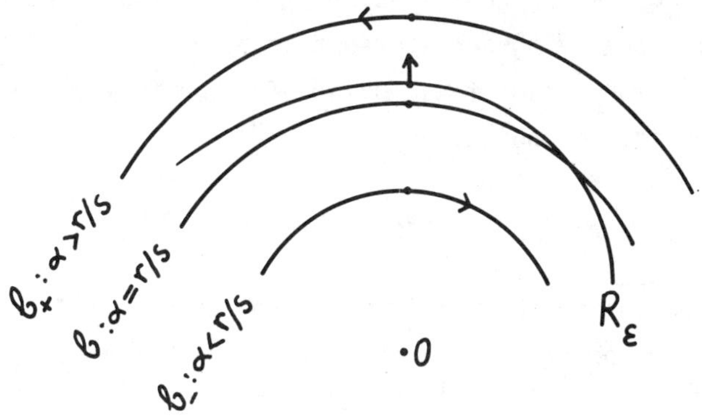

Fig. 42

Under the perturbed mapping T_ε^s these relative twists are preserved if ε is small enough. Thus, on any radius from 0 there must be one point whose angular coordinate is unchanged by T_ε^s. These "radially mapped" points make up a curve R_ε, close to C. Any of the sought-for fixed points of T_ε^s must lie on R_ε. Applying T_ε^s to R_ε generates another curve $T_\varepsilon^s R_\varepsilon$ (fig. 43). This image curve *must intersect* R_ε, because it must have the same area as R_ε and also enclose 0. Ignoring degenerate

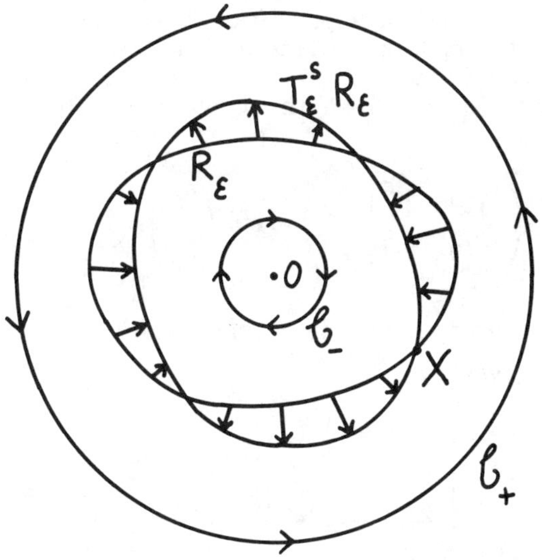

Fig. 43

cases of measure zero in which R_ε touches $T_\varepsilon^s R_\varepsilon$, we find that there must be an *even number* of these intersections.

Each intersection X is a fixed point of T_ε^s. However the orbit of X under T_ε consists of points

$$X,\quad T_\varepsilon X,\quad T_\varepsilon^2 X \ldots T_\varepsilon^s X = X,\quad T_\varepsilon^{s+1} X = T_\varepsilon X, \ldots \tag{7.8}$$

so that *all* the points on this orbit are fixed points of T_ε^s. The orbit has s distinct points on it, so that the number of intersections must be an even multiple of s: There are *2ks fixed points of T_ε^s*.

Therefore, some fixed points are preserved under perturbation. It is obvious from the method of proof that in general not all will be preserved - all but a finite number will be destroyed, so that formal techniques for calculating perturbed invariant curves (tori) must diverge. This can be rigorously proved for polynomial mappings, where TX is a polynomial function of the coordinates x_0 and y_0 of X. Then the fixed point equation for T^s, namely

$$T^s X = X, \tag{7.9}$$

consists of two polynomial equations in x and y, which by a theorem of Bézout have only a finite number of roots (fixed points).

The fixed points we have found are of two basic types, and we now examine these. The type of a fixed point is defined by the form of the nearby invariant curves. To examine these it is sufficient to *linearize* the mapping equations near the fixed point. Without loss of generality we can take the fixed point as the origin O of the mapping plane (q,p). Then the linearized mapping T must take the form

$$\begin{pmatrix} q_1 \\ p_1 \end{pmatrix} = \begin{pmatrix} T_{11} q_0 + T_{12} p_0 \\ T_{21} q_0 + T_{22} p_0 \end{pmatrix} \equiv (T) \begin{pmatrix} q_0 \\ p_0 \end{pmatrix}. \tag{7.10}$$

The nature of O is determined by the eigenvalues λ_1, λ_2 of the matrix T, which are given by

$$\det \begin{vmatrix} T_{11} - \lambda & T_{12} \\ T_{21} & T_{22} - \lambda \end{vmatrix} = 0 . \tag{7.11}$$

Because the mapping is area preserving, det T is unity and

$$\lambda_2 = \lambda_1^{-1} \ . \tag{7.12}$$

Thus the eigenvalues are either *real numbers*, λ and λ^{-1}, or *complex conjugates on the unit circle* (because T is a real mapping); we examine these cases separately.

If λ_1 and λ_2 are *complex*, we can write

$$\lambda_1 = e^{i\alpha} \ , \ \lambda_2 = e^{-i\alpha} \ , \tag{7.13}$$

and T can always be reduced, by a linear change of coordinates, to the form

$$\begin{pmatrix} q_1 \\ p_1 \end{pmatrix} = \begin{pmatrix} q_0 \cos \alpha - p_0 \sin \alpha \\ q_0 \sin \alpha + p_0 \cos \alpha \end{pmatrix} . \tag{7.14}$$

This is just a simple rotation by constant angle α, an obvious special case of the twist mapping (7.4); the invariant curves are circles. In the general case where the λ's are complex, the invariant curves are ellipses (fig. 44), and the fixed point 0 is said to be of *elliptic type*.

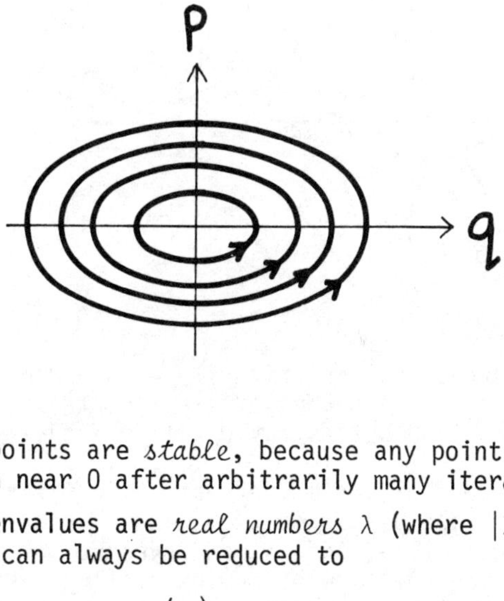

Fig. 44

Elliptic fixed points are *stable*, because any point ("orbit") near to 0 will remain near 0 after arbitrarily many iterations of T.

If the eigenvalues are *real numbers* λ (where $|\lambda|>1$) and λ^{-1}, the effect of T can always be reduced to

$$\begin{pmatrix} q_1 \\ p_1 \end{pmatrix} = \begin{pmatrix} \lambda q_0 \\ \frac{1}{\lambda} p_0 \end{pmatrix} \tag{7.15}$$

in which the invariant curves are hyperbolae p = const./q (fig. 45). Thus, this kind of fixed point is said to be of *hyperbolic type*. Actually there are two sorts of hyperbolic fixed point: the *ordinary hyperbolic point*, where $\lambda>0$ and the iterates of any point remain on one branch:

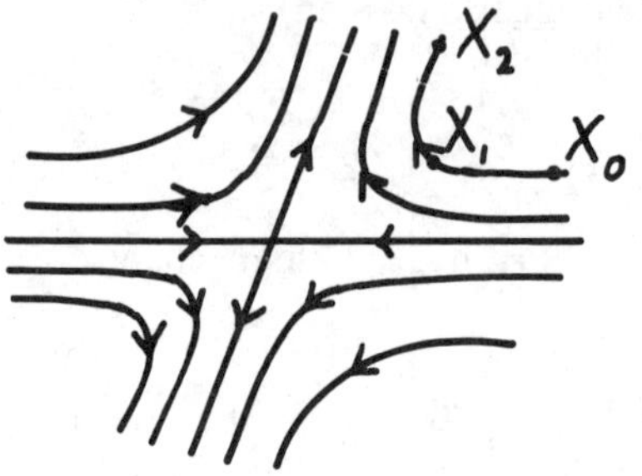

Fig. 45

And the *hyperbolic fixed point with reflection*, where $\lambda<0$ and the iterates jump back and forth between opposite branches (fig. 46).

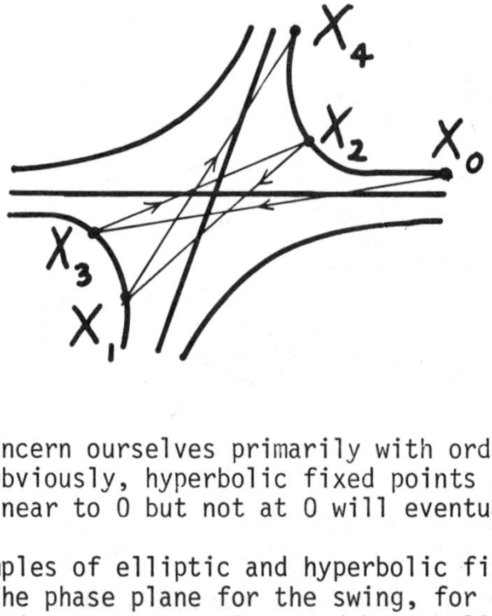

Fig. 46

We shall concern ourselves primarily with ordinary hyperbolic fixed points. Obviously, hyperbolic fixed points are *unstable*, because any point near to 0 but not at 0 will eventually map far away from 0.

Simple examples of elliptic and hyperbolic fixed points occurred in Section 2. The phase plane for the swing, for example, contains elliptic fixed points at p=0, $q=2n\pi$, and hyperbolic fixed points at p=0, $q=(2n+1)\pi$ (fig. 47). Another interesting case is that of plane motion under central force specified by a potential V(r). We take the surface of section as the "radial" plane (r,p_r) through the angle $\Theta=0$. For fixed energy E, the orbit of the moving mass is restricted by the constancy of angular momentum I_2 (Eq. 2.31) to lie on the invariant curves

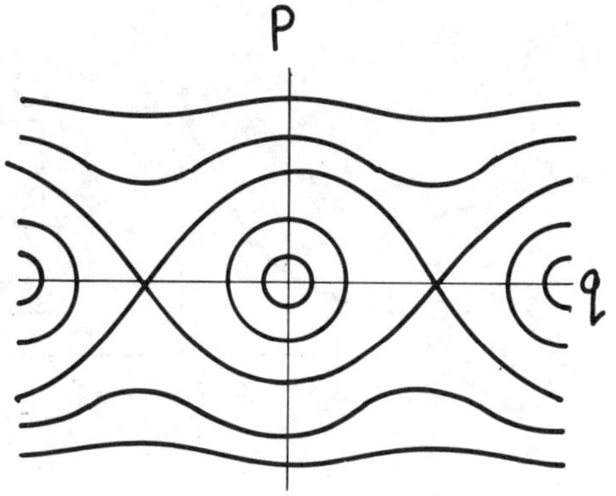

Fig. 47

$$p_r = \pm \left[2m(E-V(r) - \frac{I_2^2}{2mr^2}\right]^{1/2} \qquad (7.16)$$

(cf. 2.33). The form of these curves is determined by

$$V_{eff}(r) = V(r) - \frac{I_2^2}{2mr^2}, \qquad (7.17)$$

and of course by V(r), which we take to be of "Lennard-Jones" type (fig. 48). The system of invariant curves depends on E (fig. 49). The hyperbolic point corresponds to so-called *orbiting*, i.e. positive-energy spiral scattering.

Between the elliptic and parabolic cases lie the special, non-generic, set-of-measure-zero, infinitely-improbable-unless-you-deliberately-set-out-to-create-them *parabolic fixed points*, where $\lambda_1=\lambda_2=\pm 1$; we consider only the "ordinary" type, where $\lambda=+1$. Then T can always be reduced to

$$\begin{pmatrix} q_1 \\ p_1 \end{pmatrix} = \begin{pmatrix} q_0 + Cp_0 \\ p_0 \end{pmatrix} \qquad (7.18)$$

where C is any constant. The invariant curves are straight lines (fig. 50). In the language of fluid mechanics, the parabolic case corresponds to simple shear flow, the elliptic case to a flow with vorticity-dominated stagnation point (the limiting case being a centre of pure rotation), and the hyperbolic case to a strain-rate-dominated flow (the limiting case being pure shear).

In the parabolic case there can actually be a *line* of fixed points (p=0 in fig. 50) . This is familiar! It occurs

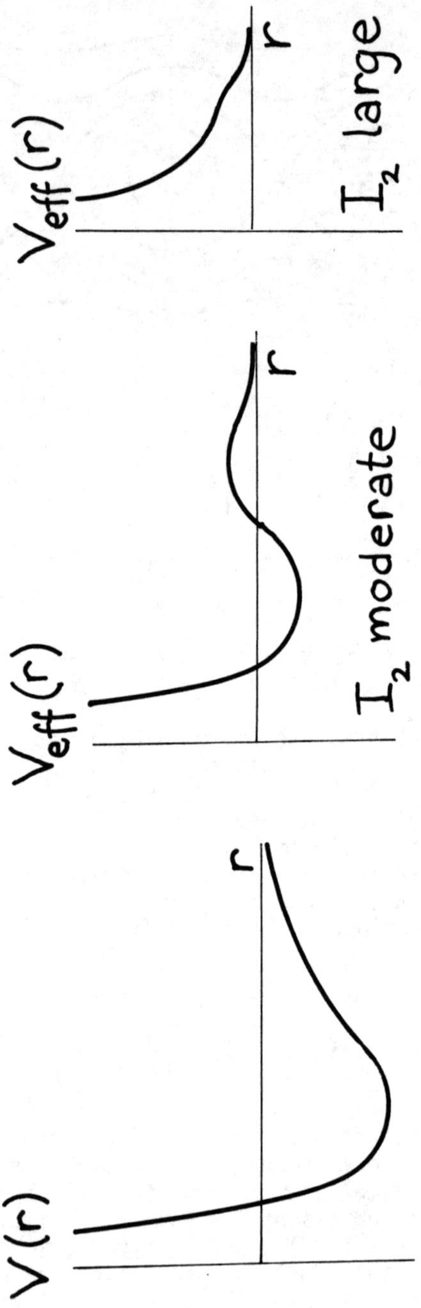

Fig. 48

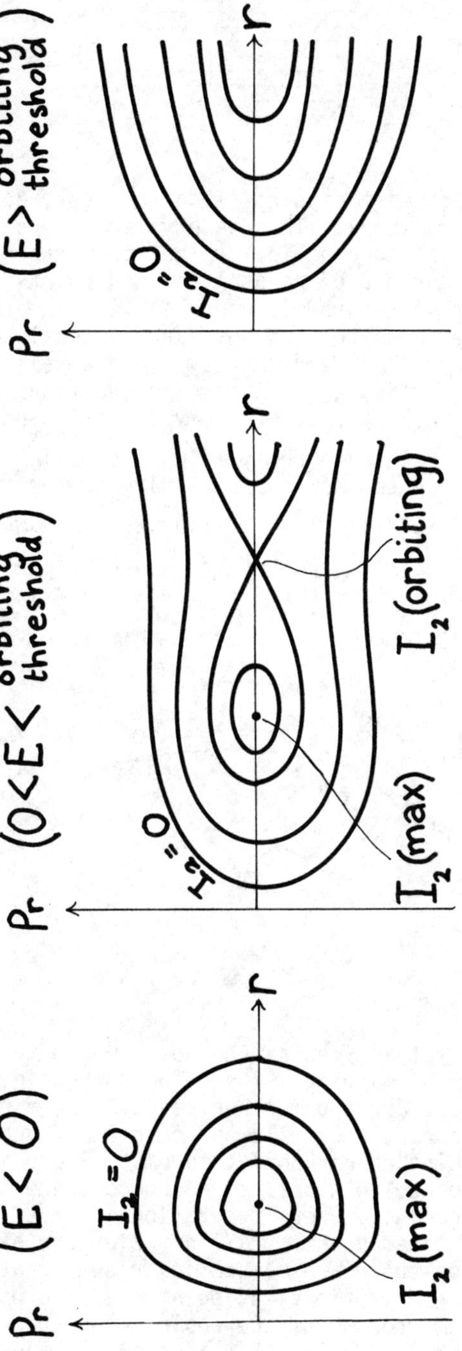

P_r
(E < 0)
$I_2 = 0$
I_2(max)
r
P_r
(0 < E < orbiting threshold)
$I_2 = 0$
I_2(max)
I_2(orbiting)
r
P_r
(E > orbiting threshold)
$I_2 = 0$
r

Fig. 49

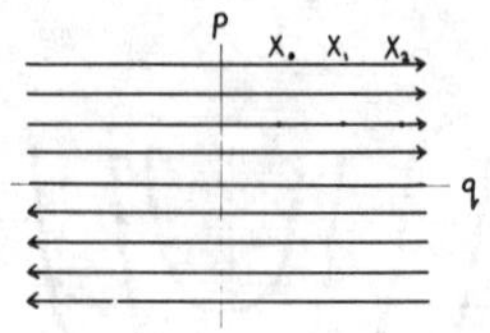

Fig. 50

in the unperturbed twist mapping (7.4) for every curve C whose rotation number is rational. This observation makes the Poincaré-Birkhoff fixed point theorem look rather inevitable, because under perturbation the curve C of parabolic fixed points "generifies" into a finite set of hyperbolic and elliptic fixed points whose eigenvalues λ are close to, but not exactly equal to unity. Hyperbolic *and* elliptic? Yes, and, moreover, in equal numbers. It is obvious from the continuity of arrows in the figure on fig. 43 that the fixed points on R_ε are alternately elliptic and hyperbolic (fig. 51). Because λ is close to unity, the hyperbolic points are *ordinary*. Therefore, we can now add something to the fixed point theorem: *of the 2ks fixed points of* T_ε^s *that remain after the break-up of the curve with rotation number r/s, precisely ks are elliptic, and ks hyperbolic, the two types forming an alternating sequence.*

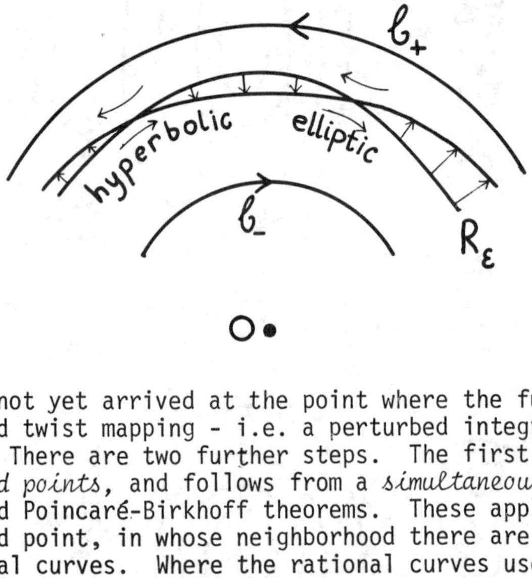

Fig. 51

We have not yet arrived at the point where the full structure of a perturbed twist mapping - i.e. a perturbed integrable system - is apparent. There are two further steps. The first concerns the *elliptic fixed points*, and follows from a *simultaneous* application of the KAM and Poincaré-Birkhoff theorems. These apply to *every* elliptic fixed point, in whose neighborhood there are closed invariant irrational curves. Where the rational curves used to be, a new structure of fixed points, half of which are elliptic, and in whose neighborhood there are closed invariant irrational curves, surrounded by *more* elliptic fixed points, and so on. Each elliptic fixed point is a microcosm of the whole, down to arbitrarily small scales. Schematically, this is shown on Fig. 52.

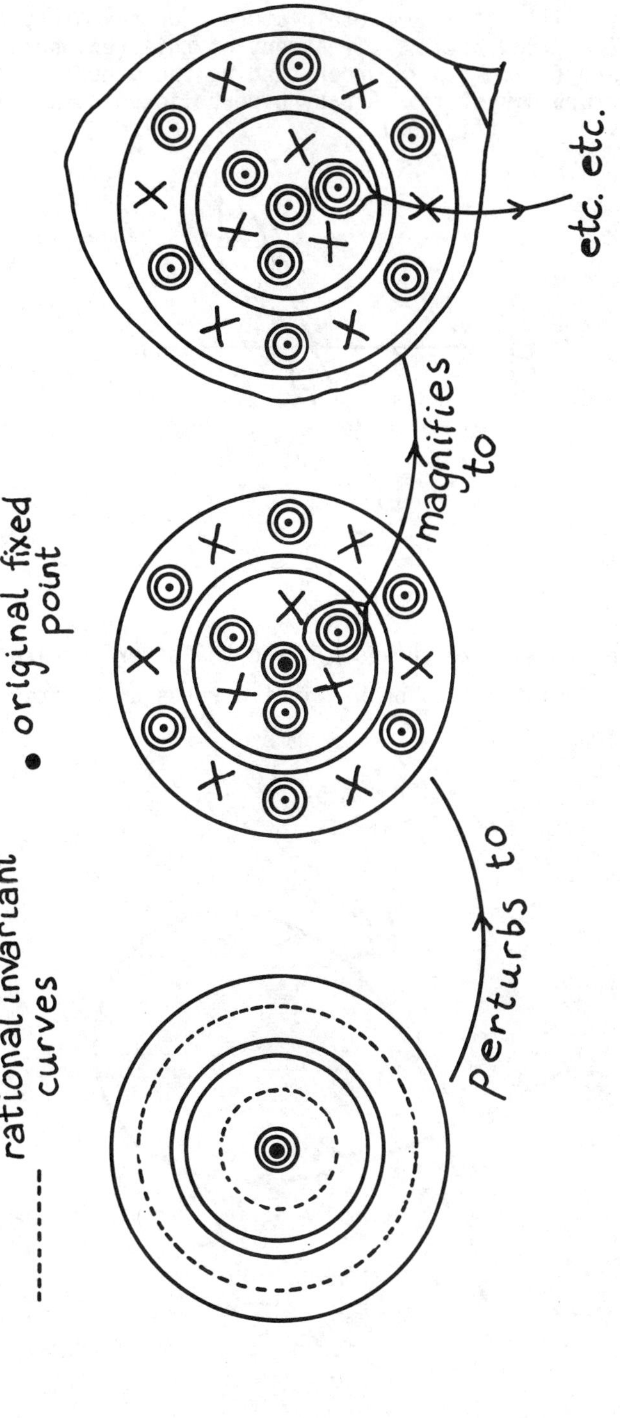
rational invariant curves
original fixed point
perturbs to
magnifies to
etc. etc.

Fig. 52

This is still not a complete picture: we have left gaps around the *hyperbolic fixed points*. Treatment of this region constitutes the second and final step in understanding the generic structure of these perturbed systems. At any hyperbolic point H, four invariant curves meet (fig. 53).

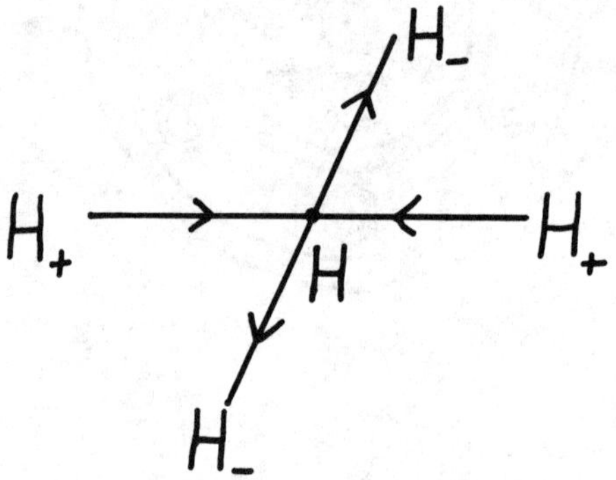

Fig. 53

Two of these are *ingoing* curves H_+, and the other two are *outgoing* curves H_-. A point X lies on H_+ if it arrives at H after infinitely many iterations of T, i.e.

$$T^s X \to H \text{ as } s \to \infty , \quad \text{if } X \text{ is on } H_+ . \tag{7.19}$$

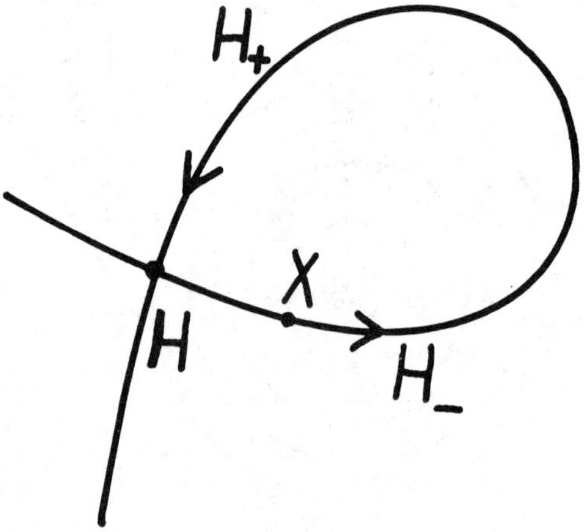

Fig. 54

Similarly, a point X is on H_- if it was at H infinitely many iterations ago, i.e.

$$T^{-s}X \to H \text{ as } s\to\infty \ , \qquad \text{if X is on } H_- \ . \tag{7.20}$$

Points on H_+ approach H infinitely slowly as $s\to\infty$, and points on H_- receded from H infinitely slowly at $s=-\infty$, as can be seen from the "standard form" (7.15) where H_+ was the axis p and H_- the axis q: as $s\to+\infty$ a point with $q_0=0$ maps onto

$$q_s = 0 \ , \qquad p_s = p_0/\lambda^s = p_0 e^{-s\ell n\lambda} \to 0 \ . \tag{7.21}$$

What happens as we follow the arcs H_+ and H_- away from H? For integrable systems the arcs join smoothly, as on figs. 47 and 49. Any point X on this invariant curve (fig. 54) can be thought of as having started out at H, mapped out along H_- and homed in back to H along H_+ after a double infinity of iterations of T. More generally, when H is a fixed point of T^s rather than T, H is one member of a set of s hyperbolic fixed points corresponding to an unstable closed orbit, and the outgoing curve H_- from H joins smoothly with the ingoing curve H_+ belonging to one of H's neighboring hyperbolic points (fig. 55).

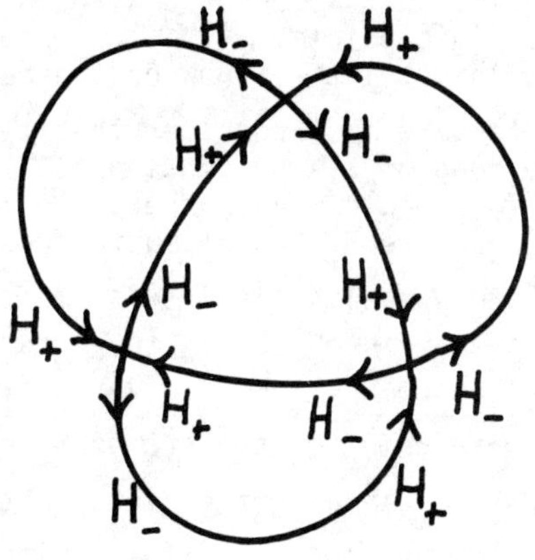

Fig. 55

But it is intuitively obvious (though hard to prove!) that this smooth joining is exceptional, nongeneric. So what happens generically? One thing that cannot happen is for any arc, say H_+, to intersect itself. For if such an intersection could occur, at X, say, then its image TX, and the image TX' of a neighboring point X', must lie close together (fig. 56). But the image TX" of the point X" shown on the sketch, cannot lie near to TX or TX', because it must be preceded by the image of the whole arc X" X'. This contradicts the continuity of the mapping (nearby points map to nearby points) and hence is impossible. Q.E.D.

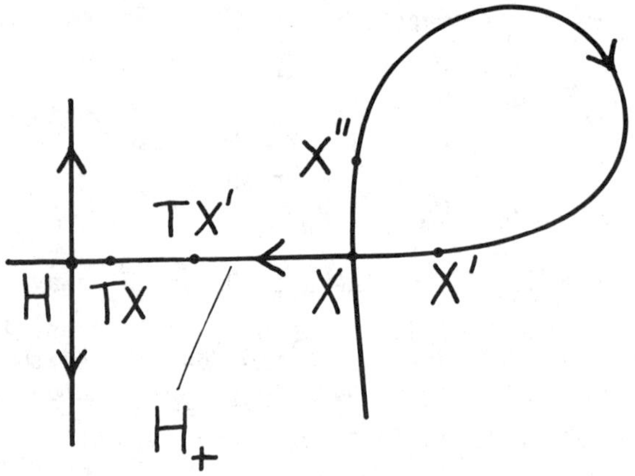

Fig. 56

However, it can and does happen that the arcs H_+ and H_- intersect *one another*. Points X where this occurs are called *homoclinic* if the arcs belong to the same fixed point H or to different points of the same unstable closed orbit, and *heteroclinic* if the arcs belong to two fixed points not associated with the same closed orbit. We consider only homoclinic points.

Consider a homoclinic point X. How do the curves $H_\pm$ continue beyond X? To answer this we must consider the iterates T^s $(-\infty < s < +\infty)$ of the neighborhood of X. By continuity, these iterated neighborhoods must all resemble one another, and in particular the neighborhood of X. Therefore H_+ and H_- must cross in all these neighborhoods: just one homoclinic point is impossible, and the *existence of one implies an infinity of others!* Thus H_+ forms a series of loops intersecting H_-, and vice versa (fig. 58). More than this is true. *Every point* of the arc of H_- between two intersections 1 and 2 of H_+ is a further intersection. This follows from the area-preserving property of T: without further inter-

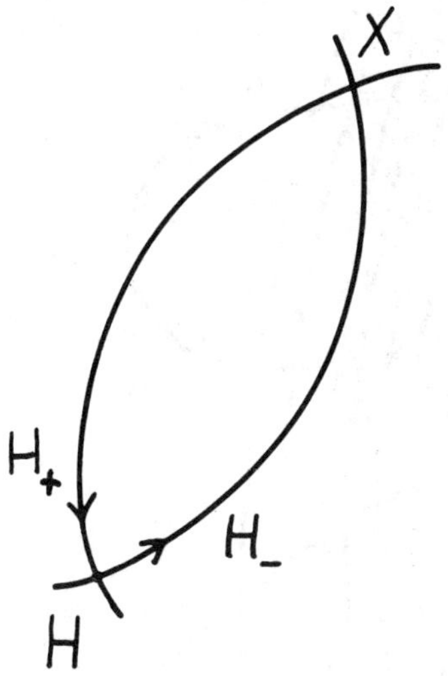

Fig. 57

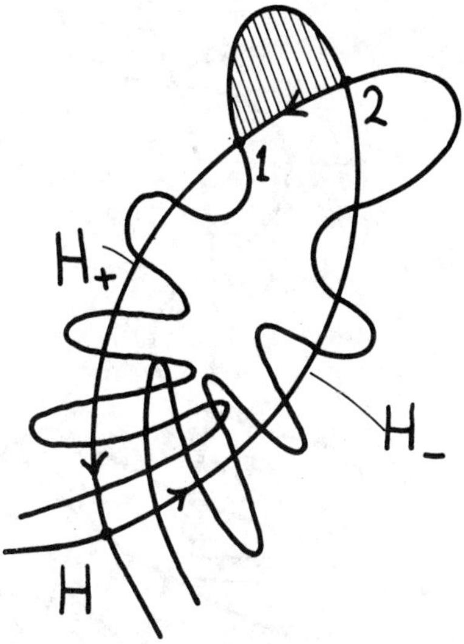

Fig. 58

sections the shaded area would be mapped onto each succeeding loop without change of area, and this could not happen infinitely many times in a finite region. All this was known to Poincaré, who wrote:

"The intersections form a kind of lattice, web or network with infinitely tight loops; neither of the two curves (H_+ and H_-) must ever intersect itself, but it must bend in such a complex fashion that it intersects all the loops of the network infinitely many times.

One is struck by the complexity of this figure which I am not even attempting to draw. Nothing can give us a better idea of the compexity of the three-body problem and of all problems in dynamics where there is no holomorphic integral and (the canonical perturbation) series diverge".

If we do try to draw what happens (following Arnol'd, Moser and others) the result is Fig. 59.

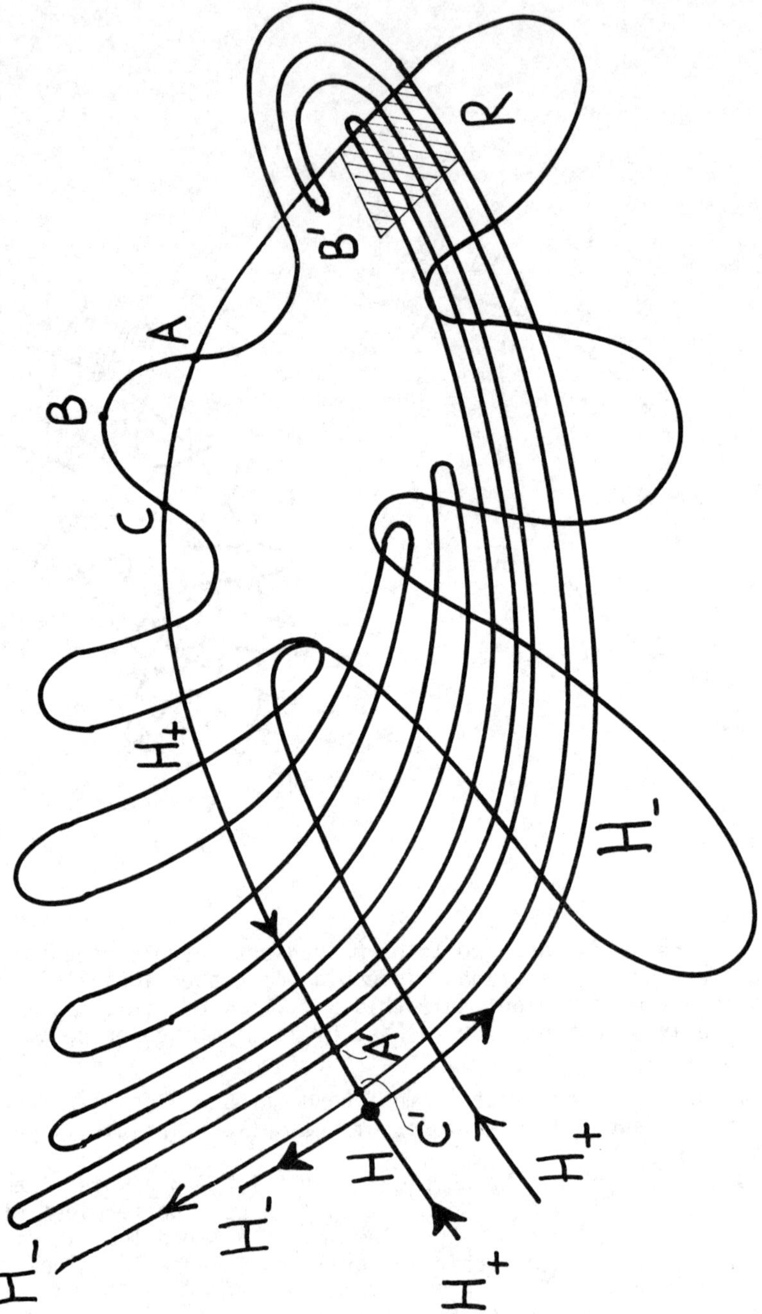

Fig. 59 (After Moser)

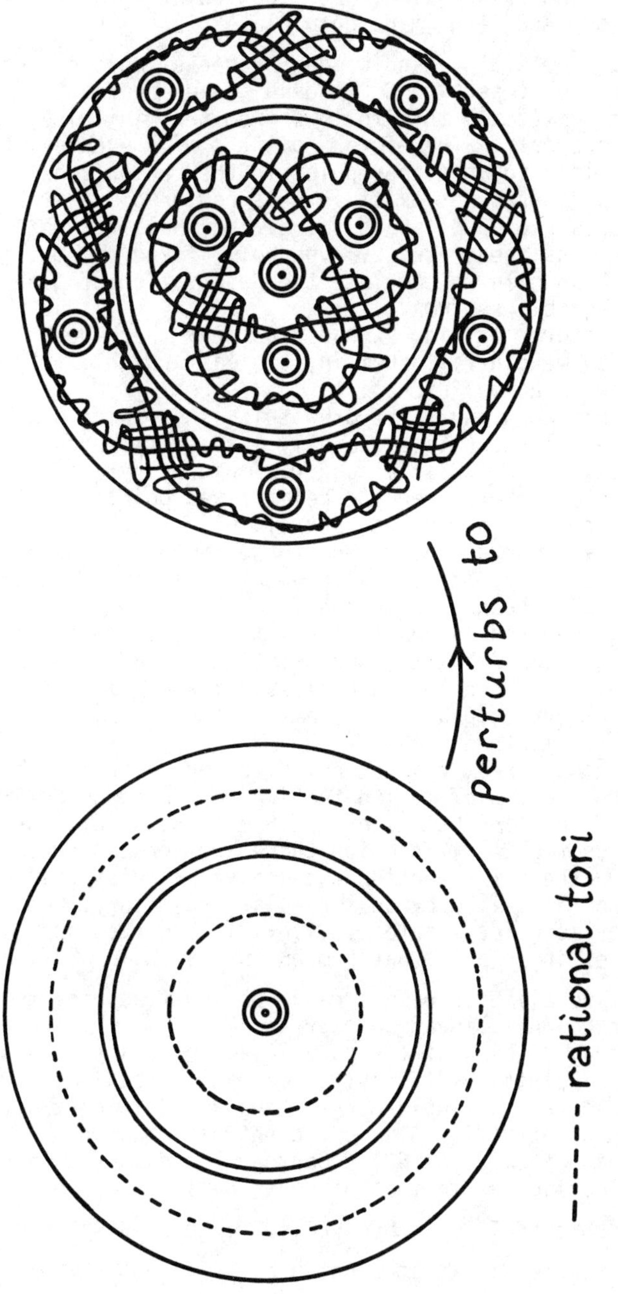

Fig. 60

This revelation of the marvelous complexity of behavior near a generic hyperbolic fixed point marks the entry of a new *stochastic* element into our discussion. For each point on an "early" loop (eg. the loop ABC of H_-)maps onto "late" loops that are ever more convoluted (eg. A'B'C') and wander over ever more extensive regions of the surface of section *S*. Indeed these invariant curves are in a sense *area-filling;* this follows from the "infinitely many intersections" property. Therefore a point X will eventually map arbitrarily close to any other point in the region considered. Smooth invariant curves do not exist in this region of *S*. Tori do not exist in this region of phase space! We shall return to these stochastic regions, generated by hyperbolic fixed points, in the last section. Meanwhile we complete the description of generically perturbed integrable systems.

The pictures on Fig. 52 had gaps near the hyperbolic fixed points. Now we can fill them in, at least roughly (fig. 60): they are dense with homoclinic points.

I do not know who first drew this astonishing picture; but even the detail shown is a woefully inadequate approximation to the true situation. What a wonderful hierarchy! Near each rational invariant curve there are hyperbolic fixed points with associated chaotically wandering curves, and elliptic fixed points surrounded with invariant curves which repeat the whole structure ad infinitum (or ad ħ - see Section 9) - a lacework of intimate intermixing of integrable and stochastic motions.

It must be emphasized that all this is in no sense pathological. It is the *generic situation* for solutions of Hamilton's equations. (For systems that are not both classical and Hamiltonian, however - e.g. dissipative or quantal systems - some of this richness of structure is smoothed away.)

There is a great deal of numerical evidence for the correctness of the picture of motion that we have been describing. The studies by Hénon and Heiles of the Hamiltonian (6.9) are a good example. On the E=1/12 section there are three hyperbolic fixed points. On the E=1/8 section these have all disappeared, and their neighborhood is filled by an irregular trajectory. At the limit of computer accuracy, one of the invariant curves surrounding the rightmost elliptic point has broken into a chain of five "islands"- that is, five elliptic points representing fixed points of T^5, each surrounded by *their* invariant curves.

To study details of the convoluted invariant arcs near hyperbolic fixed points is difficult, because it requires a whole curve to be mapped, not a single point. *Any* arc joining two (or the same) hyperbolic points (fig. 61) must eventually map onto an invariant curve joining them (fig. 62). Contopoulos has used this technique on a problem similar to that of Hénon-Heiles. There is a closed unstable orbit of T^5 with hyperbolic fixed points $X_0 \ldots X_4$ (fig. 63). He starts with a straight arc $X_0 \to X_1$, and then iterates.

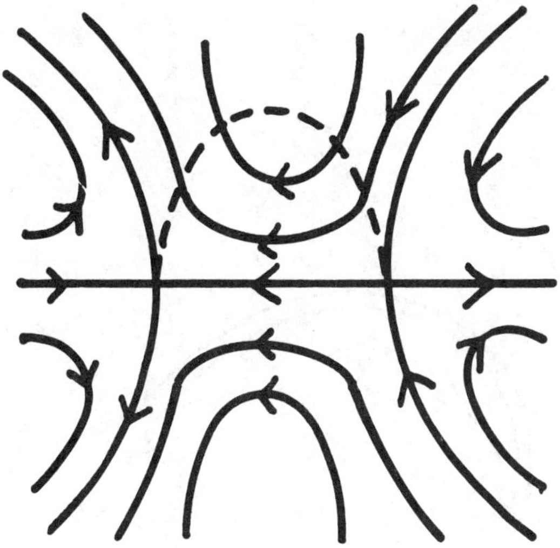

Fig. 61

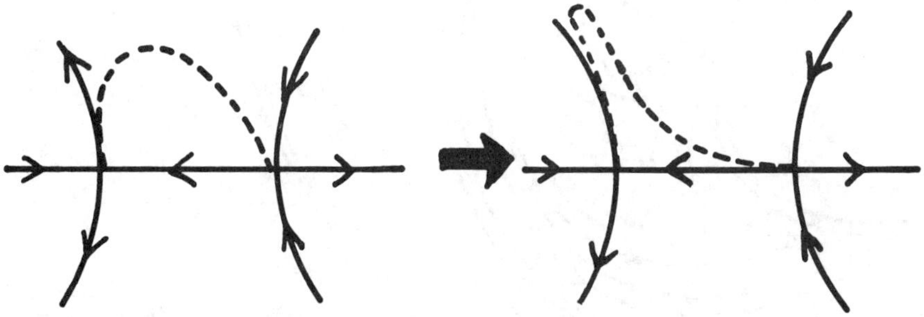

Fig. 62

The complicated outcome, even after just one cycle of iteration, is fully apparent. (Some invariant curves around elliptic fixed points are shown dotted.)

To study the hierarchy of elliptic fixed points is very time-consuming if the twist mapping is generated via a Hamiltonian, because the system's trajectory between intersections with S_x must be determined by solving the equations of motion. It is much easier to employ an *algebraic mapping*, where TX_0 is a simple explicit function of X_0. The simplest perturbed twist mapping is a quadratic addition to a rotation, and Hénon showed that no essential generality is lost by taking T as the following area-preserving mapping:

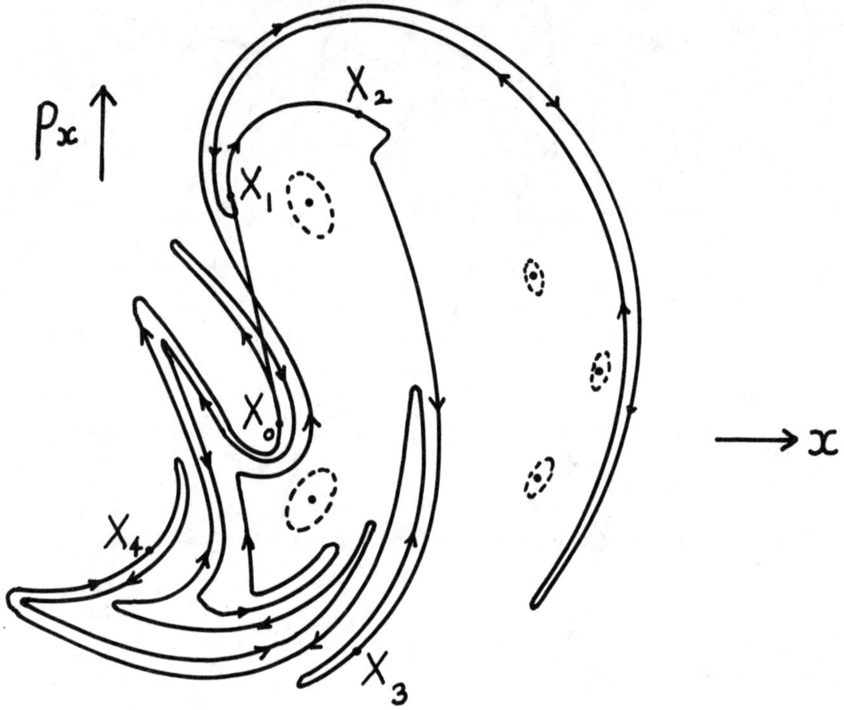

Fig. 63 (after Contopoulos)

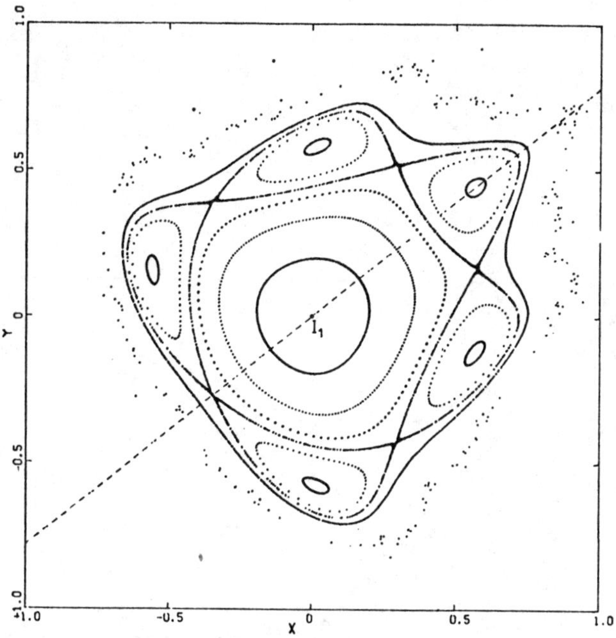

Fig. 64a

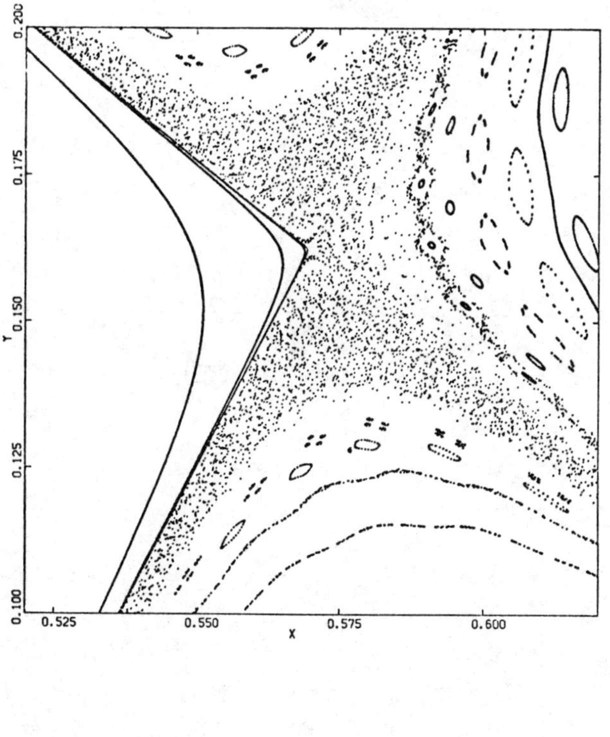

Fig. 64b

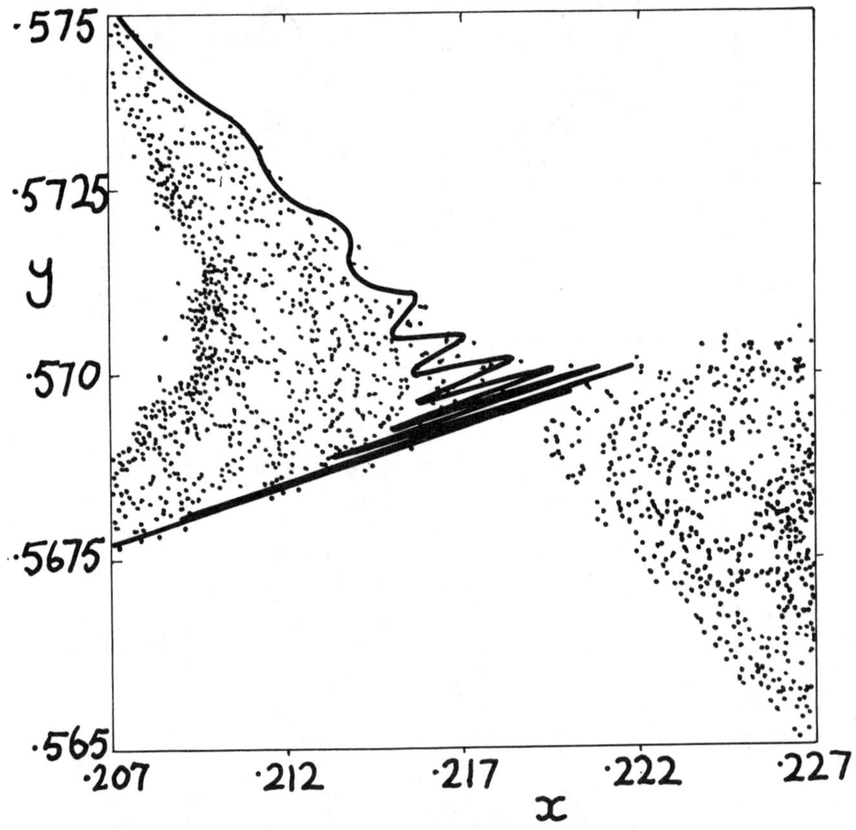

Fig. 65

$$\begin{pmatrix} x_1 \\ p_1 \end{pmatrix} = T \begin{pmatrix} x_0 \\ p_0 \end{pmatrix} = \begin{pmatrix} x_0 \cos\alpha - (p_0 - x_0^2)\sin\alpha \\ p_0 \sin\alpha + (p_0 - x_0^2)\cos\alpha \end{pmatrix}. \qquad (7.22)$$

This astonishingly simple transformation shows all the complexity we have been discussing. The angle α is fixed - it is a parameter of the mapping. The perturbation terms in x_0^2 are small near the origin 0, which is an elliptic fixed point of the unperturbed mapping. According to the KAM theorem, the perturbed mapping will also have closed invariant curves near 0. Far from 0 the perturbation is large and it is not hard to show that all points X escape exponentially fast to infinity under iterations of T.

The interesting region lies at moderate distances from 0. Fig. 64a shows Hénon's calculation of the mapping plane for $\alpha=76.11^0$. Within the non-escaping region there is a chain of 5 elliptic islands around 0, interlaced with 5 hyperbolic points, all these being fixed points of T^5, exemplifying the Poincaré-Birkhoff theorem. The hyperbolic fixed points, which we expect to be the nuclei of irregular motion, look a bit fuzzy. Fig. 64b is Hénon's magnification of the region near the right most hyperbolic point. Not only is "area-exlporing" chaos nearly visible (the dots are 50,000 iterations of a single point!), but several stages of the hierarchy of islands can be seen as well. If the central fixed point is taken as zero-order, then first-, second-, and third-order islands can be seen in these beautiful pictures.

The Hénon mapping (7.22) also displays convoluted invariant curves in the irregular regions near hyperbolic fixed points. Using $\alpha=66.42^0$ Cuthill (unpublished) mapped a short line segment emanating from one of the six hyperbolic points. After 146 iterations the line stretched into the irregular region near the next hyperbolic point and had begun to oscillate, as fig. 65 shows. (The line segment was of length 0.012 and it was necessary to include 55000 points on it in order for the line not to separate into dots when stretched by the mapping. The distance between neighboring hyperbolic points is about 0.5.)

We end this section with a few disconnected remarks. After introducing hyperbolic points with reflection (in whose neighborhoods iterates of a point zig-zag between opposite invariant curves), we never mentioned them again. They can arise for large perturbations by the "conversion" of elliptic fixed points with rotation angles near π (i.e. eigenvalues near -1), and give rise to strongly irregular behavior in generic cases.

Likewise, we have not considered heteroclinic points (intersections of invariant curves through fixed points of different closed orbits). These must always occur along with the homoclinic points that we found to be the source of irregular behavior, because there were infinitely many other rational tori within the destroyed zone near a low-order rational torus. In fact Chirikov has analyzed irregularity in physical terms precisely by considering these "overlapping resonances" and the associated heteroclinic points.

Next, here is an imperfect analogy that might help in understanding the structure in phase space whose sections the perturbed twist mappings represent. Imagine winding a cable starting from a "primary" single loop of thin wire (fig. 66). Cover it with concentric sheaths of plastic (tori). Interrupt this sheathing to find

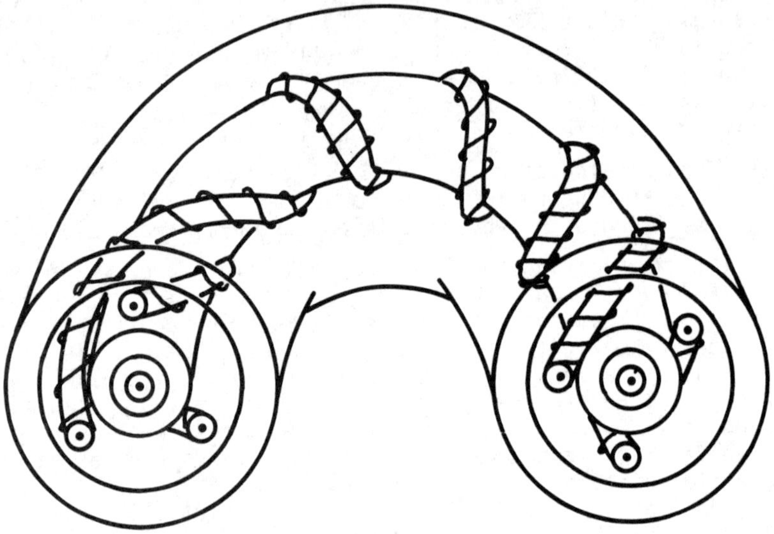

Fig. 66

a secondary sheathed loop in a spiral about the primary, to close after a few windings. On this secondary loop are tertiary, quaternary, ... windings. Continue the interrupted primary sheathings to surround the secondaries. Repeat ad infinitum. When this process has been completed, there will be some vacant spaces. Fill each with an infinitely long, tangled wire. Mathematicians have recently begun to study such structures, called, not surprisingly, "solenoids".

Finally, remember that the last two sections have dealt in detail only with systems with two degrees of freedom. These have a special property: the (two dimensional) tori stratify the (three-dimensional) energy surface E. Therefore the irregular orbits, which wander through regions where rational tori have been destroyed, are trapped between remaining irrational tori, and can explore a region of E which, while three-dimensional, is nevertheless restricted, and in particular, disconnected from other irregular regions in E (fig. 67). For more degrees of freedom, however, the tori do not stratify E. When N=3, for example, the tori are three-dimen-

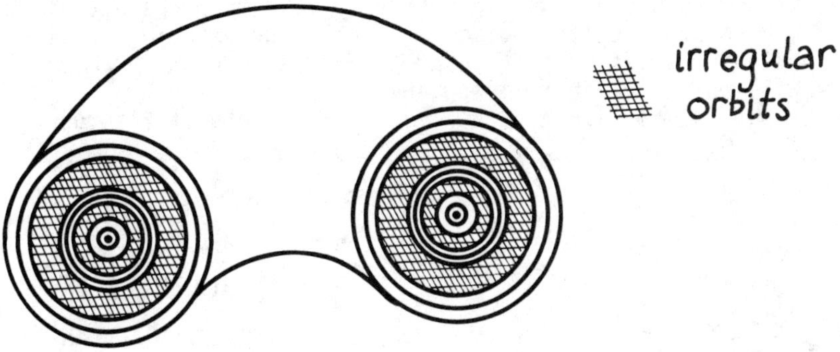

Fig. 67

sional while E is five-dimensional. Then the gaps form one single connected region, and it is conceivable, although unproven, that one single irregular orbit might cover them all. The tori are then a bit like lines in three dimensions (fig. 68). The possibility of so-called "Arnol'd diffusion" of the irregular orbits means that for $N>2$ the existence of invariant tori for perturbed motion is no guarantee of stability of motion, since irregular wandering orbits, that are not trapped, exist arbitrarily close to tori.

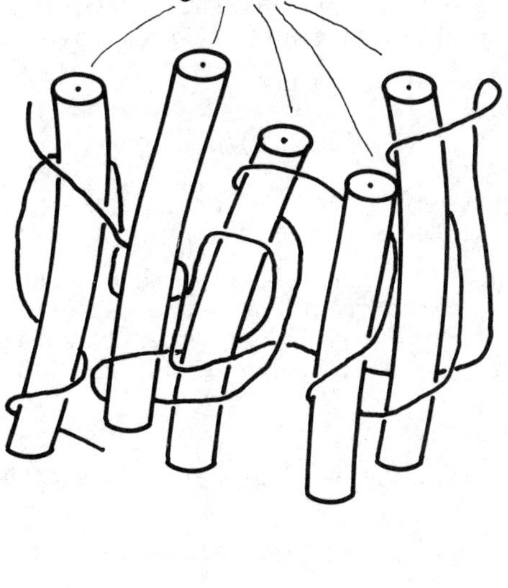

Fig. 68

8. STRONGLY IRREGULAR MOTION*

We have traced the origin of irregular motion in dynamical systems to hyperbolic fixed points of associated area-preserving mappings. If, therefore, we want models for strongly irregular motion, it is obviously sensible to try to find mappings *all* of whose fixed points are hyperbolic. (Clearly such mappings will not be perturbed twist mappings, and the corresponding dynamical systems will not be close to integrable.)

One such example is *Arnol'd's cat map* on the unit 2-torus (fig. 69).

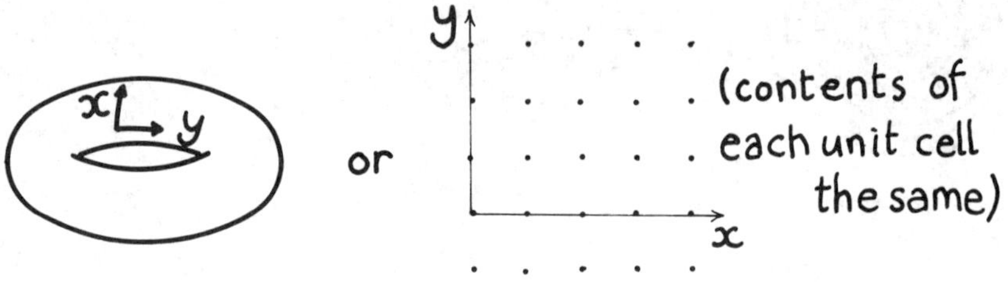

Fig. 69

The map is

$$\begin{pmatrix} x_1 \\ y_1 \end{pmatrix} = \begin{pmatrix} 1 & 1 \\ 1 & 2 \end{pmatrix} \begin{pmatrix} x_0 \\ y_0 \end{pmatrix} \equiv (T) \begin{pmatrix} x_0 \\ y_0 \end{pmatrix} . \tag{8.1}$$

This is a special case of "rational linear automorphisms of the torus", where the matrix can have any integral coefficients. The eigenvalues of T^n and λ^n and λ^{-n}, where

$$\lambda = (3 + \sqrt{5})/2 , \tag{8.2}$$

so that any fixed points of T^n (closed orbits) must be of hyperbolic type. Any point on the torus for which x_0 and y_0 are rational fractions is a fixed point of T^n for some n - the n's becoming larger with the denominators of the fractions - and these rationals are the *only* fixed points (because T has integer coefficients). Thus (0,0) is the fixed point of T, and (2/3, 1/5) and (3/5, 4/5) are fixed points of T^2.

---

* See also Sec. 3 of Ref. 10, Ch. 3 of Ref. 17, and Ch. 1-3 of Refs. 2, 6 and 7.

The mapping (8.1) shears each unit cell as shown in fig. 70.

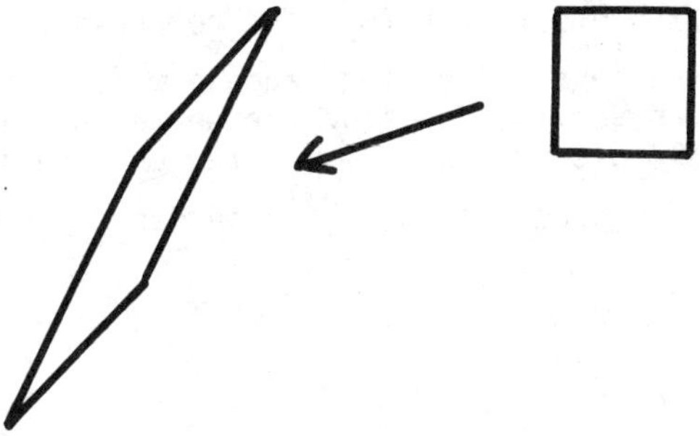

Fig. 70

What this does to a cat on the torus is shown in fig. 71. After just two iterations, the cat is wound around the torus in complicated filaments. Any small area element will ultimately (after T^{∞}) wrap densely round the torus, because its behavior under T repeats in microcosm that of the unit cell: it stretched by λ in one direction, and contracts by λ in a perpendicular direction (the angle, 58.3^0, made by the stretch axis with Ox, has the golden section $(\sqrt{5} + 1)/2$ as its tangent). The disintegration of the cat arises from the unstable, hyperbolic, nature of T, which causes initially close points to map far apart.

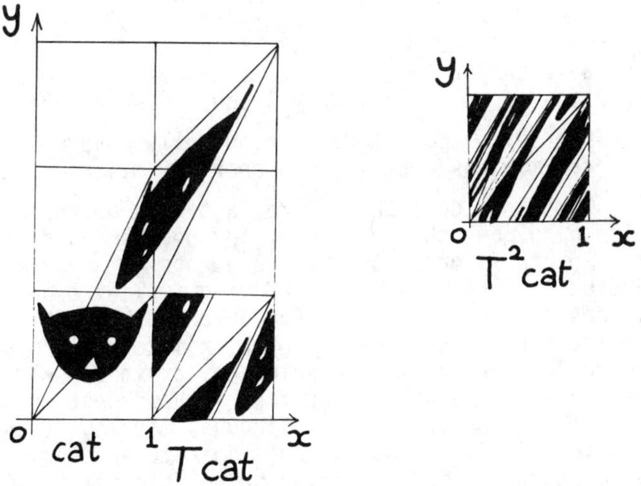

Fig. 71 (After Arnol'd and Avez).

Arnol'ds cat map has *homoclinic points* - intersections of the ingoing and outgoing curves H_+ and H_- for a hyperbolic fixed point - and we found earlier that these are associated with irregular behavior. To find the homoclinic points of the fixed point (0,0) of T, simply draw the axes of stretch (H_-) and compression (H_+) from (0,0). These are irrational directions and so wrap densely round the torus (fig. 72), never intersecting themselves but intersecting one another infinitely often. They also intersect the invariant curves of the other fixed points (of $T^{n\neq 1}$) in densely distributed *heteroclinic* points.

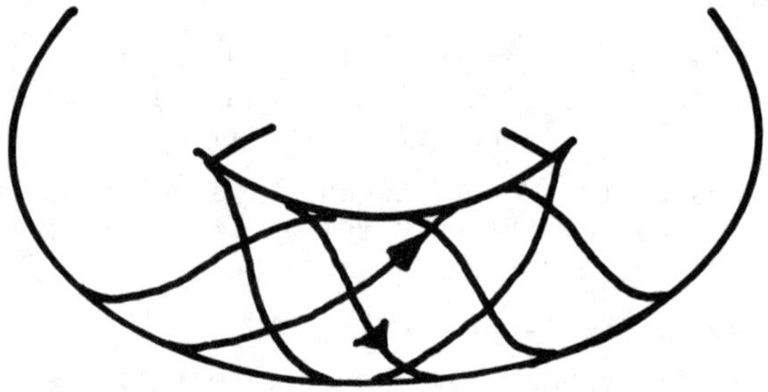

Fig. 72

This behavior under $T^{n\to\infty}$ is clearly *ergodic:* "time" averages over the iterates of any point (x_0, y_0) equal "space" averages over the torus, since any set of iterates eventually covers the torus. However, the cat map has a stronger property: *mixing*. This means that after T^{∞} not only does the entire past of a point cover the space, but so also does the present of any neighborhood of the original point. In other words, the area elements get ever more drawn out and eventually cover the space. Mixing implies ergodicity, but ergodicity does not imply mixing. For an example of this, recall the phase-space tori of Section 2, on which integrable systems live. If the frequencies are incommensurable, any orbit densely fills the torus - the orbit is ergodic on the torus (though not on the whole energy surface!). However, the (continuous) mapping of any point on the torus (eq. 2.35) is such as to *translate* the whole torus rather than distort it, so that neighboring points map together and this system is non-mixing (fig. 73).

It is not only mappings of the plane that exhibit mixing; "real" dynamical systems show it too. Perhaps the most important

$$a \sim \frac{GM\ell}{D^3} . \tag{8.3}$$

This will render 1's position of impact on 2 uncertain by

$$\Delta S \sim \frac{a\ell^2}{V^2} , \tag{8.4}$$

and its angle of reflection uncertain by

$$\Delta\Theta \sim \frac{\Delta s}{r} = \frac{GM\ell^3}{D^3V^2r} . \tag{8.5}$$

This in turn will render the reflection angle after succeeding collisions uncertain by $(\ell/r)\ \Delta\Theta$, $(\ell/r)^2\ \Delta\Theta$, $(\ell/r)^3\ \Delta\Theta$ ..., so that the number, n, of collisions after which the determinacy is lost (angle uncertain by 1 radian) is

$$n \sim \frac{\ell n(1/\Delta\Theta)}{\ell n\ \ell/r} = \frac{\ell n\ (\frac{D^3V^2r}{GM\ell^3})}{\ell n\ \ell/r} . \tag{8.6}$$

Now let the fluid be oxygen at NTP and let M be an electron at the limit of the observable universe, i.e. $D \sim 10^{10}$ light years. Then we get $n \sim 56$! If the spheres are billiard balls ($r \sim 3$cm, $\rho \sim 30$cm, $V \sim 1\text{ms}^{-1}$) and M is someone in the billiard-room ($M \sim 50$kg, $D \sim 1$m), then we get $n \sim 9$! Unstable indeed! It would be amusing to work out the effects of these perturbations on the electrons in computers doing molecular dynamics calculations. (The uncertainty arising from quantum mechanics gives n=0 for oxygen - i.e. even the first collision cannot be accurately "aimed" - and n=15 for billiard balls.)

It is worth emphasizing that Sinai's proof that hard spheres form a mixing system is valid not only in the "thermodynamic limit" (exceeding one!). The simplest case is two discs, moving on a closed two-dimensional surface with Euclidean metric; this has to have the topology of a torus. Even if one disc is fixed (a hole in the torus) the motion of the other is mixing (fig. 75).

Another mixing system is a mass point moving on a geodesic on a closed surface whose Gaussian curvature (product of two principal curvatures) is negative everywhere. (This is rather unimaginable in that a closed negatively-curved surface cannot be embedded in a Euclidean 3-space, but in its four-dimensional form it may have cosmological implications.) The essential point is that on such a surface two geodesics that are close and parallel at some point will separate exponentially for past and future times (fig. 76): this

Fig. 73

of these is the *hard-sphere fluid*, whose mixing was finally rigorously established by Sinai about ten years ago. Because of the infinite contact potential this is obviously not a perturbation of any simpler system (e.g. non-interacting particles). As with Arnol'd's cat, the mixing arises from the unstable nature of the motion, and in this case the instability is the result of collisions between the spheres' convex surfaces.

The instability is well illustrated by a calculation due to Chirikov, which also shows how unpredictable these mixing systems are, *even in principle*. No hard-sphere fluid is isolated, for it is impossible to screen out at least gravitational perturbations. Therefore let the fluid (sphere radius r, mean separation ℓ, mean particle speed V) be perturbed by a mass M a distance D away (fig. 74).

Consider a molecule 1 moving to collide with 2. Both 1 and 2 will fall towards M, but at different rates, and the tidal (i.e. relative) acceleration will be (order of magnitude)

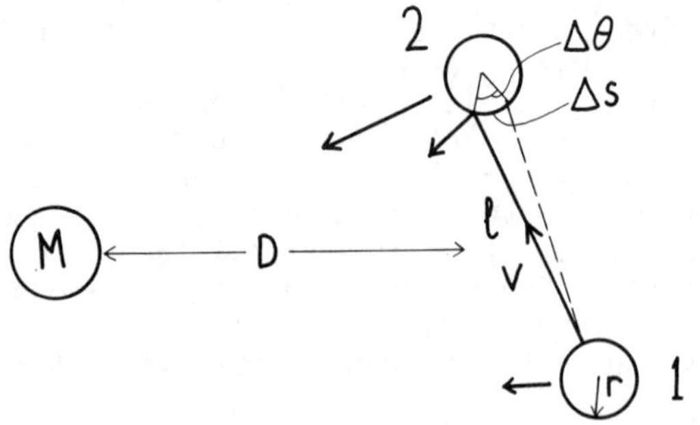

Fig. 74

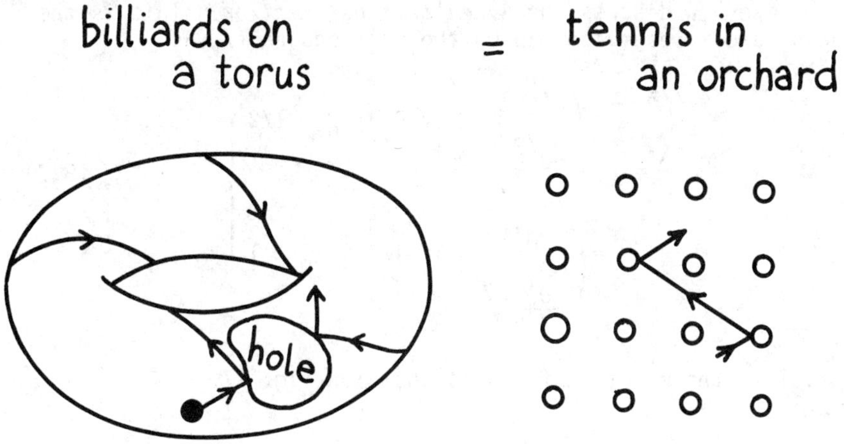

Fig. 75

contrasts with the convergence that occurs on positively-curved surfaces, e.g. spheres. This behavior can be derived from the fact that a geodesic is a space curve, lying in the surface, whose principal normal always coincides with the surface normal. The "torus billiard table" is actually a special case of this negative-curvature system.

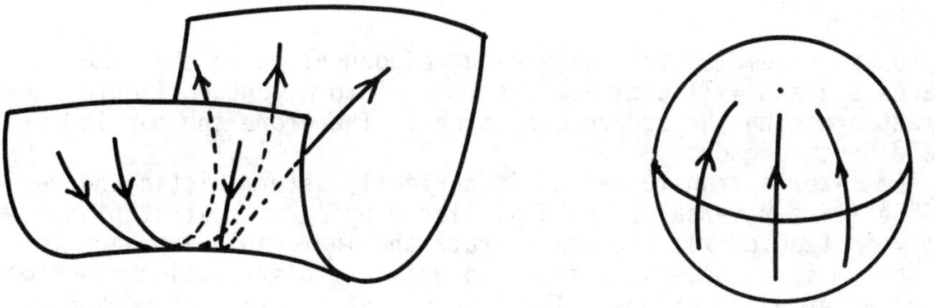

Fig. 76

Now we understand, at least in outline, how ergodicity, mixing and loss of determinacy arise. We have yet to consider the *approach to equilibrium*, and *randomness*. As a model for these two kinds of behavior, we now discuss the *Baker's transformation*. This is the following area-preserving map on the unit square (x,y):

$$\begin{pmatrix} x_1 \\ y_1 \end{pmatrix} = \begin{pmatrix} 2x_0 \\ y_0/2 \end{pmatrix} \quad \text{if } 0 \le x_0 < 1/2$$
$$= \begin{pmatrix} 2x_0 - 1 \\ \frac{y_0+1}{2} \end{pmatrix} \quad \text{if } 1/2 \le x_0 < 1 \qquad (8.7)$$

Pictorially, the map transforms as shown on fig. 77.

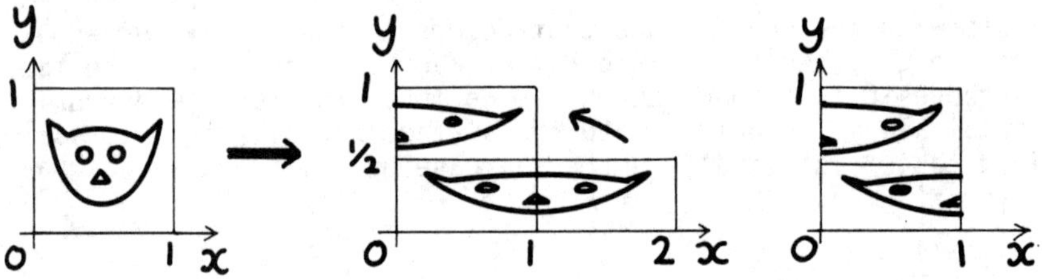

Fig. 77

The process resembles the rolling-out of dough; hence the name. Any area element will ultimately stretch into a long horizontal filament crossing the square many times. Therefore the map is mixing and hence ergodic.

The Baker's transformation is perfectly deterministic and reversible (time-reversal gives a similar mapping with stretching along y instead of x). To demonstrate the inexorable approach to equilibrium it is necessary first to define a distribution function $f_n(x,y)$ after n iterations. The "Liouville" equation of motion for $f_n(x,y)$ is, from (8.7),

$$f_n(x,y) = f_{n-1}(\tfrac{x}{2}, 2y) \quad \text{if } 0 \le y < 1/2$$
$$= f_{n-1}(\tfrac{x+1}{2}, 2y-1) \quad \text{if } 1/2 \le y < 1 . \qquad (8.8)$$

Next we must coarse-grain f_n to remove some of its fine-scale information. We do this by defining

$$W_n(x) \equiv \int_0^1 dy\ f_n(x,y)\ . \tag{8.9}$$

This corresponds to integrating out uninteresting phase-space variables in a mechanical problem. From (8.8) we get W's equation of motion:

$$W_n(x) = \int_0^{1/2} dy\ f_{n-1}(\tfrac{x}{2},\ 2y) + \int_{1/2}^{1} dy\ f_{n-1}\ (\tfrac{x+1}{2},\ 2y-1)\ , \tag{8.10}$$

i.e.

$$W_n(x) = 1/2\ [W_{n-1}(\tfrac{x}{2}) + W_{n-1}(\tfrac{x+1}{2})]\ . \tag{8.11}$$

If $f_n(x,y)$ is normalized to unity, so is $W_n(x)$, and this equation preserves that normalization under iteration.

The "equilibrium" solution $W_n(x)=1$ obviously satisfies (8.11). Moreover, *any* initial $W_0(x)$ will tend to unity after an infinity of iterations, since one iteration replaces the value of W at x by the mean of its values at two points (x/2 and x+1/2) surrounding x - i.e. the mapping has a smoothing effect. It is amusing to verify this by showing how iteration must destroy all Fourier components $e^{2\pi i\ell x}$ (ℓ=1,2,3...) describing the variation of $W_0(x)$. From (8.11),

$$e^{2\pi i\ell x} \to \frac{(1+e^{\frac{2\pi i\ell}{2}})e^{\frac{2\pi i\ell x}{2}}}{2} \to \frac{(1+e^{\frac{2\pi i\ell}{2}})}{2}\,\frac{(1+e^{\frac{2\pi i\ell}{4}})}{2}\,e^{\frac{2\pi i\ell x}{4}} \to \tag{8.12}$$

$$\ldots \prod_{n=1}^{\infty} [\frac{1+e^{\frac{2\pi i\ell}{2^n}}}{2}]\ .$$

A bracket n in this product must always vanish, since this requires

$$\frac{2\pi\ell}{2^n} = (2m+1)\pi\ ,\ \text{i.e.}\ \ell=2^{n-1}(2m+1)\ , \tag{8.13}$$

for some m, so that any odd-ℓ components vanish on the first iteration, and any even ℓ must be 2^n times an odd number for some n. Thus any initial $W_0(x)$ does indeed iterate to equilibrium.

Equation (8.11) for $W_n(x)$ resembles the "rate equation" for a random walk process (steps to x from x/2 and (x+1)/2), and we now show that the Baker's transformation (8.7) does indeed have a stochastic character. Let x_0 and y_0 each be represented by a binary "decimal":

$$x_0 = .a_1a_2a_3a_4 \;\dots \qquad \text{(a's and b's all 0 or 1).}$$

$$y_0 = .b_1b_2b_3b_4 \;\dots \tag{8.14}$$

Put these "back to back":

$$\{\dots\dots\; b_4b_3b_2b_1 \cdot a_1a_2a_3a_4 \;\dots\dots\} \,. \tag{8.15}$$

The Baker's transformation corresponds to a shift of the decimal point one place to the right! This is because such a shift doubles x and halves y, and automatically takes care of the conditions arising if $x_0 \gtrless 1/2$. Let us agree to label the orbit of (x_0,y_0) a sequence of 0's and 1's according to whether the iterated points have $x<1/2$ or $x>1/2$. Thus the sequence of numbers is just the sequence of first binary digits of the x-values of the iterated points, a doubly infinite sequence $-\infty < n < +\infty$. Because the Baker's transformation is isomorphic to the shift of decimal point in (8.15), and the first binary digit of x is the number immediately to the right of the decimal point, the sought-for sequence is simply

$$\{\dots\dots\; b_4b_3b_2b_1a_1a_2a_3a_4 \;\dots\dots\} \,. \tag{8.16}$$

Now, "almost all" initial points (x_0,y_0) are *irrational*, so that the decimals (8.14) are nonterminating, nonrepeating sequences that could have been obtained by tossing a coin (0 = heads, 1 = tails).

Therefore almost all orbits in the Baker's transformation, although perfectly deterministic, can be made to generate a set of "random" numbers. (The exceptional orbits are the set-of-measure-zero closed orbits generated by rational initial points (x_0,y_0).)

The central mathematical tool that is used here is the correspondence between the system considered (Baker's transformation) and the so-called *Bernouilli shift on infinitely many symbols* (moving the · in 8.15). In recent years a number of such correspondences have been found, and it seems as if this idea goes right to the heart of the problem of finding randomness in deterministic systems. Most important, it has been shown that *near any homoclinic point of a mapping* another mapping can be found which corresponds to a Bernouilli shift. To describe this map, refer to fig. 59. In any region R (e.g. the shaded quadrilateral) there will be a dense set of points P that eventually map back into R. The map $\tilde{T}$ in question is from P to its point of first return to R, $\tilde{T}P$.

Such homoclinic points, we learned, occur for generic Hamiltonian systems and not just for abstract algebraic mappings. Therefore, in real mechanical systems there are orbits describable by random sequences of integers, and we end this section with three examples of such systems. All involve a change in the unperturbed motion as energy varies, from bounded to unbounded.

The first example is *Sitnikov's case of the three-body problem.* Two equal "primary" masses M move in ellipses about their centre of mass O. The test mass, $m \to 0$, in whose motion we are interested, moves along the line OZ perpendicular to the plane in which the primaries move (fig. 78). Its orbit is z(t). The perturbation ε is the eccentricity of the primaries' ellipse. With suitable scaling (G=1, M=1/2, stretching of t coordinate) we have the distance r(t) of either primary from O as

$$r_\varepsilon(t) = \frac{1}{2}(1 - \varepsilon \cos 2t) + O(\varepsilon^2) . \qquad (8.17)$$

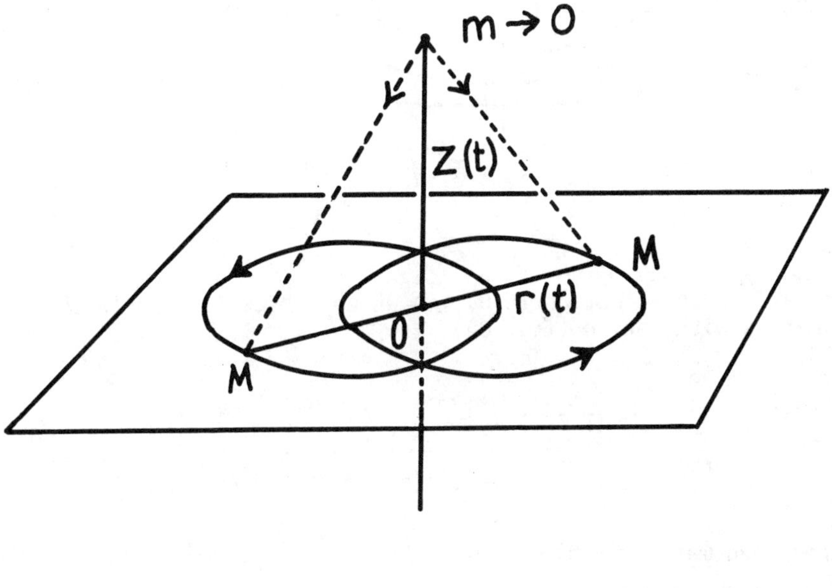

Fig. 78

Under the primaries' gravitation m's equation is

$$\frac{d^2z}{dt^2} = -\frac{z}{(z^2+r_\varepsilon^2(t))^{3/2}} , \tag{8.18}$$

which comes from a Hamiltonian

$$H = \frac{p^2}{2} - \frac{1}{\sqrt{z^2+r_\varepsilon^2(t)}} . \tag{8.19}$$

For the unperturbed motion ($r_\varepsilon \to r_0 = 1/2$), H(=E) is a constant of motion, and the phase plane z,p for this one-dimensional system is as shown in fig. 79.

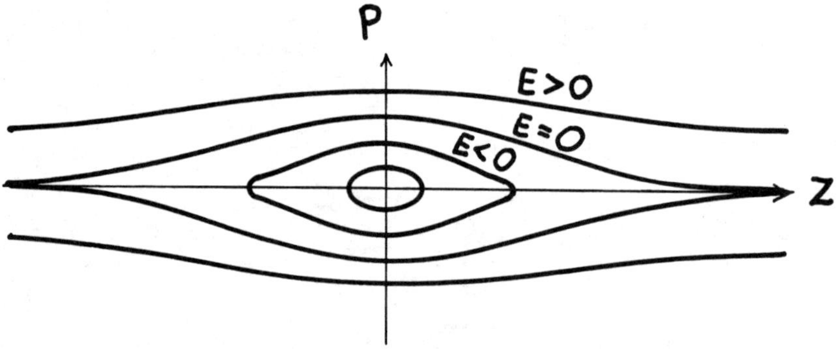

Fig. 79

E=0 separates bounded motion (E<0), in which m oscillates between the primaries with period (fig. 80)

$$T(E) = \frac{2\pi}{\omega(E)} = 4 \int_0^{(1/E^2-1/4)^{1/2}} \frac{dz}{\{2(E+(z^2+1/4)^{-1/2}\}^{1/2}} \quad -2<E<0 , \tag{8.20}$$

from unbounded motion (E>0) in which m moves from $z = \pm\infty$ to $z = \mp\infty$. We can define a discrete mapping on $S = (z,p)$ by plotting the position and momentum of m at the discrete times

$$t_n = \{.....2,-1,0,1,2,3, ...\} , \tag{8.21}$$

that correspond to periods of the primaries.

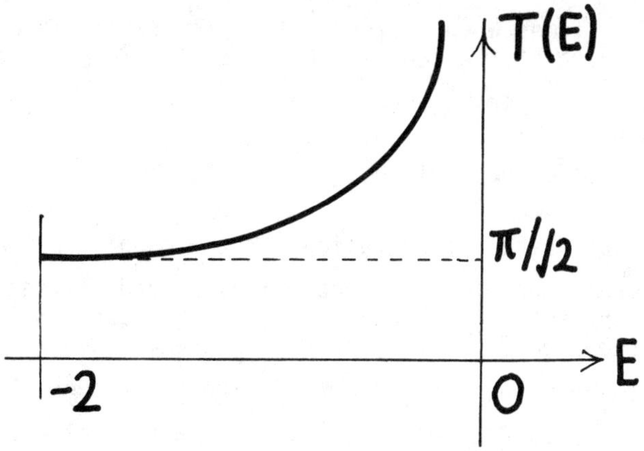

Fig. 80

For ε=0 the iterates of any point X = (z,p) move around one of the E-invariant curves (fig. 81), making ever smaller fractions of a revolution as E→0 from below. The line E=0 is an invariant curve joining two unstable fixed points at |z| = ∞, p=0, each of which is a degenerate kind of hyperbolic point. When the eccentricity ε is switched on, H is no longer a constant of the motion, but the KAM theorem tells us that closed invariant curves fill most of the region near the origin p=0, z=0, and we also know that there will be small irregular regions near unperturbed orbits with rational period T (remember the perturbation has period unity). These irregular orbits are bounded however - the resonance between M's and m's motion can never catapult m to infinity - because the irregular motion is trapped (on S) between smooth KAM curves.

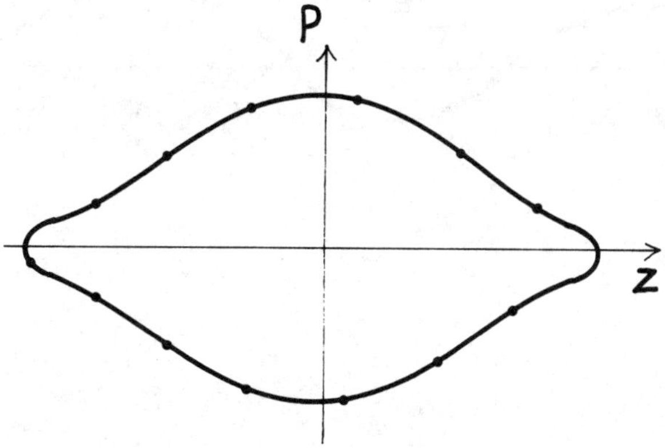

Fig. 81

The situation is however very different near E=0, where the unperturbed motion is itself unstable. We define an integer sequence from our mapping as follows: let t_k be the times of zeroes of any solution z(t) (i.e. times when m crosses the primaries' plane). Without loss of generality we can take $0 < t_0 < 1$. Then we define the doubly infinite integer sequence $\{S_k\}$

$$S_k \equiv \text{integer part of } t_{k+1} - t_k, \; z(t_k) = 0 \; . \qquad (8.22)$$

Thus S_k measures the number of iterations of the mapping between successive zero-crossings of z(t). What Sitnikov and Alexseev showed was this: given any small ε, a motion z(t) can be found that corresponds to *any* sequence $\{S_k\}$ provided all S_k exceed some number $m(\varepsilon)$. As $\varepsilon \to 0$, $m(\varepsilon) \to \infty$ so that the erratic orbits (random $\{S_k\}$) whose existence this result implies (fig. 82) are concentrated in narrow regions near E=0.

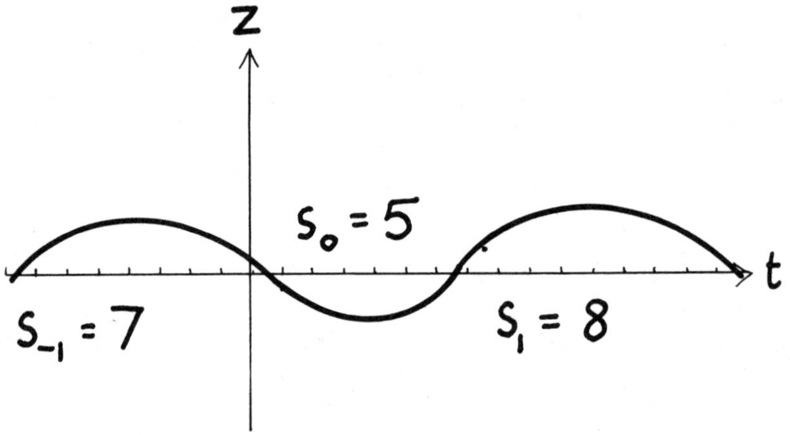

Fig. 82

An important feature of this result follows from the existence of sequences $\{S_k\}$ with one or two S_k equal to infinity. These correspond to "escape" orbits, which oscillate infinitely often before z becomes infinite (fig. 83),

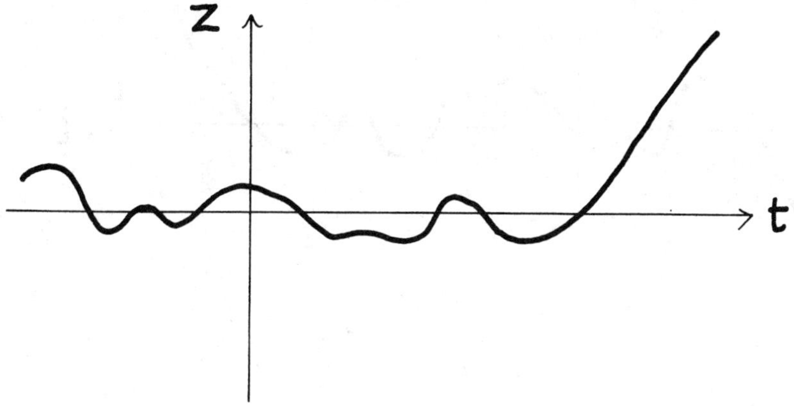

Fig. 83

or "capture" orbits, which fall in from infinity and then oscillate infinitely often (fig. 84),

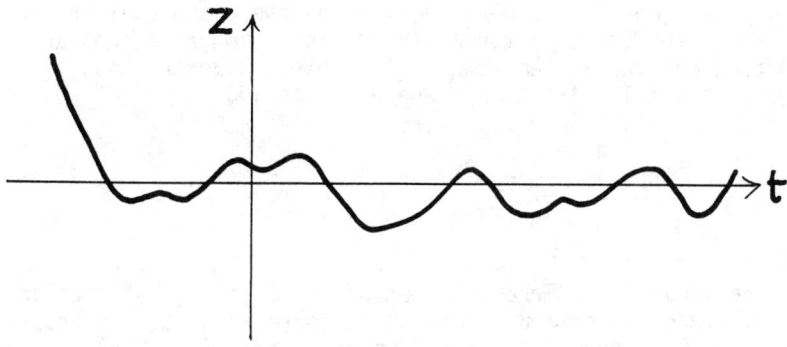

Fig. 84

or "capture and escape" orbits, where m falls in from infinity, oscillates arbitrarily often and then escapes (fig. 85).

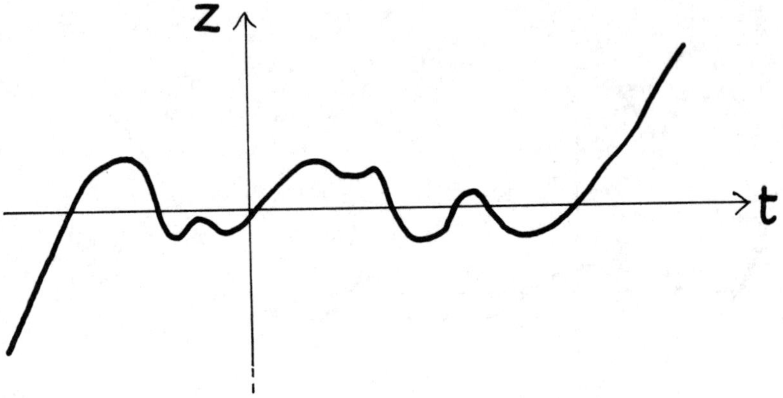

Fig. 85

Of course, the S_k can be totally *random* (provided $S_k > m(\varepsilon)$), so once again we have "stochasticity within determinism". These results apply also for large perturbations ε.

Our second example is the *swing*. The unperturbed case, and associated invariant curves, was considered in section 2. The perturbed case has the swing periodically excited by varying the length ℓ with period unity:

$$\ell \to \ell_0(1 + \varepsilon \cos 2\pi t) . \tag{8.23}$$

The map on the plane $S = \{q,p\}$ is defined by the p and q at integer times t. When $\varepsilon \neq 0$ the Hamiltonian (2.20) is no longer a constant of the motion and the system is nonintegrable. Nevertheless, invariant curves still exist except where the period

$$T = 2\mathrm{Re} \int_{-\pi}^{\pi} \frac{dq}{\sqrt{\frac{2}{m\ell^2}(E+mg\ell \cos q)}} , \quad E < mg\ell , \tag{8.24}$$

is nearly rational. The irregular motions in these "gaps" are limited in amplitude except near $E=mg\ell$ which corresponds to the unperturbed orbit with a hyperbolic fixed point at $q=\pm\pi$ (fig. 86). Here it is possible to define arbitrary sequences $\{S_k\}$ as in (8.22) with q replacing z. Thus S_k measures the integral part of the time in-

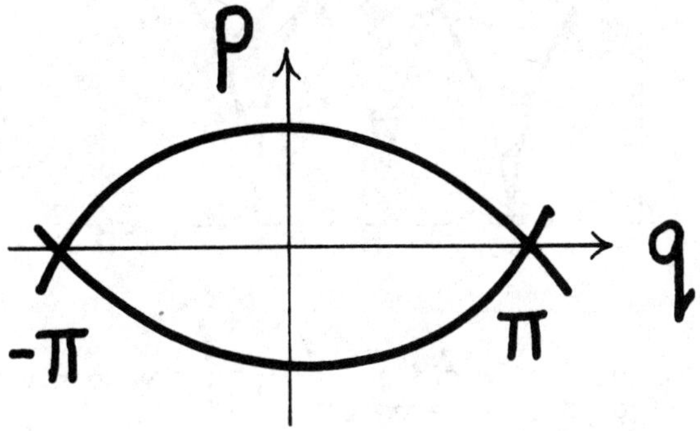

Fig. 86

terval between t_k and t_{k+1} when the swing is momentarily at rest. By analogy with the previous case it is likely that any sequence $\{S_k\}$ with $S_k > m(\varepsilon)$ corresponds to an orbit. In this problem, however, the topology of motion is different and suggests defining a new sequence $\{r_k\}$ by

$$r_k = \text{integer part of } |q(t_{k+1})-q(t_k)|/2\pi \; ; \qquad (8.25)$$

then r_k gives the number of rotations per libration. For the unperturbed swing $r_k = 0$ if $E < mg\ell$ and $r_k = \infty$ if $E > mg\ell$; it would be interesting to know whether r_k can take on a "stochastic" range of values when $\varepsilon \neq 0$, for energies near $mg\ell$.

These strongly irregular motions occur for E near $mg\ell$ - the unstable case. It is amusing to look at the perturbed swing motion *near equilibrium* ($E = -mg\ell$), where the unperturbed motion of q(t) is like a harmonic oscillator. Then (2.22) becomes, when linearized

$$\ddot{q}(t) + \frac{g}{\ell_0}(1-\varepsilon \cos 2\pi t)q(t) = 0. \qquad (8.26)$$

This is Mathieu's equation, better known to physicists as Schrödinger's equation for an electron in a one-dimensional solid with sinusoidal potential with unit spatial period. The "wave function" is

$$q(t) = e^{ikt} \times \text{periodic function of } t,$$

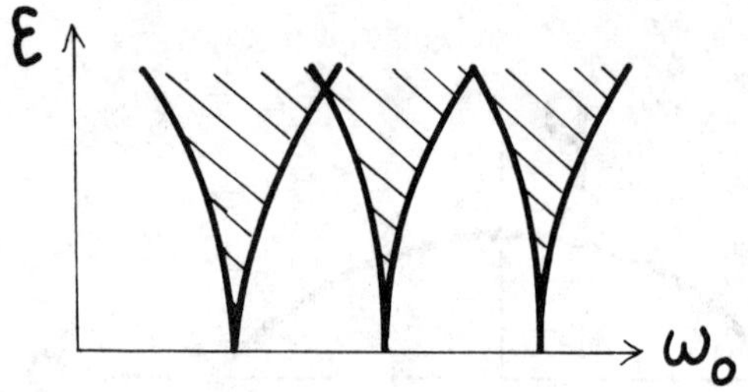

Fig. 87

where k is real except for "energies" g/ℓ_0 in the nearly-free-electron "band gaps" (fig. 87) centered on

$$\frac{g}{\ell_0} \equiv \omega_0^2 = (\frac{n}{2} \times 2\pi)^2 \ . \qquad (8.28)$$

In these gaps, k is imaginary and e^{ikt} can grow exponentially - the swing's equilibrium is *unstable* and its deflection grows exponentially. What has happened is that for these unperturbed equilibrium frequencies ω_0 the elliptic fixed point at the origin of (q,p) has eigenvalues near ± 1, i.e. is nearly *parabolic*, and so is easily converted into a hyperbolic fixed point. For small ε the real, nonlinear problem has invariant curves near q=p=0 so that the exponential instability is soon quenched - the swing's frequency is no longer ω_0 and the resonance is lost. Children don't know this, but they automatically adjust the frequency (of altering the length ℓ) to suit the "local" unperturbed frequency, and this nongeneric perturbation beats the KAM theorem! A dramatic example of this "adaptive pumping" occurs at the shrine of Santiago de Campostella in Spain, where according to H. Pomerance, pilgrims get incense to burn by swinging a brazier hanging from the ceiling, increasing the amplitude to about 180° by shortening and lengthening the supporting rope (fig. 88).

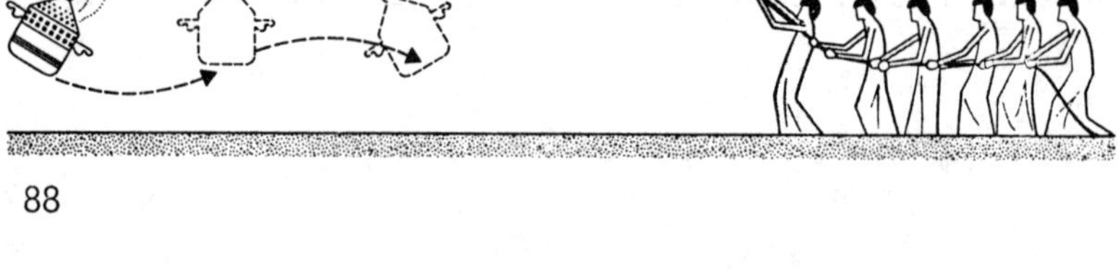

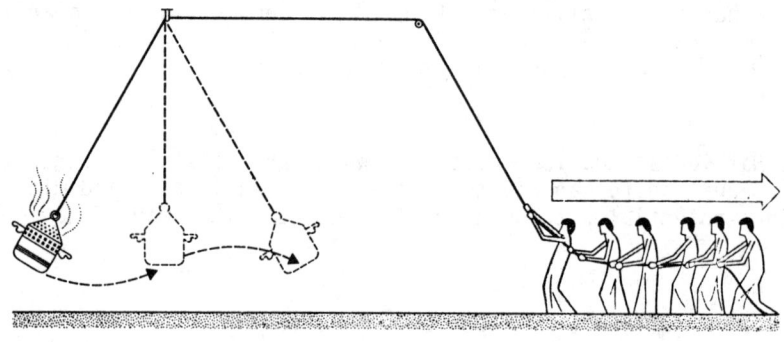

Fig. 88

Our final example is the *collision of a rigid-rotor molecule with a surface*. The molecule is a dumbbell with length d and two equal masses m/2, with centre of mass at height z above the surface, with which each mass interacts according to a Lennard-Jones type of potential U(z) (fig. 89).

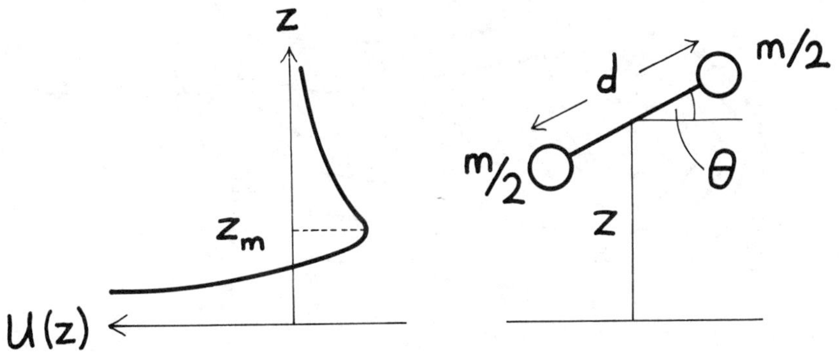

Fig. 89

This problem has two degrees of freedom, z and the dumbell's angle Θ. Consideration of the forces and torques acting leads to the following Hamiltonian for small d:

$$H = \frac{p_z^2}{2m} + 2U(z) + \frac{2p_\Theta^2}{md^2} + \varepsilon \frac{U''(z)d^2}{4} \sin^2\Theta . \qquad (8.29)$$

The last term is the perturbation, describing the torque exerted by the surface. $\varepsilon\to 0$ describes the gradual switching-off of this torque (e.g. by sphericising the dumbbell). We take the surface of section as $S_z = \{a, p_z\}$ defined by $\Theta=0$ (molecule parallel to the reflecting surface). H(=E) is always a constant of the motion, fixed on S_z.

When $\varepsilon=0$ (unperturbed case), p_Θ is a constant of the motion too (unhindered rotation), and S_z is covered by invariant curves whose equation is

$$P_z = \pm \{(E - 2U(z) - 2P_\Theta^2/md^2)2m\}^{1/2} , \qquad (8.30)$$

sketched in fig. 90 for some positive E.

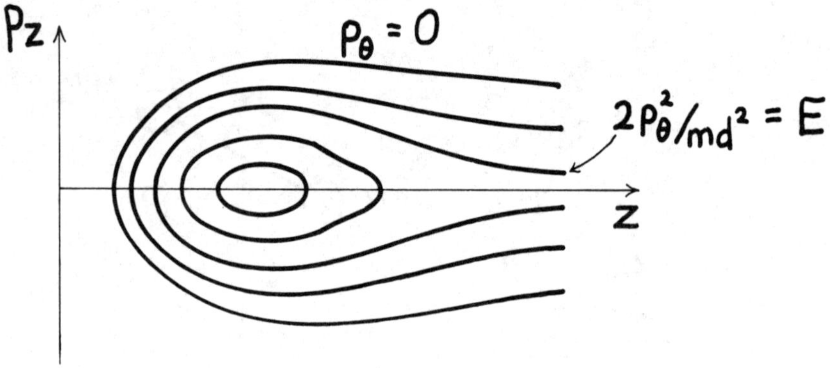

Fig. 90

Once again there is a quasi-hyperbolic fixed point at infinity (cf. Sitnikov's problem), when $2p_\Theta^2/md^2 = E$. The "rotation number" $\alpha(P_\Theta)$ of this mapping is (cf. 7.3)

$$\alpha(P_\Theta) = \frac{\omega_z}{\omega_\Theta} = \frac{\pi\ md^2}{2p_\Theta \mathrm{Re} \int_0^\infty dz/\{\frac{2}{m}(E - 2U(z) - 2p_\Theta^2/md^2)\}^{1/2}} , \qquad (8.31)$$

which vanishes at the "escape" critical angular momentum $|p_\Theta| = d\sqrt{mE/2}$ (fig. 91).

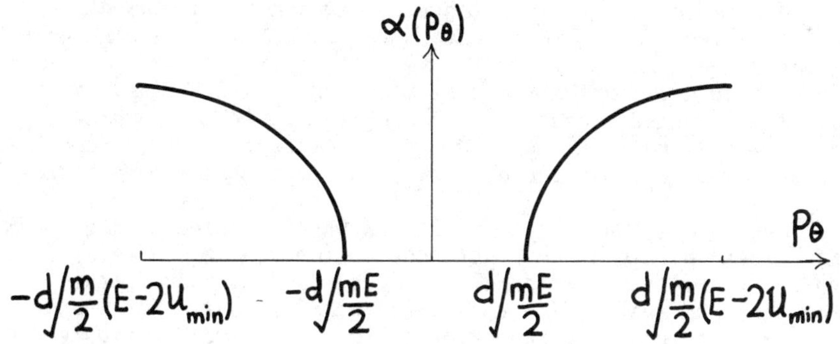

Fig. 91

When $\varepsilon \neq 0$ we once again expect strongly irregular behavior near the p_Θ corresponding to escape for the given E. An interesting integer sequence is again $\{r_k\}$ of (8.25), with q replaced by θ and t_k denoting the k'th crossing of the potential minimum z_m by the molecule. Then an infinity in the sequence $\{r_k\}$ would correspond to escape (capture) of the molecule by the surface, preceded (succeeded) by oscillatory trapping in the well of U(z). Rigorous statements about $\{r_k\}$ and similar sequences would be very useful in surface physics, where questions of capture and escape are studied largely by computations which, in view of the probably pathological dependence on initial conditions, are difficult to interpret.

Perhaps the most striking aspect of these modern developments in mechanics is the detailed understanding of the way in which stochastic elements enter into the motion of systems governed by causal equations. It is instructive to end this section by quoting from an essay written by Maxwell in 1873, showing how sophisticated was his philosophical thinking on these matters.

"*It is a metaphysical doctrine that from the same antecedents follow the same consequents. No one can gainsay this. But it is not of much use in a world like this, in which the same antecedents never again concur, and nothing ever happens twice...*

The physical axiom which has a somewhat similar aspect is 'that from like antecedents follow like consequents.' But here we have passed from sameness to likeness, from absolute accuracy to a more or less rough approximation. There are certain classes of phenomena

... in which a small error in the data only introduces a small error in the result ... The course of events in these cases is stable.

"There are other classes of phenomena which are more complicated, and in which cases of instability may occur, the number of such cases increasing, in an extremely rapid manner, as the number of variables increases...

"....Every existence above a certain rank has its singular points: the higher the rank, the more of them. At these points, influences whose physical magnitude is too small to be taken account of by a finite being, may produce results of the highest importance ...

"If, therefore, those cultivators of physical science from whom the intelligent public deduce their conception of the physicist... are led in persuit of the arcana of science to the study of the singularities and instabilities, rather than the continuities and stabilities of things, the promotion of natural knowledge may tend to remove that prejudice in favor of determinism which seems to arise from assuming that the physical science of the future is a mere magnified image of that of the past."

9. SEMICLASSICAL QUANTUM THEORY FOR NONINTEGRABLE SYSTEMS

In a general context, all of the complicated classical behavior that we have described must be regarded as the limiting behavior of the corresponding quantal system when Planck's constant $\hbar$ is negligible. Now in this last section we shall discuss some of the largely unsolved problems arising when $\hbar$ is not negligible but is small enough (in comparison with classical quantities of the same physical dimension) for us to hope that the quantal behavior can be understood in terms of the classical behavior. In other words we intend to discuss "semiclassical mechanics". For simplicity the treatment here will be restricted to *bound* quantum and classical systems, where the main problem is the determination of semi-classical energy levels. This is not a problem that can easily be left to a computer, because of the interaction between numerical noise and the increasingly fine scale of oscillation of wave functions as $\hbar \to 0$.

For *integrable systems* the problem is well understood and the levels are given explicitly by a quantum condition best expressed with the action-angle formalism explained in Section 2, as follows. Let the energy levels in an N-dimensional system be labelled by N quantum numbers $\underline{m} \equiv (m_1 \ldots m_N)$. Then the $\underline{m}$'th bound state is associated with a *particular torus* $\underline{I}_{\underline{m}}$, and its energy $E_{\underline{m}}$ is given by the Hamiltonian (2.14) expressed in action variables:

$$E_{\underline{m}} = H(\underline{I}_{\underline{m}}) \ . \tag{9.1}$$

The quantized tori $\underline{I}_{\underline{m}}$ lie on the points of a lattice in the N-dimensional $\underline{I}$ space whose unit cells have side length $\hbar$. The only subtlety is that the origin of the lattice is usually not at the origin of $\underline{I}$ space, so that

$$\underline{I}_{\underline{m}} = (\underline{m} + \underline{\alpha}/4)\hbar \ , \tag{9.2}$$

where $\alpha \equiv (\alpha_1 .. \alpha_1 ... \alpha_N)$ describes this displacement. The numbers α_i are integers equal to the number of "turning points" of the projection onto $\underset{\sim}{q}$ space of the i-th irreducible circuit γ_i of the torus $\underline{I}_{\underline{m}}$, i.e. the number of places on γ_i where the torus is "normal" to the $\underline{q}$ space. On the schematic fig. 92, for example, $\alpha_1 = 0$ ("rotation") and $\alpha_2 = 2$ ("libration").

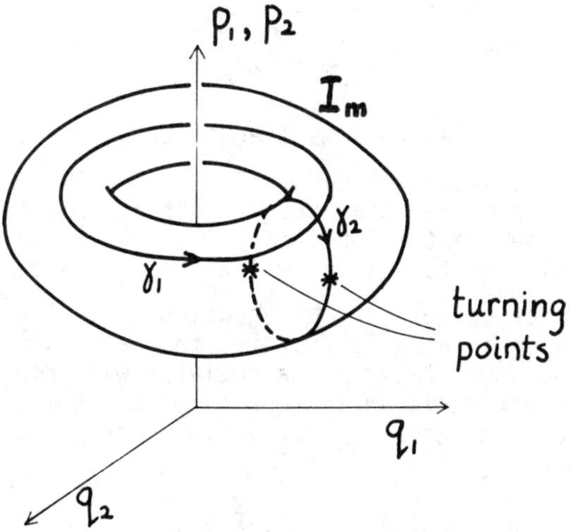

Fig. 92

The quantum condition (9.1)-(9.2) generalizes the old rules of Bohr and Sommerfeld. Perhaps the easiest way to obtain it is by demanding single-valuedness of the simplest W.K.B. wave function obtained by solving the time-independent Schrödinger equation to lowest orders in $\hbar$. This gives a travelling wave at any point q and hence corresponds to just one of the possible intersections in phase space of the manifold $\underline{q}$=constant with the torus $\underline{I}$ to which the wave function corresponds. This "local" W.K.B. wave is

$$\psi(\underline{q},\underline{I}) = C\left(\det \frac{\partial^2 S(\underline{q},\underline{I})}{\partial q_i \partial I_j}\right)^{1/2} e^{\frac{i}{\hbar} S(\underline{q},\underline{I})} \tag{9.3}$$

where $S(\underline{q},\underline{I})$ is the action integral (2.6) with the constants $\underline{f}$ chosen as $\underline{I}$, and C is a constant. After going around γ_i the action S in (9.3) has acquired an increment of $2\pi I_i$ (cf. equation 2.13). However this is not the only source of change in ψ, because the determinant becomes infinite at turning points on γ_i. To see this, realize that (2.15) implies

$$\det \frac{\partial^2 S}{\partial q_i \partial I_j} - [\text{Jacobian}(\frac{\partial \underline{q}}{\partial \underline{\Theta}})]^{-1} \tag{9.4}$$

The Jacobian vanishes at turning points on γ_i, and if the α_i such zeroes are simple each contributes a factor $e^{i\pi}$, so that the total phase increment of ψ round γ_i is

$$\frac{2\pi I_i}{\hbar} - \frac{\alpha_i \pi}{2} . \tag{9.5}$$

For ψ to be single valued this must equal $2\pi m_i$, where m_i is an integer, and (9.2) follows at once.

It must be emphasized that the frequencies $\underline{\omega}$ (equation 2.36) play no part in the quantum conditions. In particular, *closed orbits*, corresponding to tori $\underline{I}$ for which the ω_i are commensurable, will in general not be selected by equation 9.2. Therefore it is false (at least for integrable systems) to claim as some authors have done that quantum states are associated with closed orbits around which the action is an integer times $\hbar$. However, the closed orbits do play a most interesting role in determining the density of states function

$$n(E) \equiv \sum_{\underline{m}} \delta(E-E_{\underline{m}}) . \tag{9.6}$$

This can be transformed, using the quantum condition, into a representation of n(E) as a sum over all topologically different closed orbits (i.e. all "rational" tori). But each closed orbit gives not a set of levels but an oscillatory contribution to n(E); the more complicated closed orbits (high-order rational tori) give faster oscillations. As more and more closed orbits are included in this "topological sum" sharp peaks begin to appear and eventually turn into the delta functions corresponding to the energy levels.

What if the quantum system is classically *nonintegrable?* There will of course still be energy levels, and it is not hard to show that on the average, each level occupies a volume h^N in phase space, so that the average density of states $\overline{n}(E)$ is

$$\bar{n}(E) \equiv \lim(\Delta E \to 0) \lim(\hbar \to 0) \int_{-\Delta E/2}^{\Delta E/2} dE' n(E+E') = \frac{1}{h^N} \iint d\underline{q}\, d\underline{p}\, \delta(E - H(\underline{q},\underline{p})). \tag{9.7}$$

But the levels can no longer be located by (9.1) and (9.2) because in nonintegrable systems the whole basis of the quantum conditions breaks down. This is because in the "irregular" regions of phase space, near unstable closed orbits, tori $\underline{I}$ do not exist, and therefore the quantum numbers $\underline{m}$ cannot be defined. As long ago as 1917 Einstein realized that semiclassical quantum mechanics must be very different for integrable and nonintegrable systems. He saw a contradiction in theoretical physics as it then existed, that the (integrable) systems which could be quantized at that time, and the (nonintegrable) systems to which statistical mechanics could be applied, fell into two mutually exclusive classes. He expressed the hope, soon to be justified, that a properly formulated quantum mechanics would remove this contradiction. These prescient remarks seem to have been ignored until 1973, when Percival pointed out that in the light of the much deeper understanding of classical mechanics provided by the KAM theorem etc. it was time to return to the problem recognized by Einstein.

Percival's suggestion was that the quantum levels in the regions of phase space occupied by irregular trajectories will form an *irregular spectrum*, with properties very different from the *regular spectrum* arising from those regions of phase space filled with KAM tori providing a basis for quantization according to (9.2). The two sorts of spectra would be distinguished by their behavior under perturbation - for example by an electromagnetic wave if the system is a nonsymmetrical molecule. Such a perturbation strongly couples together levels of the regular spectrum with similar quantum numbers $\underline{m}$; these coupled levels have energy differences of order $\hbar$. By contrast, under perturbation *all* levels in a given irregular region would be weakly coupled; these have the much smaller energy spacing h^N, so that the irregular spectrum is much more sensitive to perturbation than the regular spectrum, and under poor resolution might be confused with a continuous spectrum.

Apart from some exploratory computations indicating that nonintegrable systems do indeed have some energy levels that are very sensitive to perturbation, practically nothing is known about the irregular spectrum. However, it is possible on the basis of the KAM theorem to arrive at what seem to be reasonable conjectures about the way that regular and irregular regions are distributed in systems whose departure from integrability is described by a perturbation parameter ε. Let us confine the discussion to two degrees of freedom, and recall the arguments of Sections 2, 3 and 4, especially the crucial equation (4.12) giving the widths of the resonance zones, near rational tori, in which irregular orbits exist. Figure 93 illustrates the lowest-order resonance zones in the "unperturbed" $\underline{I}$ space which has been quantized according to equation (9.2) [the 1/1 tori for example, inhabit the locus of points in $\underline{I}$ space where the normals to the contours $H(\underline{I})=E$ lie at 45° to the I_1 axis].

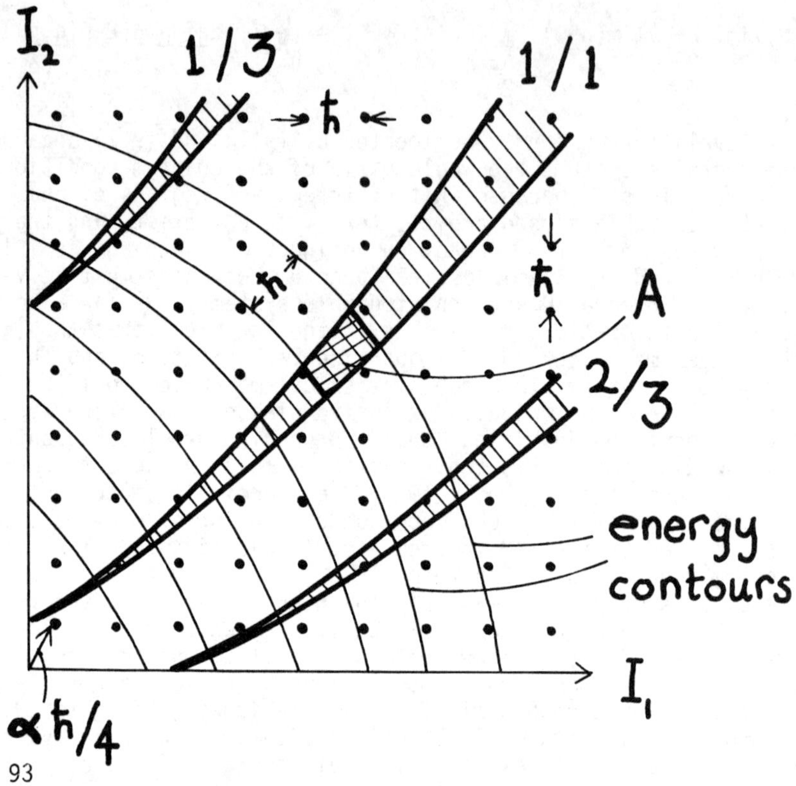

Fig. 93

I claim that the approach to the classical limit is non-uniform in ε. For fixed (small) ε there appear to be *three semi-classical regimes* as $\hbar\to 0$. These regimes are distinguished by values of a parameter β that will now be defined. Surrounding any point $\underline{I}$ is the area $\hbar^2$ corresponding to a quantum state. This area will be crossed (fig. 93) by infinitely many resonance zones, the widest of which has the frequency ratio r/s with smallest s (equation 4.12). Between two energy contours whose perpendicular separation $|\Delta \underline{I}|$ is $\hbar$ this widest resonance will occupy an action area A (fig. 93). Then β is defined as

$$\beta \equiv A/\hbar^2 . \tag{9.8}$$

When β is small it is a measure of the proportion of the quantum area $\hbar^2$ occupied by irregular trajectories. Large β indicates that resonant zones near $\underline{I}$ contain many quantum states. Elementary geometry and use of (4.12) give the estimate

$$\beta \simeq \frac{\bar{h}\text{x width of widest resonance}}{\bar{h}^2}$$

$$\sim \frac{\bar{h}\text{x } |\underline{I}|\text{x angular width of widest-resonance}}{\bar{h}^2}$$

$$\sim K(\varepsilon)\ |\underline{I}|/\bar{h}s^{2.5}\ .$$

Therefore β is large in the semiclassical limit ($\bar{h}$ small), and also for large perturbations ε, low order resonances (s small) and high excited states ($|\underline{I}|$ large).

In the *first semiclassical regime* $\bar{h}$ is small enough for a semiclassical treatment of the unperturbed system to be valid, but ε is so small that $\beta \ll 1$ for all $\underline{I}$ in the energy region of interest, even those crossed by lowest order resonances (s=1). The irregular regions occupy only a small fraction of the quantum area $\bar{h}^2$ and so do not affect the form of the quantum states $|\psi\rangle$. In effect Planck's constant $\bar{h}$ blurs all the pathology of the classical orbit structure. Under these circumstances quantization by tori based on equations (9.1) and (9.2) can be employed to locate the perturbed quantum levels $E_{\underline{m}}$, the actions $\underline{I}$ being approximately calculated by means of a perturbation scheme such as that discussed in Section 2.

Marking $\bar{h}$ smaller leads to the *second semiclassical regime*, in which β is of order unity for states whose actions $\underline{I}$ lie in the lowest resonance zones. Then there will be a few states whose quantum area is dominated by irregular trajectories. The energies of these states will still be given approximately by (9.1) with $\underline{I}$ obtained by interpolation from tori near the irregular region. However, the wave function $\psi(\underline{q})$ will no longer be given by the WKB expression (9.3) because the torus on which it is based no longer exists.

So what does such an "irregular state" look like? Since it is in phase space rather than in real space that the irregularity associated with nonintegrability manifests itself, it seems sensible to study a quantum object defined on phase space. Such an object is the Wigner function $\psi(\underline{q},\underline{p})$, defined as

$$\psi(\underline{q},\underline{p}) \equiv \frac{1}{h^{2N}} \int d\underline{Q} \int d\underline{\Pi}\, e^{-\frac{i}{\bar{h}}(\underline{p}\cdot\underline{Q}+\underline{q}\cdot\underline{\Pi})} \langle\psi| e^{\frac{i}{\bar{h}}(\hat{\underline{q}}\cdot\underline{\Pi}+\hat{\underline{p}}\cdot\underline{Q})} |\psi\rangle \quad , \qquad (9.10)$$

where ^ denotes an operator. It is well known that this can be written in the following unsymmetrical form involving the wave function $\psi(\underline{q})$:

$$\Psi(\underline{q},\underline{p}) = \frac{1}{(\pi\bar{h})^N} \int d\underline{X}\, e^{-2i\underline{p}\cdot\underline{X}/\bar{h}}\, \psi(\underline{q} + \underline{X})\psi^*(\underline{q}-\underline{X}) \ . \qquad (9.11)$$

For states in the "integrable" parts of phase space filled with tori the WKB wave function can be employed to take the classical limit of Ψ, with the pleasant result that the Wigner function for such a state with quantum number $\underline{m}$ condenses as $h\to 0$ onto a delta function on the torus $\underline{I}_{\underline{m}}$, i.e.

$$\Psi_{\underline{m}}(\underline{q},\underline{p}) \xrightarrow{\hbar\to 0} \frac{(\underline{I}(\underline{q},\underline{p})-\underline{I}_{\underline{m}})}{(2\pi)^N} . \tag{9.12}$$

(When h increases from zero $\Psi_{\underline{m}}$ develops "fringes" about the torus $\underline{I}_{\underline{m}}$ that have a characteristic "Airy function" form).

This makes it natural to conjecture that the Wigner function for an "irregular state" spreads over the corresponding irregular region in phase space, and a surface of section for the energy E of the quantum state might show a series of *randomly distributed maxima and minima* of Ψ. Figure 94 is a sketch of this conjectured behavior, to be compared with say, the inner irregular region on fig. 60. At this state we can only guess what sort of randomness

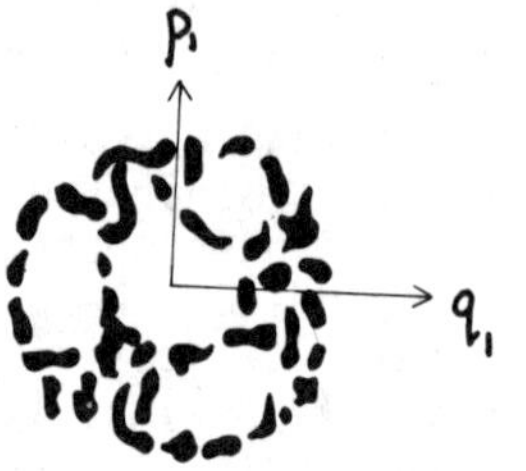

Fig. 94

Ψ will display. Probably it will be the "Gaussian" randomness studied in noise theory. It is also likely that the wave function $\Psi(\underline{q})$ in coordinate space is also of Gaussian random type for irregular states, with the mean intensity falling to zero at 'anticaustics' on the boundary of the region in $\underline{q}$ space explored by the orbit. This morphology of Ψ contrasts strongly with that for regular states, which have strong patterns of maxima and minima near intense caustics at the classical boundaries.

Further diminishing $\hbar$ leads to the *third semiclassical regime* which is the semiclassical limit proper. Now β is large for $\underline{I}$ in all lower-order resonance zones. The corresponding irregular regions in phase space will be densely populated with quantum states; in other words the lattice spacing in fig. 93 gets so small that each zone contains many lattice points. The group of states in each irregular region cannot now be individually labelled with quantum numbers although they may be said to share a "vague quantum number" corresponding to the destroyed region $\underline{I}$ in unperturbed action space.

Quantization by tori cannot now be applied in any sense. The Wigner function Ψ for any single state will presumably spread over the whole irregular region of the energy shell, and the surface of section is conjectured to resemble fig. 94 but with a much finer granularity in the randomness.

This does not exhaust the description of the generic structure of the third semiclassical regime, because there will be points $\underline{I}$ in high-order resonance zones where β is of order unity, and points $\underline{I}$ in still higher-order zones where β is small. Therefore along with the groups of "irregular states" just described there will also be states of the type described for the first and second regimes. What seems to be happening is that the smaller values of $\hbar$ expose more of the *infinite heterogeneity of the classical orbit structure* so that the quantum states become more varied in nature as well as more numerous.

When ε is zero this heterogeneity of structure is absent, because the system is integrable and there are no irregular regions; Wigner's function Ψ for every state is a "fringed torus". When ε is large this heterogeneity is also absent, because the resonant zones have expanded and eaten away all the tori and all motions are irregular; Wigner's function for every state should now be disordered and spread all over the energy shell in phase space. Fig. 95 summarizes this picture of the generic structure of the semiclassical limit.

My opinion is that the full elucidation of the nature of irregular states and of the mingling of regular and irregular states as ε and $\hbar$ vary will require the development of new conceptual and mathematical tools. Perhaps Wilson's celebrated "renormalization group" technique recently developed to study disorder on all scales in statistical mechanics might play some part.

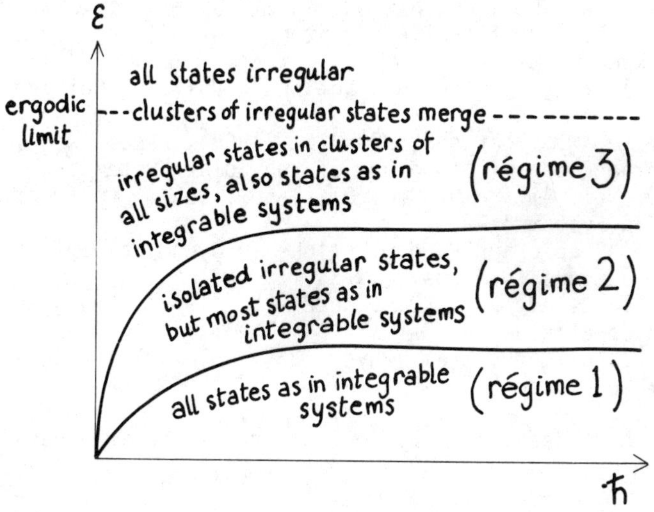

Fig. 95

REFERENCES

1. ARNOL'D, V. I., Russian-Mathematical Surveys 18, No. 6, 85-191 (1963), "Small Denominators and Problems of Stability of Motion in Classical and Celestial Mechanics".
2. ARNOL'D, V. I. and AVEZ, A., Ergodic Problems of Classical Mechanics (Benjamin, W., (1968)).
3. BERRY, M. V., Phil. Trans. Roy. Soc. (London), A287, 237-71 (1977), "Semiclassical Mechanics in Phase Space: a Study of Wigner's Function" J. Phys. A10, 2083-2091 (1977) "Regular and Irregular Semiclassical Wave Functions".
4. BORN, M., The Mechanics of the Atom (Ungar, New York, 1960).
5. BRILLOUIN, L., Scientific Uncertainty, and Information Theory, Part 2 (Academic Press, New York (1964)).
6. CAMPBELL, L. and GARNETT, W., The Life of James Clerk Maxwell (Macmillan Co., London, 1882).
7. CHIRIKOV, B. V., "Research Concerning the Theory of Non-linear Resonances and Stochasticity", Translation 71-40 CERN, Geneva, (1971).
8. EINSTEIN, A., Verh. Dt. Phys. Ges. 19, 82-92, (1917), "Zum Quantensatz von Sommerfeld und Epstein".
9. EMINHIZER, C. R., HELLEMAN, R. H. G. and MONTROLL, E. W., J. Math. Phys., 17, 121-140, (1976), "On a Convergent Nonlinear Perturbation Theory Without Small Denominators or Secular Terms".
10. FORD, J., "The Statistical Mechanics of Classical Analytic Dynamics" in Fundamental Problems in Statistical Mechanics III', edited by Cohen (North-Holland, Amsterdam 1975), 215-255.
11. FRANKLIN, F., in "The Rings of Saturn", Palluconi and Pettengill, Eds., NASA SP-343 (1974).
12. HÉNON, M., Quarterly of Applied Math. 27, 291-312, (1969), "Numerical Study of Quadratic Area-Preserving Mappings".
13. HÉNON, M. and HEILES, C., Astronomical Journal 69, 73-79 (1964), "The Applicability of the Third Integral of Motion: Some Numerical Experiments".
14. KHINCHIN, A. Ya., "Continued Fractions" (Univ. of Chicago Press, 1964).
15. MOLCHANOV, A. M., Icarus 8, 203-215, (1968), (see also, 11, 88-113) "The Resonant Structure of the Solar System".
16. MOSER, J., Mem. Am. Math. Soc. 81, 1-60 (1968), "Lectures on Hamiltonian Systems".
17. MOSER, J., "Stable and Random Motions in Dynamical Systems", Princeton University Press, (1973).
18. PERCIVAL, I. C., J. Phys. B. 6, L229-232 (1973), "Regular and Irregular Spectra".
19. PERCIVAL, I. C., Adv. Chem. Phys., 36, 1-61 (1977), "Semiclassical Theory of Bound States".

N. B. In most cases works by Soviet authors appeared earlier in Russian.

A PICTURE BOOK OF STOCHASTICITY*

Joseph Ford
School of Physics, Georgia Institute of Technology
Atlanta, Georgia 30318

ABSTRACT

Once upon a time, not so long ago, the Hamiltonian $H(Q,P) = H_0(P) + \varepsilon V(Q,P)$ was regarded as a very unpredictable fellow -- sometimes nice and integrable but more often violently stochastic. This is the story of how he got that way and why it matters. It all involves resonances in his personality and how they interact, so lets begin with them ...

INTRODUCTION

At this Christmas season conference on non-linear dynamics, it is perhaps quite appropriate to introduce the following pictorial review with the above fable-like title and abstract. But the analogy lies much deeper; for the story we present here, treating non-linear resonances as the source of chaotic trajectory behavior in Hamiltonian system, is a highly valuable but nonetheless intuitive "fable" which will eventually be replaced by a more rigorous, if less picturesque, general theory. Even so, this "fable" will likely serve as a convenient introduction to an incredibly complex subject for many years to come. Various versions[1] of this story have appeared frequently in the recent literature, and many in this audience will be quite familiar with it. For them, it is hoped that this retelling of the tale contains at least a few interesting deviations from time to time.

HAMILTONIAN SYSTEMS

In order to introduce the notation in our most general Hamiltonian systems, let us begin with the oscillator system.

$$H = \frac{1}{2}\sum_{k=1}^{N}(P_k^{\,2} + \omega_k^{\,2}Q_k^{\,2}) + V_3(Q,P) + V_4(Q,P) + \ldots, \quad (1)$$

where the Q_k and P_k denote coordinates and momenta respectively, where $\omega_k > 0$ are the positive frequencies of the harmonic approximation, and where V_3, V_4, ... denote cubic, quartic, ... polynominals in the Q_k and P_k. We now canonically change variables from the "rectangular" (Q_k,P_k) coordinates to the "polar" coordinates (ϕ_k, J_k) via

*This work supported in part by the National Science Foundation.

ISSN: 0094-243X/78/121/$1.50

$$Q_k = (2J_k/\omega_k)^{1/2} \cos \phi_k, \; P_k = -(2J_k\omega_k)^{1/2} \sin \phi_k. \tag{2}$$

In these "polar" coordinates, Hamiltonian (1) reads

$$H = H_0(J_1,\ldots,J_N) + \varepsilon\, V(J_1,\ldots,\phi_N), \tag{3}$$

where all the pure J-terms are included in H_0, where V involves only angle dependent terms, and where ε is a perturbation parameter introduced so that we may regard V as a perturbation on the obviously solvable (integrable) H_0. Independent of its origin, we regard Eq. (3) as specifying our most general conservative Hamiltonian system. Generalizing slightly, we permit V to depend periodically on the time variable $\tau = \Omega t + \tau_0$ and write

$$H = H_0(J_1,\ldots,J_N) + \varepsilon V(J_1,\ldots,\phi_N, \tau). \tag{4}$$

Expanding Hamiltonian (4) in a "double" Fourier series, we may write

$$H = H_0(J_1,\ldots,J_N) + \varepsilon \sum V_{mn}(J)\cos (m\cdot\phi+n\tau), \tag{5}$$

where $m\cdot\phi = \sum m_k\phi_k$. We shall regard Hamiltonian (5) as our most general Hamiltonian system. Equation (5) has several virtues:

1. H_0 is obviously integrable,
2. The resonant terms are "obvious",
3. H_0 plus any single angle dependent term is integrable.

The equations of motion for the unperturbed Hamiltonian H_0 read

$$\dot{J}_k = 0 \quad \text{and} \quad \dot{\phi}_k = \partial H_0/\partial J_k \equiv \omega_k(J), \tag{6}$$

where the dot superscript denotes time derivative. Equations (6) have the immediate solution

$$J_k = J_{k0} \quad \text{and} \quad \phi_k = \omega_k(J)t + \phi_{k0}, \tag{7}$$

where J_{k0} and ϕ_{k0} denote initial values and where, in general, $\omega_k = \omega_k(J_1,\ldots,J_N)$. We have thus shown that H_0 is integrable (solvable) by the simple device of directly integrating its equations of motion. More generally, any Hamiltonian which can be canonically transformed by an analytic, single-valued change of variables to read $H = H_0(J_1,\ldots,J_N)$ is said to be integrable.[2] Such a transformation will exist when a Hamiltonian $H(Q_1,\ldots P_N)$ has N independent, analytic, single-valued constants of the motion. Trajectories for such integrable systems must lie on the smooth, N-dimensional surfaces defined by the constant J_k. For bounded motion, these surfaces are tori.[2] We may regard the J_k as specifying the constant "radii" of the tori with the ϕ_k providing the angular positions. In Fig. 1, we show a cross-sectional view of a torus for the case N = 2. Quite generally for any integrable system, its trajectories lie on one or another of a set of nested tori with each torus bearing quasi-periodic (or strictly periodic) motion having the constant frequencies $\omega_k = \omega_k(J_1,\ldots J_N)$.

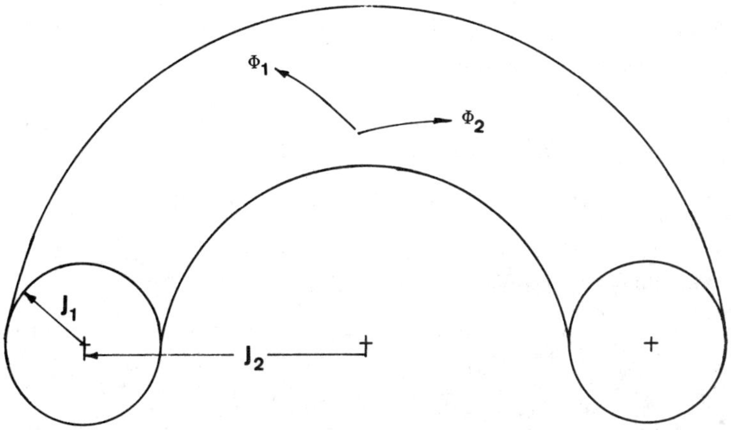

Fig. 1. A cross-sectional view of a two-dimensional torus or doughnut. The variables are those which appear in eq. (7) when N = 2.

Whenever one or another of the perturbing terms in Eq. (5) "drives" the H_0-motion at one of the natural frequencies occurring on one of its tori, then one expects a resonant response in which the affected trajectory departs the confines of its H_0-torus. But how do we recognize such a resonant term in Eq. (5)? Let us consider a single perturbing term and write

$$H = H_0(J) + \varepsilon V_{mn}(J) \cos (m\cdot\phi+n\tau) + \ldots \quad . \tag{8}$$

Then the equations of motion are

$$\dot{J}_k = \varepsilon m_k V_{mn}(J) \sin (m\cdot\phi+n\tau), \tag{9a}$$

$$\dot{\phi}_k = \omega_k + \varepsilon(\partial V_{mn}/\partial J_k) \cos(m\cdot\phi+n\tau) . \quad (9b)$$

When ε is small, we may use $J_k = J_{k0}$ and $\phi_k = \omega_k t + \phi_{k0}$ on the right hand side of Eq. (9a) to obtain

$$J_k = J_{k0} - \frac{\varepsilon m_k V_{mn} \cos(m\cdot\phi+n\tau)}{[m\cdot\omega(J)+n\Omega]} \quad (10)$$

valid to first order in ε. Thus, we immediately see that a given term cos $(m\cdot\phi+n\tau)$ is resonant at first order in ε provided that

$$[m\cdot\omega(J)+n\Omega] \lesssim \varepsilon \quad (11)$$

for some values of J, where $\omega(J) = \partial H_0/\partial J$. In particular, a term cos $(m\cdot\phi+n\tau)$ for which $n > 0$ and all $m_k > 0$ can never be resonant, at least to first order in ε.

Finally, if the sum in Hamiltonian (5) is replaced by any single one if its terms, then Hamiltonian (5) becomes integrable as we show via several examples in the following.

RESONANCE IN SIMPLE ONE DEGREE OF FREEDOM SYSTEMS

To see what all this looks like for a simple example, let us begin with the driven harmonic oscillator described by

$$\ddot{q} = -\omega_0^2 q + \varepsilon A\cos\tau , \quad (12)$$

where $\tau = \Omega t + \tau_0$. The associated Hamiltonian is

$$H = (1/2)(p^2 + \omega_0^2 q^2) - \varepsilon Aq \cos\tau . \quad (13)$$

Then using $q = (2J/\omega_0)^{1/2} \cos\phi$ and $p = -(2J\omega_0)^{1/2} \cos\phi$, we find

$$H = \omega_0 J - \frac{\varepsilon A}{2}\left(\frac{2J}{\omega_0}\right)^{1/2} [\cos(\phi-\tau) + \cos(\phi+\tau)] , \quad (14)$$

where here $H_0 = \omega_0 J$ Let us now note that the driven solution of Eq. (12) [or Hamiltonian (14)] may be written

$$\frac{\varepsilon A \cos\tau}{\omega_0^2 - \Omega^2} = \frac{\varepsilon A}{2\omega_0}\frac{\cos\tau}{(\omega_0 - \Omega)} + \frac{\varepsilon A}{2\omega_0}\frac{\cos\tau}{(\omega_0 + \Omega)} \quad (15)$$

Moreover, it is straightforward to show that $[\varepsilon A \cos\tau/2\omega_0(\omega_0-\Omega)]$ is the driven solution of the harmonic (linear) oscillator Hamiltonian

$$H = \omega_0 J - \frac{\varepsilon A}{2}\left(\frac{2J}{\omega_0}\right)^{1/2} \cos\,(\phi-\tau)\ . \tag{16}$$

while $[\varepsilon A \cos \tau/2\omega_0(\omega_0+\Omega)]$ is the driven solution for the harmonic

$$H = \omega_0 J - \frac{\varepsilon A}{2}\left(\frac{2J}{\omega_0}\right)^{1/2} \cos\,(\phi+\tau)\ . \tag{17}$$

Thus, as anticipated, Hamiltonian (16) retains all the essential resonant behavior of Hamiltonian (14).

Let us now slightly modify Hamiltonian (16) to obtain the non-linear oscillator Hamiltonian

$$H = \omega_0 J + \alpha J^2 - \frac{\varepsilon A}{2}\left(\frac{2J}{\omega_0}\right)^{1/2} \cos\,(\phi-\tau)\ , \tag{18}$$

where now $\omega(J) = \partial H_0/\partial J = (\omega_0 + 2\alpha J)$ depends on J when $\alpha \neq 0$. Then, by introducing the time-dependent, canonical change of variables $\mathcal{J} = J$ and $\Theta = \phi-\tau$, we may obtain the time-independent Hamiltonian

$$\mathcal{H} = (\omega_0-\Omega)\mathcal{J} + \alpha\mathcal{J}^2 - \frac{\varepsilon A}{2}\left(\frac{2\mathcal{J}}{\omega_0}\right)^{1/2} \cos\Theta\ . \tag{19}$$

In Eq. (19), $\mathcal{H}$ is a constant of the motion and we may graph its trajectories just as we graph the elliptical orbits for $H = (1/2) \times (p^2 + \omega_0^2 q^2)$. In this way, we may easily visualize certain differences between linear and non-linear resonances.

When $\alpha = 0$, we typically obtain the graph shown in Fig. 2, where $Q = (2\mathcal{J}/\omega_0)^{1/2} \cos\Theta$ and $P = -(2\mathcal{J}\omega_0)^{1/2} \sin\Theta$. Here we note that the driving resonance displaces the unperturbed $\varepsilon = 0$ orbits laterally along the Q-axis. The equilibrium point on the Q-axis corresponds to the periodic driven solution of Eq. (15). As Ω tends to ω_0, this equilibrium point tends to infinity. Thus at precise resonance $\Omega = \omega_0$, all the orbits diverge as vertical straight lines to infinity. Moreover even near resonance $\Omega \approx \omega_0$, the orbit initially passing through the origin $\mathcal{J}(0) = 0$ $[Q(0) = P(0) = 0]$ is unstable and departs arbitrarily far from the origin as Ω tends toward ω_0.

For $\alpha \neq 0$, the non-linear dependence of ω on $\mathcal{J}$, given by $\omega = \omega_0 + 2\alpha\mathcal{J}$, stabilizes this resonance, and the orbit passing through $J(0) = 0$, for example, is always bounded even at $\Omega = \omega_0$.

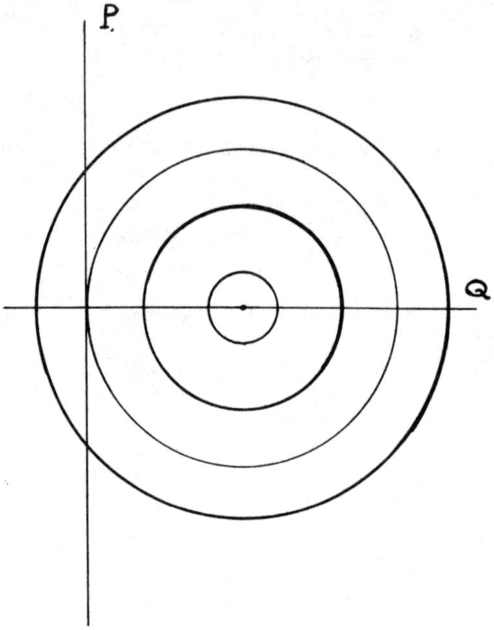

Fig. 2. A graph of orbits for Hamiltonian (19) with $\alpha = 0$. Here $\Omega \neq \omega_0$.

Indeed, set $\Omega = \omega_0$ in Eq. (19). Then for $J(0) = 0$, we have that $H \equiv 0$. Solving Eq. (19) for J then yields the bounded orbit

$$J^3 = \frac{\varepsilon}{4\alpha}\left(\frac{2}{\omega_0}\right)[1 + \cos 2\Theta] , \tag{20}$$

where we have discarded the solution $J = 0$. But the dependence of ω on J has another equally significant effect, for now we always have a resonant J-value near which $\omega(J) \approx \Omega$. In particular when $\Omega \neq \omega_0$ and $|\Omega - \omega_0| \gg 1$, we have the picture shown in Fig. 3. Here, we note that the non-linearity has introduced two new equilibrium points on the Q-axis (one stable and one unstable) in addition to the "linear" equilibrium point near the origin. Had α been zero here, we would have had only ovals essentially centered on the origin; but for $\alpha \neq 0$, we find the bounded non-linear resonant zone shown in Fig. 3 which has a finite ΔJ width. Indeed, as α decreases or ε increases, this width can become rather large.

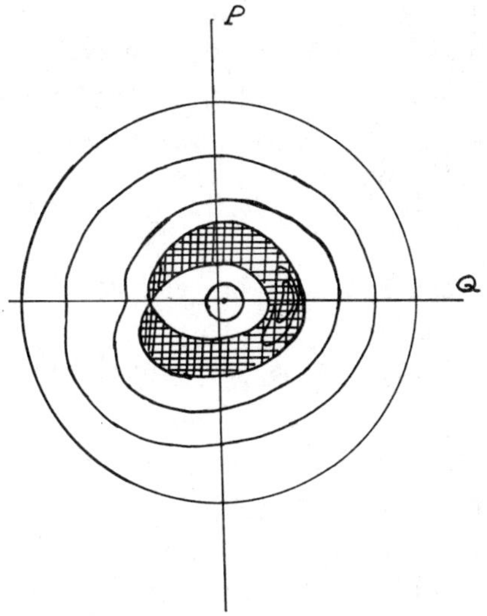

Fig. 3. A graph of orbits for Hamiltonian (19) with $\alpha \neq 0$ and $|\Omega-\omega_0|>>1$. The zone of non-linear resonance is cross-hatched.

RESONANCE IN SIMPLE, CONSERVATIVE TWO DEGREES OF FREEDOM SYSTEM

In order to obtain some typical pictures of resonant zones for conservative systems with two degrees of freedom, let us first consider the system

$$H = J_1+J_2-J_1^2-3\,J_1J_2+J_2^2+\varepsilon\,J_1J_2\cos 2(\phi_1-\phi_2), \quad (21)$$

where $H_0 = J_1+J_2-J_1^2-3\,J_1J_2+J_2^2$. Recall now that the H_0 orbits lie on two-dimensional tori, one of which is drawn in Fig. 1. Each torus bears orbits having frequencies given by $\omega_1 = 1-2\,J_1-3\,J_3$ and $\omega_2 = 1-3\,J_1+2\,J_1$. We thus expect the small perturbation $\cos(2\phi_1-2\phi_2)$ to resonantly distort a zone of unperturbed tori "centered" about the torus having $\omega_1(J_1,J_2) = \omega_2(J_1,J_2)$.

We now note that Hamiltonian (21), itself a constant of the motion, has the additional constant $I = J_1+J_2$. Thus Hamiltonian (21) is integrable, yielding motion lying on perturbed tori. We thus wish to determine here the distortion of the $\varepsilon = 0$ tori when $\varepsilon \neq 0$. If we use $I = J_1+J_2$ to eliminate J_1 from Eq. (21), we obtain

$$H = I-I^2+[2-\varepsilon\cos 2(\phi_1-\phi_2)]J_2^2-[1-\varepsilon\cos 2(\phi_1-\phi_2)]IJ_2, \quad (22)$$

which defines the two-dimensional perturbed tori lying in the three-space (J_2,ϕ_1,ϕ_2). If we now set $\phi_1 = 3\pi/2$ in Eq. (22), we obtain the intersection of the tori with the plane $\phi_1 = 3\pi/2$. In short, we obtain a two-dimensional, cross-sectional view of the perturbed tori. In the (J_2,ϕ_2) plane, the cross-sectional curves are given by

$$H = I-I^2+[2+\varepsilon \cos 2\phi_2]J_2{}^2-[1+\varepsilon \cos 2\phi_2]IJ_2\,. \tag{23}$$

In Fig. 4, we present a typical cross-sectional view of the perturbed tori, where $Q_2 = (2J_2)^{1/2} \cos \phi_2$ and $P_2 = -(2J_2)^{1/2} \sin \phi_2$. Here we note that the unperturbed $\varepsilon = 0$ tori (given by J_2 = constant) are only slightly distorted except in the crescent shaped resonant zones for which $\omega_1(J_1,J_2) \approx \omega_2(J_1,J_2)$. Also, let us note that the positions and widths of these resonant zones vary as ε or the energy $H = E$ change.

Next , let us consider the integrable resonance

$$H = J_1+J_2-J_1{}^2-3J_1J_2+J_2{}^2+\varepsilon J_1J_2{}^{3/2} \cos (2\phi_1-3\phi_2) \tag{24}$$

which has the additional constant $I = 3J_1+2J_2$. Note that H_0 is the same for both Eq. (21) and Eq. (24). Equation (21) perturbs H_0 with a so-called 2-2 resonance while Eq. (24) involves the 2-3 resonance. We expect $\cos (2\phi_1-3\phi_2)$ to distort the unperturbed $\varepsilon = 0$ tori bearing frequencies $2\omega_1(J_1,J_2) \simeq 3\omega_2(J_1,J_2)$. The equation for the cross-sectional curves of perturbed tori here reads

$$H = \frac{I}{3} - \frac{I^2}{9} + \left(\frac{1}{3} - \frac{5}{9}\, I\right) J_2 + \frac{23}{9}\, J_2{}^2 + \varepsilon\left[\frac{2}{3}\, J_2{}^{5/2} - \left(\frac{I}{3}\right) J_2{}^{3/2}\right] \cos 3\phi_2\,. \tag{25}$$

Typical curves generated by Eq. (25) are shown in Fig. 5.

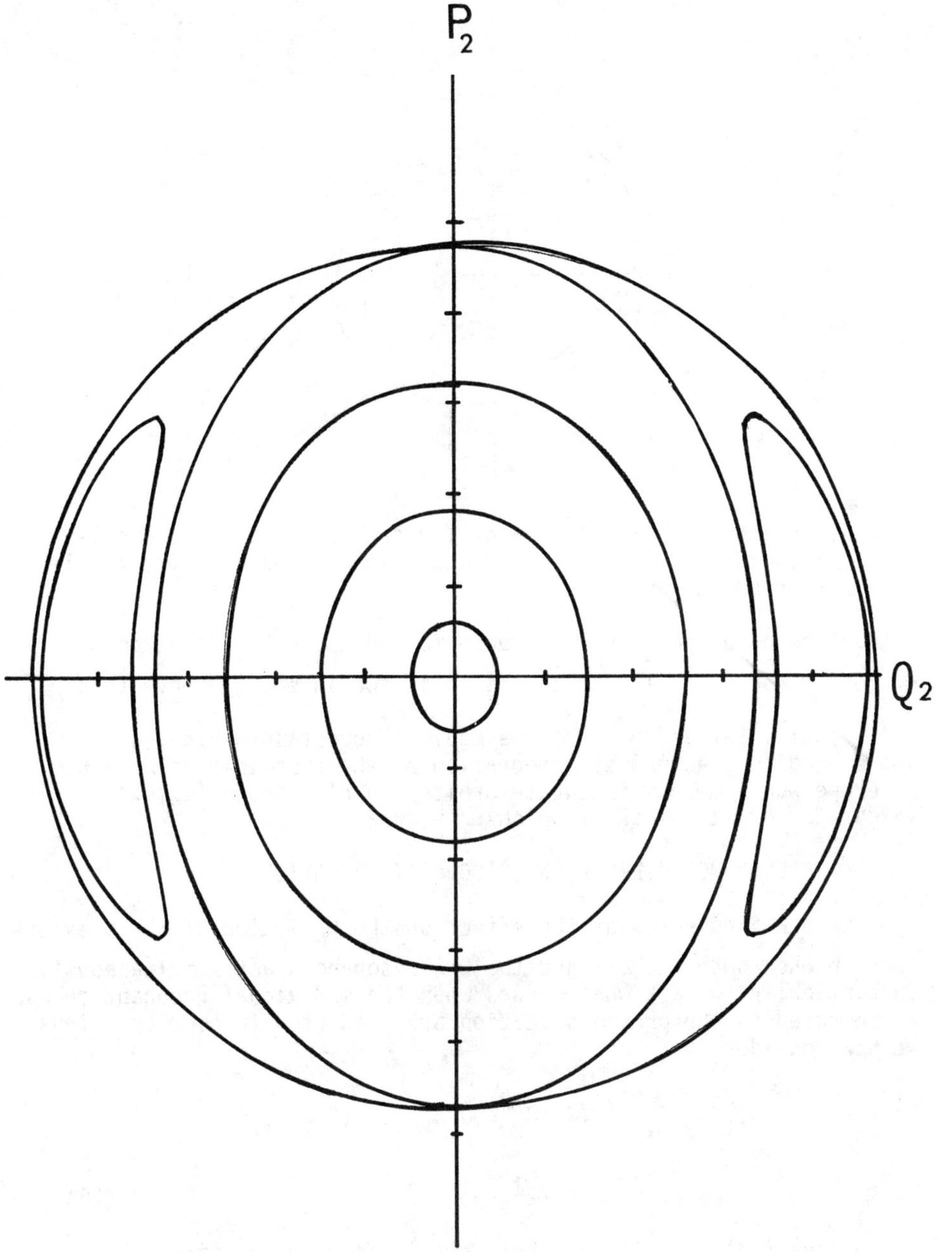

Fig. 4. A cross-sectional view of the perturbed tori for the integrable Hamiltonian (21). The crescent shaped regions are the resonant regions.

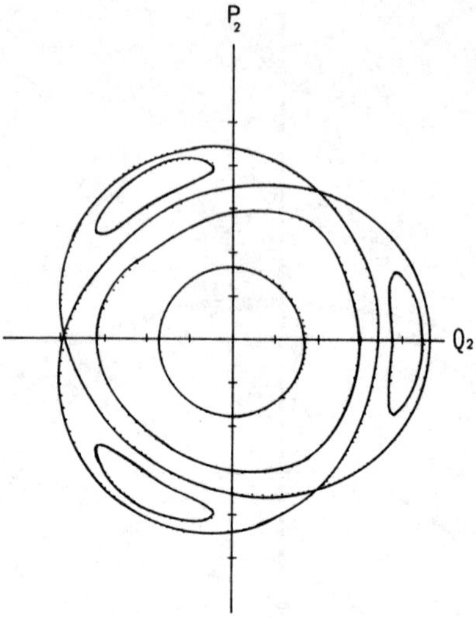

Fig. 5. A cross-sectional view of the perturbed tori for Hamiltonian (24). The 2-3 resonant zone appears as the triple crescent region.

It must be noted here that the unperturbed $2\omega_1 = 3\omega_2$ torus does not occur for $E < 0.16$; thus, Fig. 5 is for an energy above this value.

Finally let us note that we could have obtained Figs. 4 and 5 by direct numerical integration of the equations of motion. Here one would merely integrate orbits and plot those (J_2, ϕ_2) values for orbit points at which $\phi_1 = 3\pi/2$.

RESONANCE OVERLAP AND STOCHASTIC BEHAVIOR

We now inquire about the effect on the H_0 motion of the previous section when both the 2-2 and the 2-3 resonances act simultaneously. In particular, we ask what occurs when the individual resonant zones as computed in the previous section are predicted to overlap. Thus, we now consider

$$H = J_1 + J_2 - J_1^2 - 3J_1J_2 + J_2^2 + \varepsilon J_1J_2 \cos(2\phi_1 - 2\phi_2)$$

$$+ \varepsilon J_1 J_2^{3/2} \cos(2\phi_1 - 3\phi_2) . \qquad (26)$$

Using Eqs. (23) and (25) of the previous section, we first obtain Fig. 6 which graphs the positive Q_2-axis intercept of the inner edge of the 2-2 resonant zone and the outer edge of the 2-3 resonant zone as a function of energy $H = E$ for fixed $\varepsilon = 0.02$. Overlap is predicted to occur at $E = 0.2095$. Using direct numerical

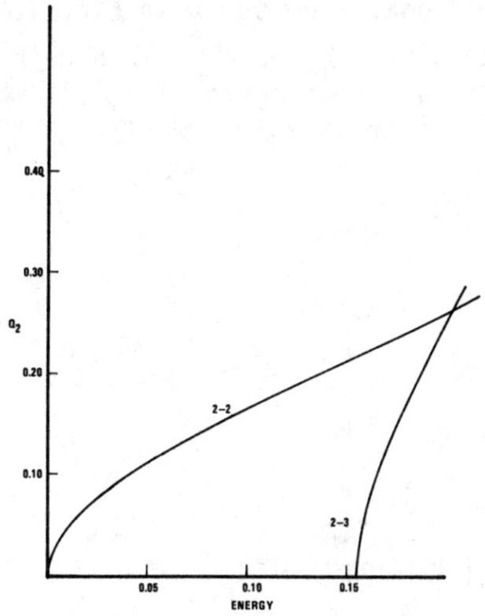

Fig. 6. Q_2-axis intercepts of the inner edge of the 2-2 zone and the outer edge of the 2-3 zone as a function of energy.

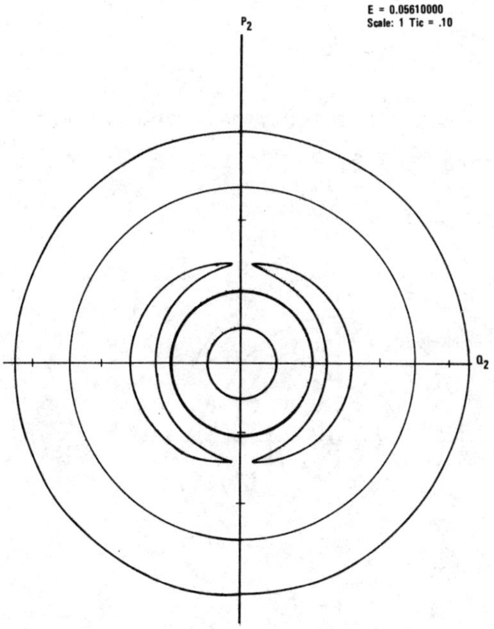

Fig. 7. Directly integrated cross-sectional view of Hamiltonian (26) at energy E = 0.056 .

integration of Hamiltonian (26) at energy E ≈ 0.056, we obtain the (Q_2,P_2)-plane cross-sectional view shown in Fig. 7. Here the 2-3 resonance zone has not yet appeared. In Fig. 8, we show the directly integrated curves at energy E = 0.18 where now the well separated 2-2 and 2-3 zones both appear. In Fig. 9, we

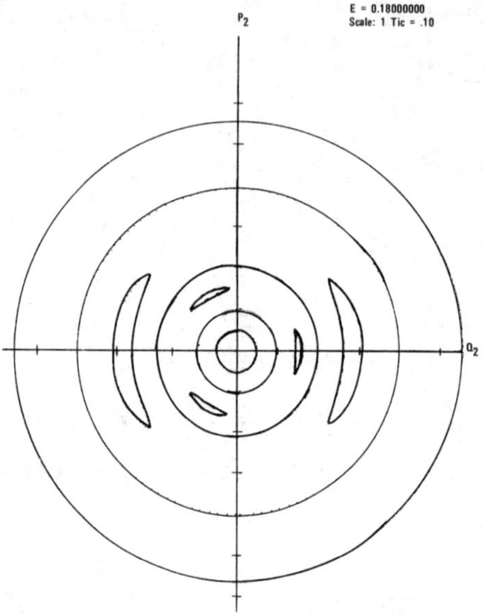

Fig. 8. A (Q_2,P_2)-plane cross-sectional view for Hamiltonian (26) at energy E = 0.18.

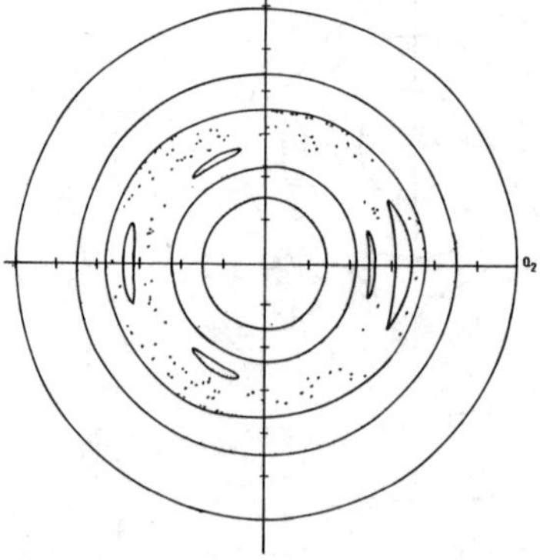

Fig. 9. The (Q_2,P_2)-plane at the energy predicted to yield resonance overlap.

show the appearance of the (Q_2,P_2)-plane cross-section at the predicted overlap energy of E = 0.2095. Here the chaotic set of dots was generated by a single orbit. The region of "overlap" thus appears to give rise to a so-called stochastic zone in which orbits are extremely erratic.

In order to gain further insight into the source of this chaotic region, let us examine Fig. 10 which shows the (Q_2,P_2)-plane at the slightly lower energy E = 0.20. Here we observe a previously unexpected narrow resonant zone containing five crescent regions which lie between the 2-2 and the 2-3 zones; in addition, a very narrow resonant zone (not shown) containing seven crescents has also been detected. These secondary resonances arise because of the interaction between the two explicitly appearing primary resonances in Hamiltonian (26). Indeed canonical

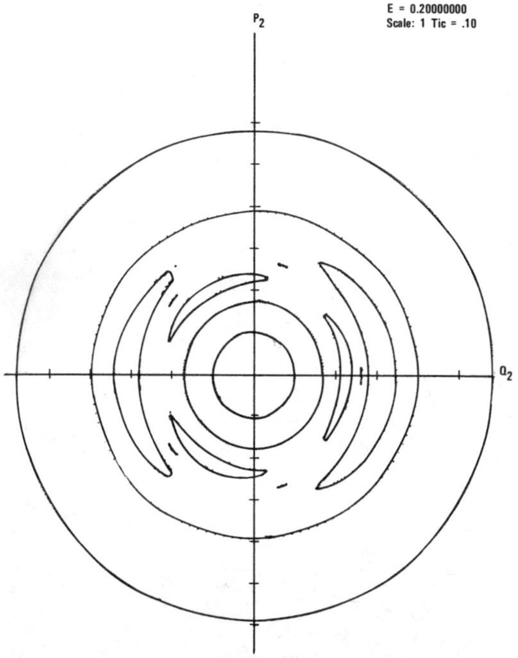

Fig. 10. At energy E = 0.20 an additional secondary resonant zone containing five narrow crescents appears between the primary 2-2 and 2-3 zones.

perturbation theory[1] may be used to show that a host of secondary and higher order resonances occur in the "overlap" region. As a consequence, orbits in this region move under the influence of many competing resonances and therefore develop acute vertigo, wandering aimlessly through phase space. Moreover as the energy increases, the size of this chaotic stochastic region increases, in many cases completely filling the allowed phase space.

If one initially starts two orbits very close together in a curve bearing region of the (Q_2,P_2)-plane, one obtains the typical linear separation shown in Fig. 11. On the other hand, two initially close orbits started in the chaotic region separate exponentially with a typical case being shown in Fig. 12. It is this sensitive exponential "forgetting" of initial conditions that leads one to label the chaotic regions as stochastic since here the final system state depends as sensitively on initial conditions as does a dice roll. Moreover, computer experiments indicate that the chaotic regions contain a dense set of unstable periodic orbits; thus one may regard the aperiodic orbits as stochastically diffusing among the dense set of scattering orbits. Further insights into the nature of these stochastic regions will be provided in later sections.

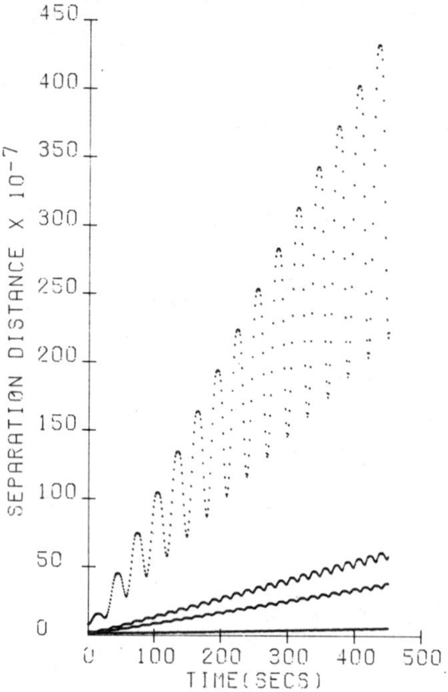

Fig. 11. This figure shows the growth of separation distance between two trajectories initially started about 10^{-7} apart in the full sphase space. Separation distance versus time is plotted for four distinct orbit-pairs. Each orbit-pair starts in a smooth level-curve region and the linear growth of separation distance with time is apparent.

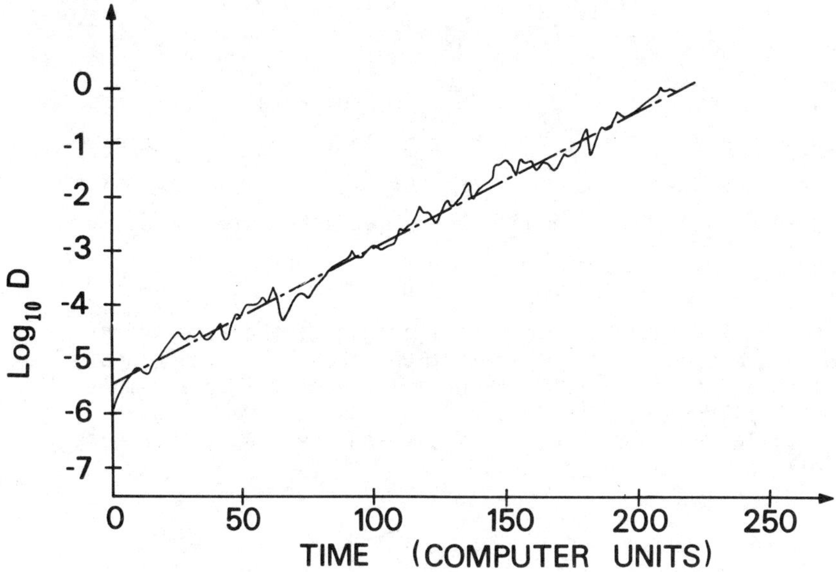

Fig. 12. Graph showing a typical curve of exponential growth in separation distance. Here, a curve of $\log_{10}D$ vs. time is plotted.

THE CHIRIKOV RESONANCE OVERLAP CRITERION

In the last two sections, we have illustrated that one may obtain an estimate of the critical parameter values at which a macroscopically visible stochastic zone first appears by computing the conditions for primary resonance overlap. Here, each primary resonance zone is computed as if that resonance were acting alone. In this section we discuss Chirikov's procedure[3] for computing this estimate. In particular, we shall illustrate his method as applied to one specific example system.

First, Chirikov computes the location and resonant width of an isolated resonance. As an example, let us consider the one degree of freedom driven oscillator

$$H = H_0(J) + \varepsilon\, V(J) \cos(\Theta - \tau) , \tag{27}$$

where $\tau = \Omega t + \tau_0$. The J-value J_r at the center of the resonant zone is determined using

$$\omega(J_r) = \partial H_0/\partial J \Big|_{J=J_r} = \Omega . \tag{28}$$

Let us now introduce the time-dependent canonical change of variables $P = J - J_r$ and $\psi = \Theta - \tau$. Hamiltonian (27) then becomes

$$H = \frac{\omega'(J_r)}{2} P^2 + \varepsilon V(J_r) \cos \psi, \tag{29}$$

where we have expanded $H_0(J_r+P)$ in powers of P retaining terms only through order P^2, where we have neglected $H_0(J_r)$, where we have retained only the lowest order term $V(J_r)$ in the expansion for $V(J_r+P)$, where the term linear in P vanishes since $\omega(J_r) = \Omega$, and where $\omega'(J_r) = \partial\omega/\partial J\big|_{J=J_r}$.

Hamiltonian (29) has placed its origin at the "center" of the resonance zone resulting in what Chirikov terms the pendulum approximation since Eq. (29) is formally identical to a simple pendulum Hamiltonian. Chirikov now assumes that Hamiltonian (29) is valid out to the edge of the resonance zone. The phase plane diagram is shown in Fig. 13.

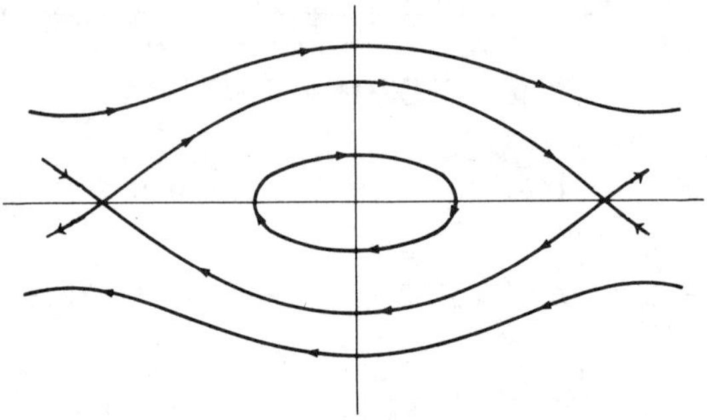

Fig. 13. The phase plane for Hamiltonian (29).

The pendulum resonant zone is bounded by the so-called separatrix curve passing through the unstable equilibrium points at $\psi = 0, 2\pi$. On the separatrix $H = \varepsilon V$, its value at $\psi = 0$. Thus, on the separatrix, we have

$$\omega' P^2/2 = \varepsilon V(1 - \cos \psi) = 2\varepsilon V \sin^2(\psi/2), \tag{30a}$$

or

$$P_s = \pm\, 2(\varepsilon V/\omega')^{1/2} \sin (\psi/2). \tag{30b}$$

In original variables, the value of J on the separatrix curve is

$$J = J_r \pm (\Delta J)_r \sin [(\Theta-\tau)/2], \tag{31}$$

where the resonance half-width $(\Delta J)_r$ is given by

$$(\Delta J)_r = 2\ [\varepsilon V(J_r)/\omega'(J_r)]^{1/2}. \qquad (32)$$

Chirikov now further approximates by taking the resonance half-width in frequency $(\Delta\omega)_r$ to be given by

$$(\Delta\omega)_r = \omega'(J_r)\ (\Delta J)_r, \qquad (32a)$$

or

$$(\Delta\omega)_r = 2\ [\varepsilon\omega'(J_r)V(J_r)]^{1/2} \qquad (32b)$$

Equation (32b) is Chirikov's estimate for the resonant half-width in terms of the frequency ω and it is valid for Hamiltonian (27).

Let us now apply these formulas to the Hamiltonian

$$H = AJ^{4/3} - (\varepsilon/2)(3\beta J)^{1/3}\ [\cos\ (\Theta-\tau_1) + \cos\ (\Theta-\tau_2)], \quad (33)$$

containing two explicit, primary driving resonances. Here A and β are constants. Now let Ω denote either Ω_1 or Ω_2. Then the pendulum approximation for either resonance acting alone yields a resonance centered at

$$J_r = \Omega^3/3\beta^4\ , \qquad (34)$$

with an ω half-width given by

$$(\Delta\omega)_r = \beta^{3/2}\ (2\varepsilon/\Omega)^{1/2}\ . \qquad (35)$$

Following Chirikov, we now define $\Delta\Omega \equiv |\Omega_1-\Omega_2|$ and we assume that $\Delta\Omega \ll \Omega_1$. In essence we are assuming that Ω_1 is close to but not identical with Ω_2. In terms of frequencies the two resonances are centered at $\Omega_1 = \omega(J_{r1})$ and $\Omega_2 = \omega(J_{r2})$; moreover the two resonant widths are approximately equal. Thus , the independent resonant zones will touch when

$$2(\Delta\omega)_r \approx \Delta\Omega, \qquad (36)$$

where $(\Delta\omega)_r$ is given by Eq. (35). Equation (36) is Chirikov's overlap criterion. Putting Eq. (35) into Eq. (36), we find

$$\frac{\Delta\Omega}{2} \approx \beta^{3/2}\ (2\varepsilon/\Omega)^{1/2}, \qquad (37)$$

where the Ω on the right is set equal to $(\Omega_1+\Omega_2)/2$. Equation (37) then predicts a critical ε value ε_c given by

$$\varepsilon_c \approx \Omega(\Delta\Omega)/8\beta^3. \tag{38}$$

For $\varepsilon \geq \varepsilon_c$, we expect that Hamiltonian (33) has a stochastic zone.

Chirikov now numerically integrates the equations of motion using $\Omega_1 = 0.217$, $\Omega_2 = 0.251$, $\Omega = 0.234$, $(3\beta J)_0^{1/3} = 0.276$, $\beta = 0.8472$, and $A = (3\beta/2\sqrt{2})^{4/3}$. Equation (38) yields $\varepsilon_c \cong 5.76 \times 10^{-5}$, while the numerical experiments described below yield $\varepsilon_c \approx 2.55 \times 10^{-5}$.

As he numerically integrates an orbit, Chirikov calculates what he calls a diffusion coefficient given by

$$D_n = \overline{[(\Delta\overline{H})^2/\Delta t]}. \tag{39}$$

Here, the total integration interval t is divided into many sub-intervals (Δt_n). The time varying total energy H is time averaged over each (Δt_n) sub-interval to yield $\overline{H}$. $\Delta\overline{H}$ is then the difference in $\overline{H}$ between any two sub-intervals (Δt_n) separated by a time interval Δt. The final average in Eq. (39) then involves averaging $[(\Delta\overline{H})^2/\Delta t]$ over all possible pairs of sub-intervals. For $\varepsilon < \varepsilon_c$, all orbits should yield D_n values which tend to zero as the total time interval t becomes large. On the other hand, for $\varepsilon \gtrsim \varepsilon_c$, an orbit started in the chaotic zone would be expected to yield a non-zero D_n due to an expected "random walk" of $\overline{H}$. In essence, Chirikov anticipates a fast exponential separation of the phase ψ for a group of initially close orbits followed by a much slower "random phase" diffusion of $\overline{H}$ itself. We shall make this "random walk" type behavior more transparent using some simple models which we discuss later. In any event, a graph of Chirikov's results is sketched in Fig. 14. Here one notes, as anticipated, an increase in D_n by many powers of ten as ε increases through ε_c.

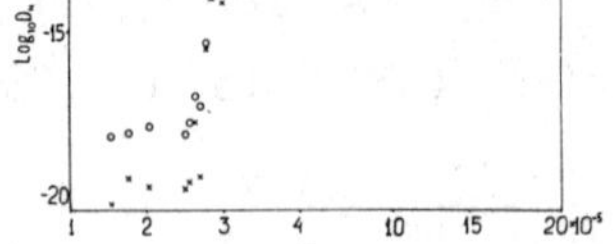

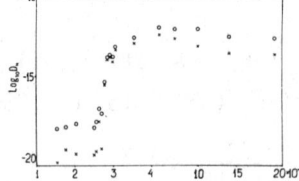

Fig. 14. A sketch of log D_n vs. ε for Hamiltonian (33).

THE BAKER'S TRANSFORMATION

We may illustrate the random walk character of the stochastic zones using a simple, rigorously ergodic[2] area preserving mapping of the unit square upon itself. Here the unit square is stretched

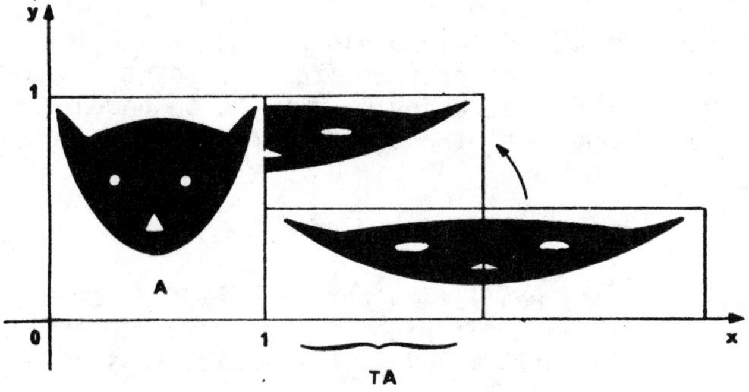

Fig. 15. A drawing of the baker's transformation.

to twice its original length and one-half its original height. This rectangle is then cut vertically and the right half placed on top of the left to reform the unit square as shown in Fig. 15. In essence, each iteration maps the point (x,y) into (2x, y/2).

Let us now develop an arithmetic representation of the baker's transformation by writing the x and y coordinates of an initial point in binary, and then writing x to the right of the decimal in the usual order, but writing y backward to the left of the same decimal. We then typically have

$$\underbrace{\cdots 11010001}_{\leftarrow y_0}\cdot\underbrace{0011101\cdots}_{x_0\rightarrow} \tag{40}$$

Now we observe that moving the decimal to the right gives the forward iterates of the point (x_0,y_0) while moving the decimal to the left provides backward iterates. Clearly each movement of the decimal to the right doubles x and halves y as required; less obviously, it also properly accounts for the cutting and folding. Initially, close points clearly separate exponentially for this model; moreover it may be shown to have an everywhere dense set of unstable periodic orbits. Thus this mapping mimics the stochastic zones previously discussed.

Now let us observe that on each forward iteration or rightward movement of the decimal in representation (40), the first digit to the right of the decimal determines whether the new iterate lies to the left or the right of x = 1/2. But in general, the sequence of zeroes and ones in these binary representations is as random as a coin toss. Thus quite clearly, the iterated points for the baker's transformation random walks between the right and left sides of the unit square. Indeed, if we define an initial probability density $W_0(x)$ on the unit square, then we obtain a random walk type diffusion equation

$$W_{n+1}(x) = (1/2)[W_n(x/2) + W_n((x+1)/2)] , \qquad (41)$$

whose derivation is immediately obvious since the point x on the (n+1) iteration can only be reached from the points (x/2) and (x+1)/2 on the previous iteration. It is to be hoped that the macroscopic consequences of the stochastic zones are now becoming clearer.

THE BOUNCING BALL MODEL

A physically more realistic "random walk" system and one closer to the Chirikov Hamiltonian of the previous section is provided by the bouncing ball model. This example was originally developed by Fermi[4] and Ulam[5] as a highly simplified model of cosmic ray acceleration. Consider a ball bouncing between two infinitely heavy walls, one fixed and one oscillating as shown in Fig. 16. The ball has instantaneous speed v and the moving wall oscillates with amplitude a, period T, and instantaneous speed $V(t)$, where $V(t)$ is a sawtooth function having the maximum value V. The minimum distance between the walls is ℓ. The exact difference equations governing the motion of this system are presented in a paper[6] by Zaslavskii and Chirikov. Following Lieberman and Lichtenberg[7], we elect to consider an approximation to these

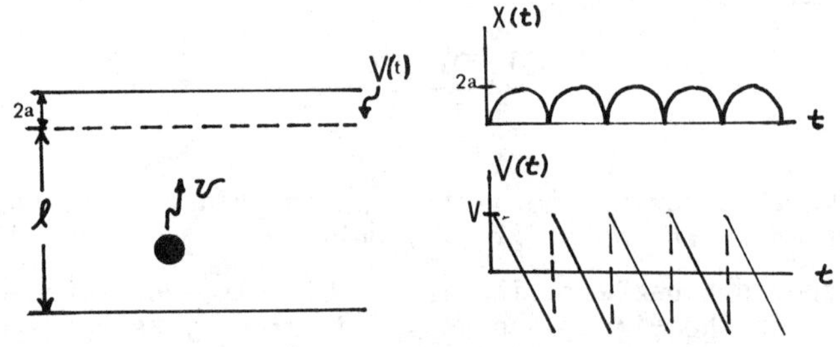

Fig. 16. Diagram of the Fermi-Ulam system used to model the acceleration of cosmic rays.

exact difference equations. The approximating equations are

$$u_{n+1} = |u_n + \psi_n - \tfrac{1}{2}| \qquad (42a)$$

$$\psi_{n+1} = \left[\psi_n + \left(\frac{M}{u_{n+1}}\right)\right] , \ (\text{mod } 1) \qquad (42b)$$

where $u_n = v_n|V$, v_n is the speed of the ball just before the nth collision with the oscillating wall, $\psi_n (0 \leq \psi_n \leq 1)$ is the phase of the oscillating wall at the nth collision, and $M = (\ell/16a)$. Equation (42) is a good approximation to the exact equations of motion provided that $M >> 1$ and $u >> 1$; however, independent of the goodness of these approximations, Eq. (42) yields the same general type of behavior as do the much more complicated exact equations. We therefore confine our attention to Eq. (42).

Our first observation is that Eq. (42) reduces our problem to the study of a plane area-preserving mapping. However, this mapping does not exhibit everywhere exponentially separating orbits throughout the (ψ,u) plane; like the systems of the preceding two sections, this system exhibits a so-called divided phase space[3]. Indeed, taking differentials of Eq. (42), we obtain

$$\begin{aligned} du_{n+1} &= du_n + d\psi_n , \\ d\psi_{n+1} &= d\psi_n - \left(\frac{M}{u^2_{n+1}}\right) du_{n+1} \end{aligned} \tag{43}$$

Now for the (ψ,u) region which shall interest us in these calculations, we have $u_n >> 1$ and $M >> u_n$; moreover, (42a) shows that $u_n >> 1$ varies slowly with n. Thus, let us approximate in (43) and set (M/u^2) equal to a constant, b say. Equation (43) may then be written

$$\begin{aligned} du_{n+1} &= du_n + d\psi_n \\ d\psi_{n+1} &= -b\, du_n + (1 - b) d\psi_n , \end{aligned} \tag{44}$$

where

$$0 < b = \frac{M}{u^2} . \tag{45}$$

A linear change of variables now permits us to write Eq. (44) in the form

$$d\zeta_{n+1} = \lambda\, d\zeta_n, \qquad d\eta_{n+1} = \lambda^{-1}\, d\eta_n , \tag{46}$$

$$\lambda = \tfrac{1}{2}(2 - b) - \tfrac{1}{2}[(2 - b)^2 - 4]^{1/2} . \tag{47}$$

From Eq. (47), we see that λ is real when $b > 4$ and is imaginary when $0 < b < 4$. Thus referring to Eq. (46), we see that iterates of Eq. (44) oscillate when $u > M^{1/2}/2$ and they exponentiate when

$u < M^{1/2}/2$. We thus expect stochastic behavior for the mapping of Eq. (42) in that (ψ,u)-plane region for which

$$u < \frac{M^{1/2}}{2} \tag{48}$$

and smooth curves for $u > M^{1/2}/2$. In the stochastic region, each small initial $(d\psi_0\, du_0)$ area-element grows exponentially in the ζ-direction and shrinks exponentially in the η-direction. Further, since the expanding ζ-direction has small but nonzero slope $(\Delta u/\Delta\psi) \approx -(1/b)$ for $b \gg 1$, both variables u and ψ locally spread exponentially, but ψ spreads more rapidly than u.

In Fig. 17, we show a composite sketch of a typical Eq. (42)-mapping based on several computer-generated figures presented by Brahic[8] and Lieberman and Lichtenberg[7]. In Brahic's paper especially, some of the mapping pictures represent a striking form of abstract art. In Fig. 17, we note that the boundary of the stochastic behavior occurs at about the predicted value of $u = M^{1/2}/2$. For larger u-values, again as predicted, stable as well as unstable fixed points appear. By direct substitution, one easily finds that the mapping T of Eq. (42) has fixed points of T itself at $(\psi,u) = (1/2, M/k)$, where k is a positive integer, and that these fixed points are stable when $u > M^{1/2}/2$. The member of this fixed point set having the largest associated stable region, as seen in Fig. 17, lies at $(1/2,M)$, and physically corresponds to the ball being reflected from the oscillating wall (at $\psi = 1/2$ when the moving wall instantaneously has zero speed) and then colliding again with the moving wall after the elapse of precisely one wall period. Fixed points of T^2, T^3, etc., can also be determined through increasingly long and tedious algebraic manipulations of Eq. (42).

For motion in the stochastic region where $u < M^{1/2}/2$, Lieberman and Lichtenberg[7] first establish that the relaxation times for the u and ψ motion differ widely. They then use this fact to obtain an irreversible rate equation which they validate using a computer. In Eq. (42), let us start with a precise initial (ψ_0,u_0) state (a definite state and not a rectangle $d\psi_0\, du_0$) for which $M \gg u \gg 1$. Then, as n increases, the sequential iterates of ψ_n will rapidly cover the whole interval $0 \lesssim \psi \lesssim 1$ in a "random" manner much before $(\Sigma|\Delta u_n|)/u_0$ becomes large. As a consequence, the fractionally small, sequential iterates Δu_n generated by Eq. (42a) will be positive or negative with about equal frequency, and u_n will perform a relatively slow "random walk" away from the initial region near u_0. Alternatively, consider an ensemble of systems with initial states spread uniformly over a small rectangle $(d\psi_0 du_0)$.

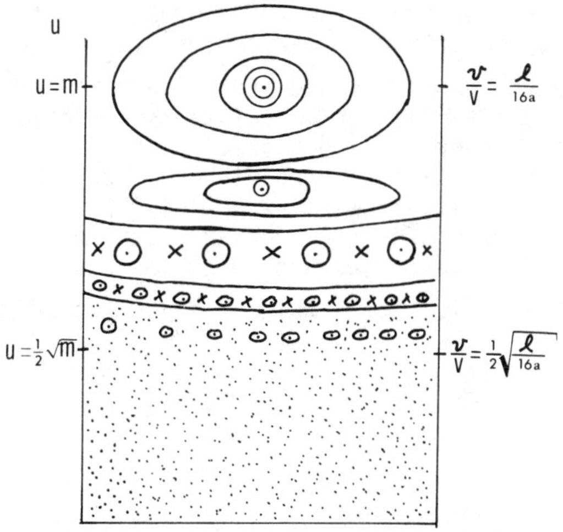

Fig. 17. Sketch of a typical mapping generated by Eq. (42).

According to the discussion following Eq. (48), the small rectangle will spread exponentially along the ζ-direction into an almost horizontal filament with $\delta\psi \sim 1$ and $(\delta u/u) \ll 1$. Although each small segment of this filament will continue locally to grow exponentially, macroscopically the next iteration of this filament will split into two or more new filaments, each having $\delta\psi \sim 1$. Moreover, since the original filament had $\delta\psi \sim 1$, Eq. (42a) ensures that half of the almost horizontal, new filaments lie slightly above (along the u-axis) the original filament and half slightly below. Similarly one more iteration splits each of these new filaments into a newer set, equally split above and below the original, new filament position. Thus in the ensemble, the system phases ψ "randomize" on an exponential time scale followed, on a much longer time scale, by a diffusive spread of the u-values.

On the basis of either of these arguments, one concludes that Eq. (42b) causes the fine-grained density $\int(\psi,u,n)$ to mix continually along the ψ-direction with exponential rapidity. Equation (42a) then ensures that the reduced probability distribution $W(u,n)$ spreads along the u-direction via a much slower random walk process which is known[9] to lead to a type of diffusion equation. One therefore expects that $W(u,n)$ satisfies the Fokker-Planck equation[9]

$$\frac{\partial W}{\partial n} = - \frac{\partial}{\partial u}(BW) + \frac{1}{2}\frac{\partial^2}{\partial u^2}(DW) \ . \qquad (49)$$

In order to verify the use of Eq. (49) for this system, B and D are calculated using Eq. (42) in the Wang-Uhlenbeck formulas[9].

The results can then be compared with the computer-calculated values for B and D. Starting from $W(u,0) = \delta(u - u_0)$, Eq. (49) predicts that the width of W should grow like $n^{1/2}$, which can also be checked against computer calculations. Finally $W(u,\infty)$ should be a constant over the stochastic region. In all cases, theory and computer experiment agree nicely. For example, in Fig. 18, adapted from Ref. 7, we show a plot of $W(u,\infty)$ versus u obtained by integrating Eq. (42) for an ensemble of systems. Here $W(u,\infty)$ is more or less constant up to the stochastic border $u = M^{1/2}/2 = 10^{3/2}/2 \approx 16$, above which W falls off quite rapidly.

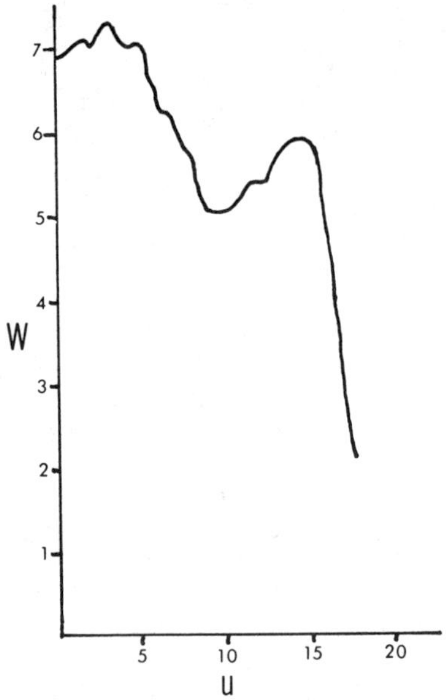

Fig. 18. A graph of the probability density $W(u,\infty)$ as a function of u. W is specified in arbitrary units.

A rigorous derivation of Eq. (49) from Eq. (42) lies in the future. Nonetheless this example and the previous one point in the direction of future progress[10] in developing rate equations for physical systems exhibiting stochastic behavior.

CONCLUDING REMARKS

Hamiltonian systems of the type $H = H_0(J) + \varepsilon V(J,\phi,\tau)$ exhibit a complete spectrum of behavior ranging from complete integrability to complete stochasticity. In this review, we have presented a pictorial, intuitive discussion of this behavior in terms of non-linear resonances and their overlap. In this view, Hamiltonian systems whose non-linear resonances have negligible (or no) overlap are either precisely integrable or else nearly integrable. On the other hand, Hamiltonian systems whose non-linear resonances increasingly overlap as some parameter is varied exhibit chaotic orbits in increasingly large phase space regions. Indeed example systems[11] have recently been discovered in which one can observe the full transition from complete integrability to complete stochasticity.

In the latter sections of this review, we have discussed example systems which provide some insight into the macroscopic consequences caused by the chaotic orbits of the underlying microscopic dynamics. In a not too remote future, one anticipates that all of equilibrium and non-equilibrium statistical mechanics will be rigorously derivable from basic dynamics, either classical or quantum. But already, practical applications are being developed in astronomy, biology, chemistry and in many areas of physics, and these developments will likely continue into the distant future.

Many people have been and are participating in this work, and the author apologizes to those whose contributions are not directly referenced here. Perhaps it suffices merely to publically state that the present author's meager understanding of this subject has been derived from a superior understanding possessed by a host of non-linear scientists so large that even a partial listing would be tediously long.

REFERENCES

1. One may find further references to the now extensive literature in J. Ford, Fundamental Problems in Statistical Mechanics, Vol. 3, edited by E. D. G. Cohen (North-Holland, Amsterdam, 1975); A. S. Wightman, Statistical Mechanics at the Turn of the Decade, edited by E. D. G. Cohen (North-Holland, Amsterdam, 1968); J. Moser, Mem. Am. Math. Soc., #81 (1968); G. Contopoulos, Bull. Astron. 2, 223 (1967); N. Saito, N. Ooyama, Y. Aizawa, and H. Hirooka, Prog. Theor. Phys. Suppl. 45, 209 (1970); G. M. Zaslavsky and B. V. Chirikov, Soviet Phys. Uspekhi 14, 549 (1972).
2. V. I. Arnold and A. Avez, Ergodic Problems of Classical Mechanics (Benjamin, New York, 1968).
3. B. V. Chirikov, Phys. Reports (to appear in 1978). The present author is appreciative for being allowed to review this paper prior to publication.
4. E. Fermi, Phys. Rev. 75, 1169 (1949).
5. S. M. Ulam, Proceedings of the Fourth Berkeley Symposium on Mathematical Statistics and Probabilities, Vol. 3 (Univ. California Press, Berkeley, 1961).
6. G. M. Zaslavsky and B. V. Chirkov, Soviet Phys. Doklady, 9, 989 (1965).
7. M. A. Lieberman and A. J. Lichtenberg, Phys. Rev. A5, 1852 (1972).
8. A. Brahic, Astron. Astrophys. 12, 98 (1971).

9. M. C. Wang and G. E. Uhlenbeck, Rev. Mod. Phys. 17, 323 (1945).
10. A. N. Kaufman, Phys. Rev. Lett. 27, 376 (1971).
11. G. Benettin and J. M. Strelcyn, Phys. Rev. A (to appear in 1978).

THEORY OF CHAOTIC MOTION WITH APPLICATION TO CONTROLLED FUSION RESEARCH

by

Yvain M. Treve
125 San Rafael Avenue
Santa Barbara, CA 93109

INTRODUCTION

The mathematical techniques that we review in this report have been selected on the basis of their relevance to at least four outstanding theoretical problems of magnetic fusion research, namely: a) ion heating; b) particle-wave interactions; c) stability of magnetic surfaces in real tokamaks; and d) strong plasma turbulence. These problems have a common feature: they all involve chaotic motions in spite of the perfectly deterministic nature of the mathematical models used for their description.

The status of our understanding about these chaotic motions differs markedly depending on the type of the mathematical models:

Regarding problems a), b), and c), their models belong to the class of the so-called non-integrable Hamiltonian systems which always possess some solutions with a chaotic behavior. To be exact, the analytical proof of this property has been given only for systems with two degrees of freedom but there is ample evidence from several theoretical and numerical computations that the property belongs generically to every non-integrable Hamiltonian system.

As to turbulence, even in hydrodynamics, there is no rigorous proof available that the Navier-Stokes equations possess turbulent solutions. This is an extremely difficult problem because the phase-space of these equations, i.e. the space of all possible velocity vector fields, has an infinite number of dimensions. Physicists and mathematicians have invoked in the past various mathematical phenomena in order to explain how the Navier-Stokes equations can possess turbulent solutions but none of these early conjectures has survived. Even the famous picture proposed by Landau in the early forties is no longer tenable since several of its implications are contradicted by experiment.

Recently, however, a new picture has been proposed by the mathematicians Ruelle and Takens which does not suffer from the defects of Landau's picture and is mathematically more plausible.

The novel and central elements of the Ruelle-Takens picture are completely new kinds of mathematical objects called strange attractors. In contrast with the well known attractors such as stable nodes, foci, or limit cycles, a strange attractor is a highly complex structure sitting in the phase-space of the dynamical equations which traps for ever all orbits passing nearby. Its most important property is that it imparts to these motions a chaotic character and an extreme sensitivity to the initial conditions.

ISSN: 0094-243X/78/147/$1.50

Although the Ruelle-Takens picture is no more than a guess at the present time it is generally believed that the actual state of affairs is not too far from it. This belief is based on preliminary mathematical results which suggest that most equations like the Navier-Stokes equations possess strange attractors in their phase-space. Also, as already mentioned, the picture does not have the defects of Landau's picture and is in qualitative agreement with several experiments specially designed to test its validity.

The important point that we want to make in this report is this: on the basis of already proven facts there is strong evidence that most realistic mathematical models of complex physical phenomena possess under certain conditions some solutions with a random, chaotic behavior. It must be emphasized that the chaotic character of these solutions is intrinsic to the model and does not require any additional assumptions about the occurence of randomness. Moreover, the set of initial conditions corresponding to these solutions has finite measure, i.e. they are observable. It appears therefore indispensable that physicists be aware of the possible existence of such solutions in order not to miss new ways of achieving some desirable effects (e.g. ion heating by electrostatic waves) and correctly interpret some experimental data. Finally, we feel that every physicist's panoply of mathematical concepts should include that of strange attractor as a new possible model of disorder in dissipative phenomena.

The kind of mathematics necessary nowadays for the in-depth study of dynamical systems is unknown to most physicists: abstract algebra, general topology, differential topology, algebraic topology, global analysis, analysis on manifolds, etc. We hope however that we have managed to write our exposé in a form readily understandable to the reader not versed in these disciplines.

PLAN

The main text is divided into two sections.

In the first section devoted to Hamiltonian systems we briefly review the essentials of the Hamilton-Jacobi theory and discuss the Kolmogorov-Arnold-Moser theorem and its implications. This gives us the opportunity to expose various standard mathematical notions and methods. In particular we present the most recent developments in perturbation theory, notably the so-called superconvergent schemes. We also discuss at some length Poincaré's method of section and area-preserving mappings for two-degree of freedom systems. In this connection, Smale's horseshoe map and its properties are described to give an idea of the methods used in symbolic dynamics for the study of random motions. This is followed by an examination of the phenomena associated with an increase in the size of the perturbation and a review of the various avenues presently being followed for the determination of the onset of stochasticity. We suggest that the modern version of an old method of Poincaré, the PCHG method (for Poincaré-Cesari-Hale-Gambill), can perhaps be used to elucidate some aspects of this problem and we indicate how to apply it to a wide class of slightly perturbed integrable systems.

The rest of the section is devoted to a review of some recent mathematical results which seem of potential value for plasma physics and to a discussion of the method of averaging and its limitations.

In section 2 we review the difficulties of the problem of turbulence and present the Ruelle-Takens picture. An example of a dynamical system with a strange attractor is constructed and the Hopf bifurcation theory is discussed. Finally we review the properties of the Lorenz model for the convective instability of an atmospheric layer which is known to have a strange attractor for sufficiently high Rayleigh numbers.

Some of the more mathematical derivations are relegated to several appendices.

References indicated by an S followed by a number are listed at the end under the heading "Sources". We give there selected sources for further reading as well as various remarks and comments.

ACKNOWLEDGEMENTS

First, I should like to express my profound gratitude to Dr. Oscar Manley for his constant support and guidance during the preparation of this report. I have also benefited greatly from conversation with Professors S. Smale, J. K. Hale, J. LaSalle, J. A. Yorke, A. Weinstein, R. B. Leipnik, and A. J. Dragt. I am particularly indebted to Professor Alan Kaufman for having read the manuscript; his constructive criticisms have been invaluable to me. Furthermore, I should like to express my appreciation to Mrs. Sarah Wilson for having typed the manuscript. This work was supported by the U.S. Department of Energy under Contract #EA-77-X-01-2865.

1. CHAOTIC MOTIONS IN HAMILTONIAN SYSTEMS

1.1 EARLIER THEORY OF HAMILTONIAN SYSTEMS

We recall that Hamiltonian systems are of the form (S-1)

$$\dot{q} = \frac{\partial H}{\partial p}, \quad \dot{p} = -\frac{\partial H}{\partial q} \tag{1}$$

where q, p are n-vectors $q = (q_1, \ldots, q_n)$, $p = (p_1, \ldots, p_n)$, the dot denotes differentiation with respect to the time t, and H, the Hamiltonian, is a given function of these variables which may also depend upon the time. The components of q and p are called the generalized coordinates and momenta, respectively. Each pair (q_i, p_i) is associated with one of the n degrees of freedom of the system.

It is always possible to convert a time-dependent Hamiltonian into a time-independent one by adding one more degree of freedom.

We shall take advantage of this possibility to restrict ourselves to time-independent Hamiltonians H(q,p). Then H is the total energy E of the system which is a constant of the motion, i.e. H(q,p) is a first integral of Eqs. (1) which can be used to reduce their order by one.

Of special importance are those Hamiltonian systems for which n single-valued independent first integrals F(q,p) are known and are such that $[F_i, F_j] = 0$ for all i, j=1, 2, 3, ..., n where the Poisson bracket of two arbitrary functions of q and p is defined by

$$[U,V] = \sum_{i=1}^{n} \left(\frac{\partial U}{\partial q_i}\frac{\partial V}{\partial p_i} - \frac{\partial U}{\partial p_i}\frac{\partial V}{\partial q_i}\right) .$$

First integrals satisfying this condition are said to be in involution.

In general, knowledge of a first integral of a differential system allows a reduction of its order by one. For Hamiltonian systems we can do better: with a known first integral different from the Hamiltonian we can reduce the order of Eqs. (1) by 2 and if we know n integrals in involution we can completely solve the problem according to the following procedure:

We seek a change of variables q,p → q', p' such that the differential equations for the new variables are derivable from a new Hamiltonian H'(q', p') as the Eqs. (1) for the old ones are derived from H(q,p), i.e.

$$\dot{q}' = \frac{\partial H'}{\partial p'} , \quad \dot{p}' = - \frac{\partial H'}{\partial q'} . \qquad (2a,b)$$

A transformation possessing this property is called canonical and from the Hamilton-Jacobi theory is derivable from a characteristic function S(q,p') according to the prescription

$$q' = \frac{\partial S}{\partial p'} , \quad p = \frac{\partial S}{\partial q} . \qquad (3a,b)$$

The importance of knowing a set of n single-valued first integral in involution is due to the fact that it is then possible to construct a characteristic function S such that the new Hamiltonian does not depend upon q' : H'(q',p') = H'(p'). Indeed this implies by Eq. (2b) that $\dot{p}'=0$, hence p' and therefore $\partial H'/\partial p' = \dot{q}' = \omega$ are constant n-vectors so that Eqs. (2) are readily integrated:

$$q'(t) = \omega t + q'_0 , \quad p' = p'_0 , \qquad (4a,b)$$

and q(t) and p(t) are obtained via the inverse transformation. The characteristic function which allows us to achieve this feat is found thus: set the n integrals F_i (or any set of n functionally independent combinations of them) equal to the components of the new momenta, i.e.

$$F(q,p) = p'$$

and solve this vector equation for $p = p^*(q,p')$. The differential form $p^*(q,p').dq$ is then the perfect differential of S which is therefore given by

$$S(q,p') = \int^{q} p^*(q,p').dq \ .$$

Once S is computed Eqs. (3) are solved for $q = q^*(q',p')$; substitution of this into (3b) gives $p = p^*(q',p')$ and the new Hamiltonian automatically does not depend upon q'. The solution of Eqs. (1) is then obtained as

$$q(t) = q^*(\omega t + \frac{\partial S}{\partial p'} \ , \ F(q_o,p_o)), \ p(t) = \frac{\partial S}{\partial q} (q(t), \ F(q_o,p_o))$$

where, in $\partial S/\partial p'$, $p' = F(q_o,p_o)$ and $q = q_o$.

Hamiltonians for which n integrals in involution are known are said to be integrable. But no systematic method is known for finding first integrals and there is to-date no analytical test to decide whether a given Hamiltonian system has single valued first integrals besides the Hamiltonian. The only approach available is via exploration of the geometrical configuration of the phase-space by numerical integration of the orbits.

It was recognized long ago that for the many integrable problems of classical mechanics the manifold M defined by $F_i(q,p)$ = constant, i=1, 2, ..., n, is an n-dimensional torus, i.e. the topological product of n circumferences. Arnold (S-2) has shown rigorously using topological arguments that it is so whenever at every point of M the gradients of the functions F are linearly independent. On a particular torus corresponding to a given value of F(q,p) the motion is quasi-periodic in general (periodic if the frequencies are commensurable) and given by Eqs. (4a) where the coordinates are defined modulo a fundamental period which is usually chosen equal to 2π. With this choice of the period, the "momenta" have the dimension of action and the coordinates are angles (S-3).

It is seen from Eqs. (4a) that depending upon the values of the frequencies ω_i and therefore on the torus the motion may be periodic or quasi-periodic (some authors say multiply periodic; the old equivalent expression "conditionally periodic" is progressively being abandoned). It is periodic for those values of q_o and p_o such that there is a time T for which $\omega_i T = k_i 2\pi$, k_i an integer, i = 1, 2, ..., n, i.e. when the frequencies are commensurable; then the orbits are closed curves on the torus. If instead all frequencies are incommensurable for a set of values of q_0, p_0 then the orbits fill the corresponding torus like a Lissajous curve. The set of such tori has positive measure while the set of tori on which the motion is periodic has zero measure since it is in one-to-one correspondence with the rational numbers.

We thus see that the behavior of integrable Hamiltonian systems is fairly well understood. In sharp contrast, the behavior of systems for which n single-valued integrals in involution do not exist is extremely complicated. Poincaré was the first to realize this fact in connection with his investigation of the restricted problem of 3 bodies. Taking advantage of the possibility of reducing this problem to a Hamiltonian system with only two degrees of freedom he obtained many important results some of which are generic, i.e. are to be found for all Hamiltonian systems of this kind (13). Following the path opened by Poincaré, Birkhoff (14) found many important properties of systems with two degrees of freedom which shed some light on the complexity of their behavior. But it is only recently that a full measure of this complexity and of its possible implications in mechanics and physics was achieved through both analytical and numerical work.

1.2 THE KOLMOGOROV-ARNOLD-MOSER THEOREM

A major landmark in the evolution of our knowledge about non-integrable systems was the announcement by Kolmogorov in 1954 of his famous theorem on the behavior of slightly perturbed Hamiltonian systems (cf ref. 7, Appendix 7). Kolmogorov gave only a sketch of the proof (15) which was latter carried out in detail by Arnold (16), (S-4) for analytic Hamiltonians. Finally, Moser (S-5) gave another proof under weaker assumptions. The theorem has thus become to be known as the KAM theorem for short.

A good understanding of the KAM theorem requires some familiarity with the theory of perturbations developed by the astronomers for the computation of the orbits of heavenly bodies. The theory deals with Hamiltonians of the general form

$$H(q,p) = H_0(q,p) + \varepsilon H_1(q,p,\varepsilon)$$

where ε is a small parameter and the Hamiltonian system associated with the unperturbed Hamiltonian H_0 is integrable in the sense indicated before. We may therefore assume that the appropriate canonical transformation has been applied to the original system so that H_0 depends upon p only. We further assume that H_1 is analytic in ε and can therefore write the Hamiltonian in the form

$$H(q,p) = H_0(p) + \varepsilon H_1(q,p) + \varepsilon^2 H_2(q,p) + \dots \quad (5)$$

Many different schemes are available (S-6) for generating approximations to the solution and they are all more or less equivalent differing only in their relative ease of implementation. We choose to present the scheme indicated by Arnold (12). This scheme is distinct from the superconvergent scheme suggested by Kolmogorov and developed by Arnold for the proof of the theorem (see ref. 16 and Appendix A).

Arnold seeks a transformation $q,p \to q',p'$ defined by a characteristic function S

$$q' = q + \varepsilon \frac{\partial S}{\partial p'} \ , \ p = p' + \varepsilon \frac{\partial S}{\partial q} \qquad (6a,b)$$

such that in the new Hamiltonian the terms depending on q' be of order ε^n with $n \geq 2$, i.e.

$$H(q,p) = H_0'(p',\varepsilon) + \varepsilon^2 H_1'(q',p') + \varepsilon^3 \ldots . \qquad (7)$$

If this can be accomplished, during a time interval $t \simeq 1/\varepsilon$ the actual solution q'(t), p'(t) will differ from the motion described by $H_0'(p',\varepsilon)$ by a quantity of order $\varepsilon^2 t$. In terms of the original q and p this will give approximate expressions for q(t), p(t) with errors $O(\varepsilon)$ over such a time interval. If we continue this process a transformation $q', p' \to q'', p''$ will bring the Hamiltonian to the form

$$H(q,p) = H_0''(p'',\varepsilon) + \varepsilon^3 H_1''(p'',p'') + \ldots$$

and the error will now be of the order of $\varepsilon^3 t$ over a time interval t, and so on.

To carry out this program we replace p in Eq. (5) by the right hand side of Eq. (6b) and Taylor expand H_0 about $p = p'$; we thus obtain

$$H(q,p) = H_0(p') + \varepsilon \frac{\partial H_0}{\partial p} \cdot \frac{\partial S}{\partial q} + \varepsilon H_1(q,p') + O(\varepsilon^2) \ .$$

We can now take advantage of the 2π-periodicity of H_1 in the q's to rewrite this as

$$H(q,p) = H_0(p') + \varepsilon \overline{H}_1(p') + \varepsilon \left(\frac{\partial H_0}{\partial p'} \cdot \frac{\partial S}{\partial q} + \tilde{H}_1(q,p') \right) + O(\varepsilon^2)$$

where $\overline{H}_1(p')$ is the mean value of H_1 and $\tilde{H}_1(q,p') = H_1(p') - \overline{H}_1(p')$ has therefore mean value zero. Clearly Eq. (7) will be obtained if we choose S such that

$$\frac{\partial H_o}{\partial p'} \cdot \frac{\partial S}{\partial q} + \tilde{H}_1(q,p') = 0 \ . \tag{8}$$

Writing the Fourier expansions of $\tilde{H}_1$ and S as

$$\tilde{H}_1(q,p') = \sum_{k\neq 0} h_k(p') \exp(i(k,q))$$

and

$$S(q,p') = \sum_{k\neq 0} S_k \exp(i(k,q)),$$

respectively ($(k,q) = \sum_{i=1}^{n} k_i q_i$), and substituting into Eq. (8) yields the Fourier coefficients for S

$$S_k(p') = i \frac{h_k(p')}{(k,\omega)} \ . \tag{9}$$

This is where we encounter the old problem of the small divisors $(k,\omega) = \sum_{i=1}^{n} k_i\omega_i$ (12): in order to be able to compute the coefficients for all k's (k,ω) must be different from zero which implies that the values of the frequencies must be incommensurable. But even if the ω's are incommensurable the procedure will not work in general because as we go to higher and higher harmonics, that is as the modulus of k increases indefinitely, some of the divisors (k,ω) will become arbitrarily close to zero. The convergence of the series for S is then in doubt.

This difficulty is encountered in all traditional perturbation methods. Yet their inventors, the astronomers, have used them with great success for the calculation of orbits over very long times. The reason for their success is that in practice the frequencies are often such that the divisors become small only for very large values of $|k|$. The corresponding values of the Fourier coefficients are then very small so that by truncating the series appropriately an excellent approximation to the solution is obtained. This will become clearer after we have discussed the KAM theorem which can be stated as follows (see Appendix B):

Suppose again that $H(q,p) = H_o(p) + \varepsilon H_1(q,p)$ is analytic in some domain G of the phase-space with ε a small parameter and H_1 2π-periodic in the q's. If in G either

$$\Delta_1 = \det\left(\frac{\partial^2 H_o}{\partial p_i \partial p_j}\right) \neq 0 \text{ or } \Delta_2 = \det\begin{pmatrix} \frac{\partial^2 H_o}{\partial p_i \partial p_j} & \frac{\partial H_o}{\partial p_i} \\ \frac{\partial H_o}{\partial p_j} & 0 \end{pmatrix} \neq 0 \ , \tag{10}$$

then G can be decomposed into two disjoint subsets: $G = G_1 + G_2$, where G_1 is invariant (i.e. the solution point q(t), p(t) belongs to G_1 for all times for any initial point q(0), p(0) belonging to G_1) and G_2 is small compared to G and tends to zero with ε. Moreover G_1 is composed of invariant n-dimensional analytical tori on which the motion is quasi-periodic with frequencies $\omega_i = \partial H_o/\partial p_i$.

In other words, most of the invariant tori corresponding to the unperturbed Hamiltonian still exist under a small perturbation. They fill a domain G_1 which occupies most of phase-space.

Note that the theorem does not tell what happens in the small domain G_2. We will come back to this question later but first we want to show that the theorem is readily applicable to a problem related to ion heating, namely the non-relativistic motion of an ion in a uniform magnetic field $B = B_o\hat{z}$ under the perturbation of an electrostatic wave.

In rectangular coordinates the Hamiltonian can be written in the form (S-7)

$$H = \frac{1}{2m}(\underline{p} + m\Omega y\hat{x})^2 + e\phi_o \sin(\underline{k}\cdot\underline{r} - \omega t) \qquad (11)$$

Without loss of generality $\underline{k}$ can be chosen parallel to the y,z - plane ($\underline{k} = k_z\hat{z} + k_1\hat{y}$). There is then a time dependent canonical transformation which brings the Hamiltonian to the time independent and dimensionless form (25)

$$H(J_1, J_2, w_1, w_2) = H_o(J_1, J_2) + \varepsilon H_1(J_1, w_1, w_2) \qquad (12)$$

where

$$H_o = J_1 + \nu J_2 + \frac{1}{2}\beta^2 J_2^2 \qquad (13a)$$

and

$$H_1 = \sin(\alpha(2J_1)^{1/2} \sin w_1 + w_2) . \qquad (13b)$$

Here J_1 and J_2 are actions conjugate to the angles w_1 and w_2, α and β are the direction cosines of k, $\nu = \omega/\Omega$, and ε measures the amplitude of the wave.

For $\varepsilon = 0$, the motion takes place on invariant tori and is periodic or quasi-periodic depending on whether the frequencies $\omega_1 = 1$ and $\omega_2 = \nu + \beta^2 J_2$ are commensurable or not. The KAM theorem applies for ε small and in the general case $\beta = k_z/k \neq 0$ since then the determinant $\Delta_2 = \beta^2 \neq 0$. Hence for ε small and $\beta \neq 0$ most of the motions take place on invariant tori and are quasi-periodic.

Note that the theorem does not apply in the situation considered by Karney and Bers (22) and Karney (23) where the wave propagates at right angle to the magnetic field since then $\beta = 0$ so that $\Delta_1 = \Delta_2 \equiv 0$.

For ε different from zero there are motions other than the quasi-periodic ones which fill a subset of phase-space of small measure and whose behavior, as we shall see, is of an entirely different nature. But first we want to make an important remark: for 2-degree of freedom Hamiltonian systems such as the one above, the phase-space is 4-dimensional but can be reduced to 3 dimensions by eliminating one of the variables using the energy integral. This considerably simplifies the study of the orbtis especially if we use the ingenious method of sections invented by Poincaré (13, Vol. 3).

Suppose γ is a closed periodic orbit and let α be a point on it. Through α let us pass a smooth surface Σ whose tangent plane makes a finite angle with the tangent to γ at α (γ and Σ are then said to intersect transversally). Now pick another point β in Σ and close to α, and consider the orbit γ' going through it. By reason of continuity we can expect that after a certain time γ' will pierce again Σ at a point β_1. If $\beta_1 = \beta$ we again have a periodic orbit; if not, we can follow γ' and mark in Σ the successive points $\beta_1, \beta_2, \ldots$ where γ' crosses Σ as t increases and we do the same as t decreases obtaining $\beta_{-1}, \beta_{-2}, \ldots$. When the system is integrable and the orbits lie on tori it is possible to find a plane of section which intersects all orbits. The process that we have just described can be viewed as a mapping $\Phi : \Sigma \to \Sigma$ which takes a point p of the plane of section into a point p' of the same plane. This mapping is called a Poincaré map. Repeated application of the mapping $\Phi(p)$, $\Phi(\Phi(p)) = \Phi^2(p)$, etc. in the case of a periodic orbit will yield a finite n such that $\Phi^n(p) = p$. Then if we let A denote the set of points $\{p, \Phi(p), \ldots, \Phi^{n-1}(p)\}$ any iterate of a point $a \varepsilon A$ also belongs to A. This property is expressed by saying that the set A is invariant under the map. There are also lines which are invariant under the map, i.e. if Γ is such a line then $\Phi(p) \varepsilon \Gamma$ for all points $p \varepsilon \Gamma$. As we shall see the existence and properties of the invariant sets of the Poincaré map are of prime importance for the understanding of the behavior of Hamiltonian systems with two degrees of freedom.

As an illustration of how such mappings work we consider the motion generated by the unperturbed Hamiltonian H_0 as given by Eq. (13a) in the case $\beta \neq 0$.

The solution is then

$$J_1 = J_1^0 \ , \ w_1(t) = t + w_1^0 \qquad , \bmod 2\pi,$$

$$J_2 = J_2^0 \ , \ w_2(t) = (\nu + \beta^2 J_2^0)t + w_2^0 \ , \ \text{mod } 2\pi,$$

where the superscript 0 refers to the initial values at t = 0.

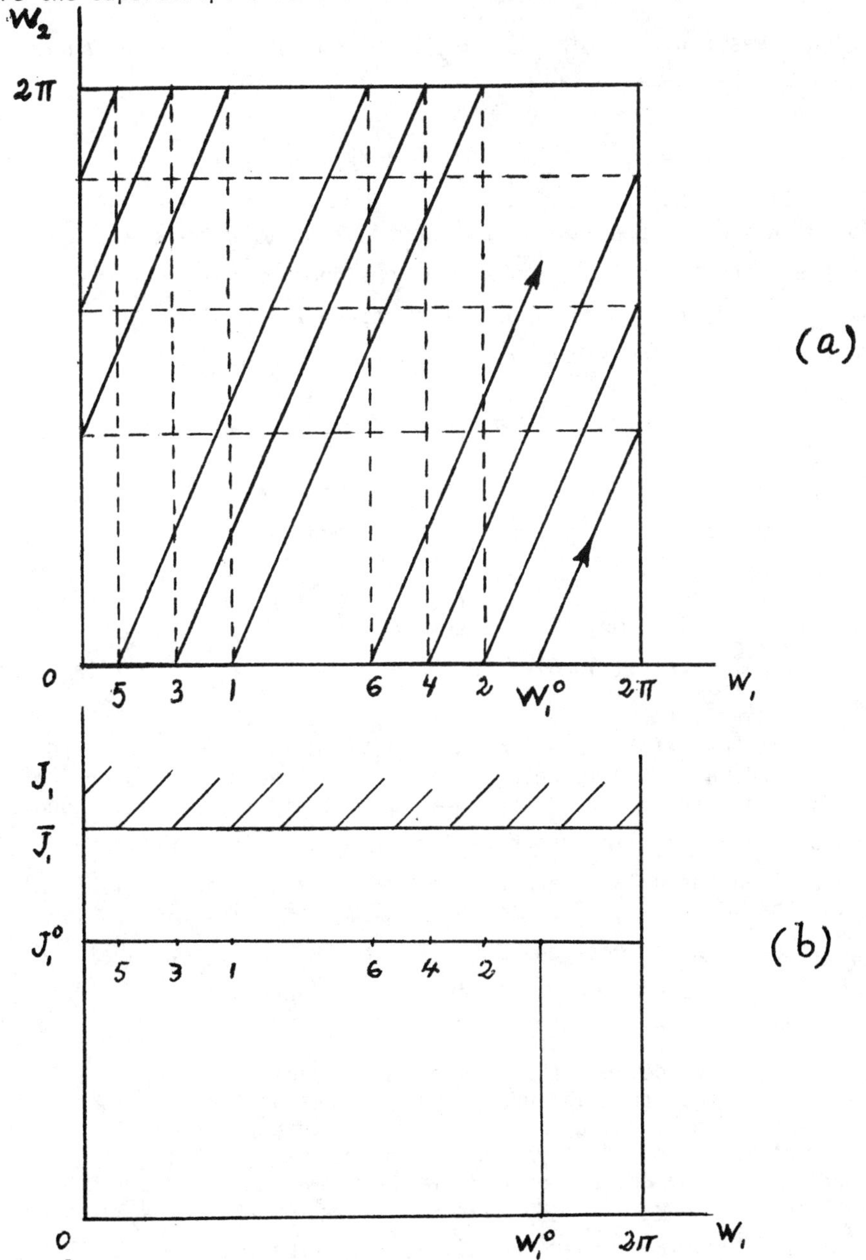

Figure 1.

We fix the energy and eliminate J_2 using

$$J_1 + \nu J_2 + \frac{1}{2}\beta^2 J_2^2 = E$$

which gives two determinations $J_2'(E,J_1)$ and $J_2''(E,J_1)$. Concentrating upon the determination

$$J_2'(E,J_1) = \frac{1}{\beta^2}(-\nu + (\nu^2 - 2\beta^2(J_1 - E))^{1/2})$$

for values of J_1 and E which make J_2' real we take the plane $w_2 = 0$ as the plane of section in the 3-dimensional phase-space of J_1, w_1, w_2. The projection of an orbit on the w_1-w_2 plane looks as in Figure (1a), the slope of the orbit being equal to

$$\omega_2(J_1^0) = \nu + \beta^2 J_2'(E,J_1^0) \ .$$

Each time the orbit crosses the plane $w_2 = 0$ we obtain a value for w_1 that we mark on the corresponding segment $J_1 = J_1^0$ in the plane of section (Fig. 1b).

Thus the study of a Poincaré map associated with an integrable system is fairly simple: the iterates $\Phi(\alpha)$, $\Phi^2(\alpha)$, ... of a point $\alpha = (w_1^0, J_1^0)$ are all located on the segment $0 \leq w_1 \leq 2\pi$ of the straight line $J_1 = J_1^0$.

Note that if the problem is formulated in terms of the ordinary coordinates and momenta the invariant lines of a 2-degree of freedom integrable system are no longer straight. Indeed, suppose that the plane $q_2 = q_2^0$ = constant is an appropriate surface of section. Then these lines are defined by the energy integral $H(q,p) = E$ and the additional first integral $F_1(q,p) = C$ in which q_2 is set equal to q_2^0, E is kept constant and C is the parameter labeling these lines. These two equations in three unknowns q_1, p_1, p_2 are equivalent to a single equation in q_1, p_1, say, which depends on C and therefore defines the one-parameter family of invariant lines. It is thus clear that in general these lines are not straight.

When a small perturbation is taken into account things become extremely complicated if the perturbed system is not integrable. Returning to the action-angle variable formulation, by the KAM theorem there are still continuous smooth lines in the w_1-J_1 plane which are invariant under the motion just like the segments $J_1 = J_1^0$

where invariant in the integrable case. These invariant lines fill most of the area available in the strip $0 \leq w_1 \leq 2\pi$ but not completely: there are gaps between them which cover a finite, although small, area.

Within these gaps one finds two kinds of motions: 1) first there are periodic motions which come from certain periodic motions of the unperturbed system, i.e. although the invariant tori corresponding to commensurable frequencies for $\varepsilon = 0$ are destroyed some of the periodic orbits that they carry survive the perturbation; 2) the rest of the motions are not periodic and display a genuine random behavior.

A heuristic explanation for the appearance of these gaps can be given:

As already indicated, in order for the series $S(q,p')$ to converge it is necessary that none of the ratios ω_i/ω_j, $i \neq j$, of the components of the frequency vector ω be a rational number. This requirement is not serious, however, for in all probability it will be fulfilled for an ω picked at random since the set of rational numbers has measure zero. Now, in expression (9) for the coefficient S_k the numerator depends solely on the Fourier coefficient h_k which, as is well known, tends to zero exponentially with $|k| \equiv \sum_1^n |k_i|$ under the analytic assumptions made about H_1. There is then the question whether there exist frequencies ω such that the small denominators (k,ω) tend to zero slowly enough so that the series for S converges. It was Kolmogorov's great contribution to point out that a classical theorem of the theory of the so-called Diophantine approximations (26) provides a positive answer to this question. This theorem states that given a frequency ω it is possible to find a number K depending on ω but not on k such that the inequality

$$|(k,\omega)| \geq K(\omega)|k|^{-(n+1)} \tag{14}$$

holds for all $k \neq 0$. Using this fact it is then easy to show that given an uncommensurable ω there is a number K for which the series for S converges.

However, the convergence of the series for S is not enough since in the perturbation scheme S is but the first of an infinite sequence of characteristic functions $S = S^{(1)}, S^{(2)}, \ldots$ producing a sequence of angle-independent Hamiltonians $H_0^{(1)}, H_0^{(2)}, \ldots$, whose calculation introduces an additional difficulty: at each level s of approximation the frequency is different since it is given by $\omega = \partial H_0^{(s)}/\partial p^{(s)}$.

Arnold found that in order for all these series to converge ω must belong to a certain set Ω of positive measure.

In the space of all admissible frequencies the complement Ω_c of Ω has a finite measure for $\varepsilon > 0$ which tends to zero with ε. What differentiates the frequency vectors in Ω from those in Ω_c is that the irrational ratios of their components, i.e. ω_i/ω_j cannot be approximated by rational numbers as well as such ratios associated with the frequency vectors belonging to Ω_c. What is meant here is that given $\mu>0$ the inequality $|(k,\omega)|\leq\mu$ can be satisfied when ω is in Ω_c for values of $|k|$ much smaller than when $\omega\varepsilon\Omega$. For instance, consider the two irrational numbers: $N \equiv (\sqrt{5} - 1)/2 = 0.618033988749...$ (the golden mean), and $\pi/5 = 0.6283185308...$. If in (k,ω) we take $k_1 = 71$, $k_2 = -113$ we get $71 - 113 \times(\pi/5)<7\times10^{-6}$ for $|k|=|71|+|-113|= 184$. On the other hand, in order to have $|k_1 + k_2N|<7\times10^{-6}$ it can be shown using the theory of continued fractions that it is necessary to go up to $|k| = 121393$, corresponding to $k_1 = 46368$, $k_2 = -75025$.

Now, to each frequency in Ω there correspond a finite number of tori $p^{(\infty)}$ obtained by inverting the equation $\omega = \partial H_0^{(\infty)}/\partial p^{(\infty)}$ and which are in one-to-one correspondence with tori $p(\omega)$ of the unperturbed system through the inverse of the canonical transformation $q, p \to q^{(\infty)}, p^{(\infty)}$. These are the tori which are preserved under the perturbation and they fill most of phase-space since the volume they occupy is proportional to the measure of Ω. On the contrary, the other tori, i.e. those corresponding in the same fashion to the frequencies in the complement Ω_c disappear under the perturbation leaving vacant a volume vanishingly small with ε like the measure of Ω_c.

Finally, because any neighborhood of a frequency in Ω contains frequencies in the complement Ω_c, there are disappearing tori in the neighborhood of any of the survivors, i.e. the volume occupied by the disappearing tori is the union of thin gaps between surviving tori which are present everywhere in phase-space. As mentioned before, these gaps are filled with orbits of a highly complicated nature, in contrast with the orbits lying on the preserved tori which give rise to quasi-periodic motions.

At the present time, the most powerful tool available for the study of these complicated orbits is Poincaré's method of section. For more than two degrees of freedom his method is difficult to apply the "surfaces" of section having 2n-2 dimensions; but for n = 2 they are usual two-dimensional surfaces so that their corresponding Poincaré map is not as difficult to study. Many fascinating generic properties of two-degree of freedom Hamiltonian systems can then be derived from those of their associated Poincaré map as we shall see.

1.3 POINCARÉ MAPS AND AREA-PRESERVING MAPPINGS

If x and y are appropriate curvilinear coordinates defined on a surface of section Σ of a two-degree of freedom, analytic Hamiltonian system, the Poincaré map Φ takes a point (x,y) into a point (x',y'), i.e.

$$\Phi : x' = f(x,y),\ y' = g(x,y) \ , \qquad (15)$$

where f and g are analytic functions. For non-integrable systems these functions cannot be determined; yet, it turns out that they possess a generic property from which a surprisingly large number of qualitative features of the system can be found. This property is that there always exists a certain function Q(x,y) such that the integral $\int Q(x,y)\,dxdy$ has the same value when evaluated over any region of Σ as over its image under Φ (27). Clearly this property can be expressed equivalently by the equation

$$Q(x,y) = Q(x',y')\,\frac{\partial(x',y')}{\partial(x,y)} = Q(x',y')\ \left(\frac{\partial f}{\partial x}\frac{\partial g}{\partial y} - \frac{\partial g}{\partial x}\frac{\partial f}{\partial y}\right)$$

that the functions f and g satisfy. It turns out that the very existence of a surface of section implies that the function Q does not vanish in the region of definition of the mapping Φ. As a consequence, there exists an invertible change of variables $(x,y) \rightarrow (X,Y)$ such that in terms of the new variables the above integral reduces to $\int dXdY$, i.e. in the plane of X and Y the Poincaré map preserves areas in the usual sense (S-8). This is why it is usually stated that the Poincaré map is area preserving although, in reality, this is rarely the case since in general the function Ω is not constant on an arbitrary surface of section. This abuse of language is not serious however since the generic properties of an area-preserving mapping of the plane X,Y are essentially topological in nature, i.e. they are preserved under the inverse transformation $(X,Y) \rightarrow (x,y)$. It is therefore sufficient - and much easier - to investigate these properties in the case of an area-preserving transformation of the plane, and this is what we are going to do in the following (S-9).

To begin with, we look at the behavior of the map Φ in a neighborhood of one of its fixed points which we assume to be at the origin x = y = 0. If z denotes the vector of components x,y we can write Eqs. (15) as

$$\Phi : z' = Az + h(z) \qquad (16)$$

where A is a constant matrix and h a series in x and y lacking constant and linear terms. The area-preserving property of Φ implies that det A = 1.

Fixed points are classified according to the nature of the eigenvalues of the matrix A. For the linearized mapping

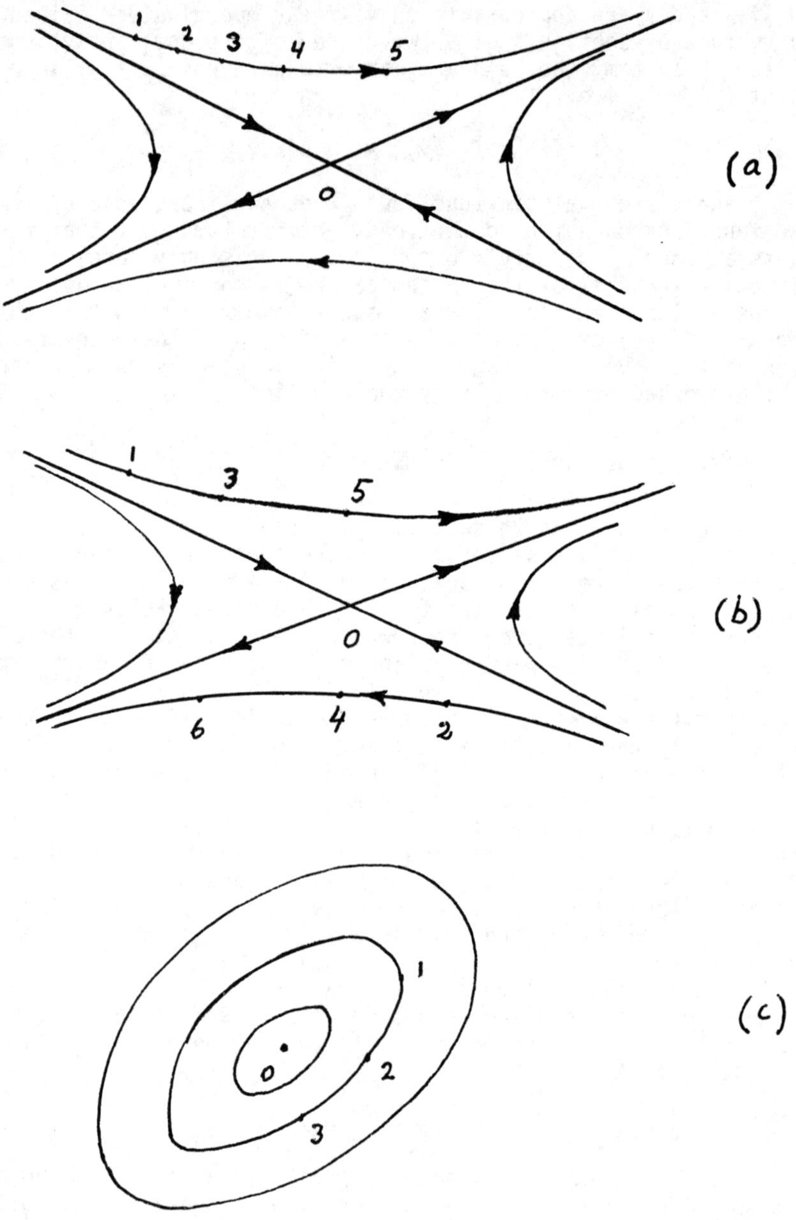

Figure 2

$$\Phi_L : z' = Az$$

an arbitrary neighborhood of the origin is filled with invariant curves whose configuration depends upon the nature of the eigenvalues λ_1, λ_2 of A, i.e. on its trace alone since the characteristic equation is of the form

$$\lambda^2 - (\text{trace } A)\lambda + 1 = 0 .$$

a) trace $A>2$: (real positive roots α, α^{-1}) the invariant curves are hyperbolae. The iterates 1, 2, ... of a point all lie on the branch of the particular hyperbola going through it. The configuration is as shown in Fig. (2a) where, as in the following figures, the numbers indicate the locations of the successive iterates under repeated application of Φ_L. The fixed point is then called hyperbolic (without reflection).
b) trace $A< - 2$ (real negative roots $-\beta$, $-\beta^{-1}$): the configuration is the same but the successive points jump alternatively from one branch of the hyperbola to the other as shown in Figure (2b). The point is called hyperbolic with reflection.
c) $- 2<$ trace $A < 2$ (complex conjugate roots $e^{\pm i\gamma}$, γ real): the invariant curves are concentric ellipses as shown in Figure (2c). The fixed point is called elliptic.
d) trace $A = \pm 2$: this is an exceptional situation in which the point is called parabolic without reflection if $\lambda_1 = \lambda_2 = 1$, and with reflection if $\lambda_1 = \lambda_2 = -1$.

Birkhoff (14) and Moser (44) have shown that the configuration of the invariant curves in a sufficiently small neighborhood of a hyperbolic point is the same for a non-linear mapping as that of its linear part. There is an analytic function $G(x,y)$ which is constant along the invariant curves. This implies that the corresponding Hamiltonian system possesses a first integral in a small region of phase-space surrounding the periodic orbit corresponding to the hyperbolic fixed point. Note that this first integral is not "isolating" in the sense that the corresponding level surfaces do not cast out regions of phase-space where orbits would remain forever. The behavior of the motions away from the periodic orbit is remarkably complicated as we are going to show now.

Among the invariant curves defined by $G(x,y)$ = constant two go through the fixed point where they are tangent to the asymptotes of the hyperbolae of the linearized map. One of them denoted W^s is stable in the sense that $\Phi^k(p) \to 0$ as $k \to \infty$ for all $p \varepsilon W^s$, while the other W^u is unstable, i.e. $\lim \Phi^k(p) \to 0$ as $k \to -\infty$. Suppose W^s and W^u intersect transversally at a point P. Then it can be shown that they intersect at an infinity of points which are the iterates P_i, $i = \pm 1, \pm 2, \ldots$ of the mapping. Now consider the sequence

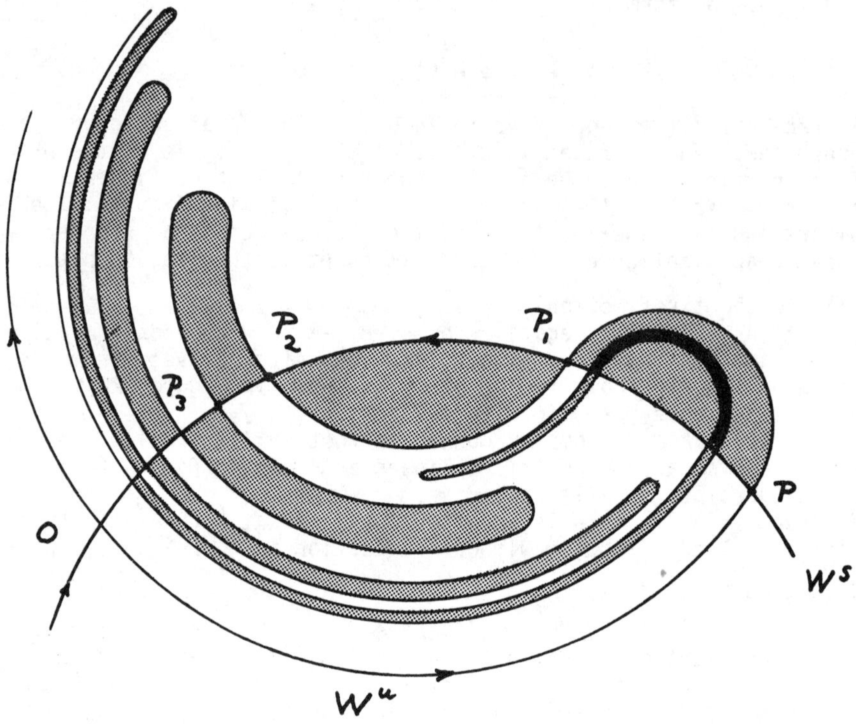

Figure 3a.

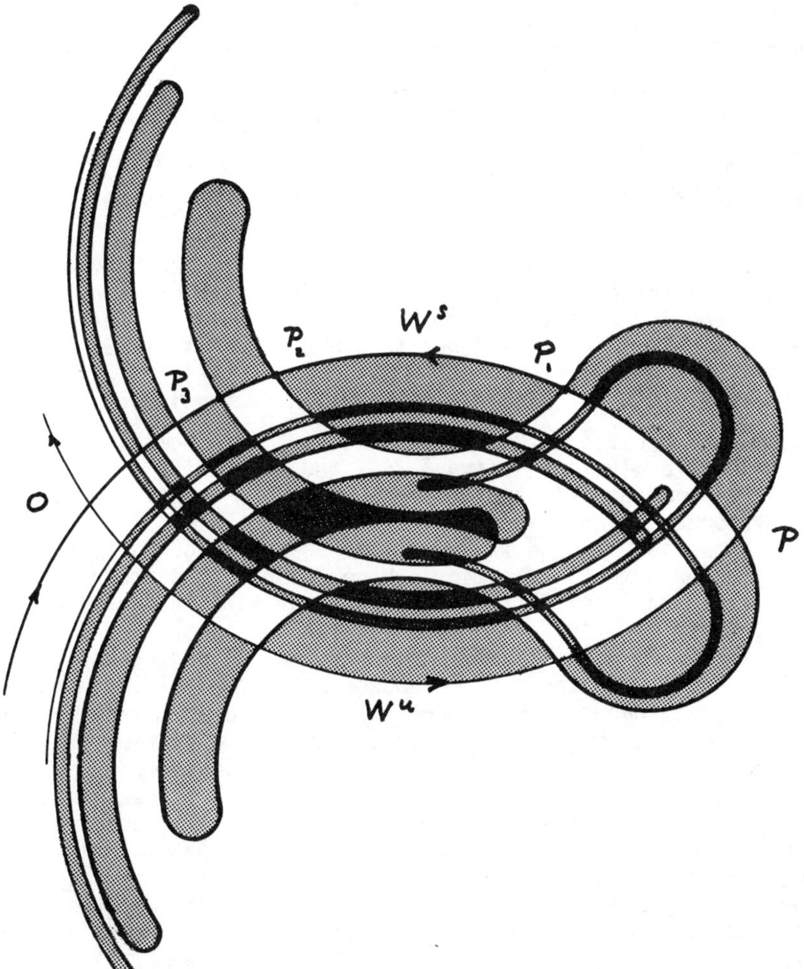

Figure 3b.

P, P_1, P_2, ...; it is infinite, lies on the arc PO of W^s, and has O as limit point. Hence the distance between two consecutive points tends to zero as i tends to infinity. Since the area between the arc P_iP_{i+1} of W^s and the corresponding arc of W^u is constant under the mapping it therefore follows that the dimension of the surface enclosed within these arcs in the direction normal to W^s must increase indefinitely producing the configuration sketched in Figure (3a) since W^u cannot intersect itself. Furthermore W^s behaves in the same way and therefore intersects W^u in an infinity of points as shown in Figure (3b). All the points where W^s and W^u intersect are called homoclinic and it is seen that the existence of one implies the existence of an infinity. Note that the orbit going through P and its iterates approaches the unstable periodic orbit going through O as the time tends to ∞ and $-\infty$. It is called a homoclinic orbit.

Because of the intricate configuration of the invariant curves W^s and W^u the behavior of the orbits close to the homoclinic orbits seems to defy analysis. Hadamard (45) and Birkhoff (14, Vol. 2, p. 653) realized however that these motions can be put into one-to-one correspondence with sequences of symbols. This was elaborated upon by Morse and Hedlund (46) who laid down the foundations of what is called symbolic dynamics. More recently this method of analysis has made considerable progress thanks to the works of Smale (47), Alekseef (48, 49) and others (cf 18).

Alekseef has very clearly explained that (48) "the basis of symbolic dynamical methods is the association of infinite words in a certain alphabet with the trajectories of a dynamical system. This is done by choosing a certain partition of a surface of section, the elements of the partition being denoted by letters. By observing those elements in which a trajectory falls as it returns successively to the surface of section, we obtain the word corresponding to the trajectory. It is clear that the fundamental mapping of the surface of section*, the shift transformation, corresponds to the shift by one letter to the left. This scheme has the following physical interpretation. To letters, i.e. elements of the partition, correspond outcomes of a certain experiment, the "macro-measurements"; to a word corresponds a sequence of results of repeated measurements in a system evolving with time".

The analysis of the properties of the Poincaré map in the vicinity of a homoclinic point is very difficult even with the help of symbolic dynamics. It is however possible to understand the essentials of the mathematical arguments involved using the close analogies which exist between the Poincaré map and certain known mappings whose properties can be derived in a heuristic way without much difficulty. The best known mapping in this category is probably Smale's horseshoe map that we proceed to describe and analyze (50).

*I.e. one forward application of the Poincaré map.

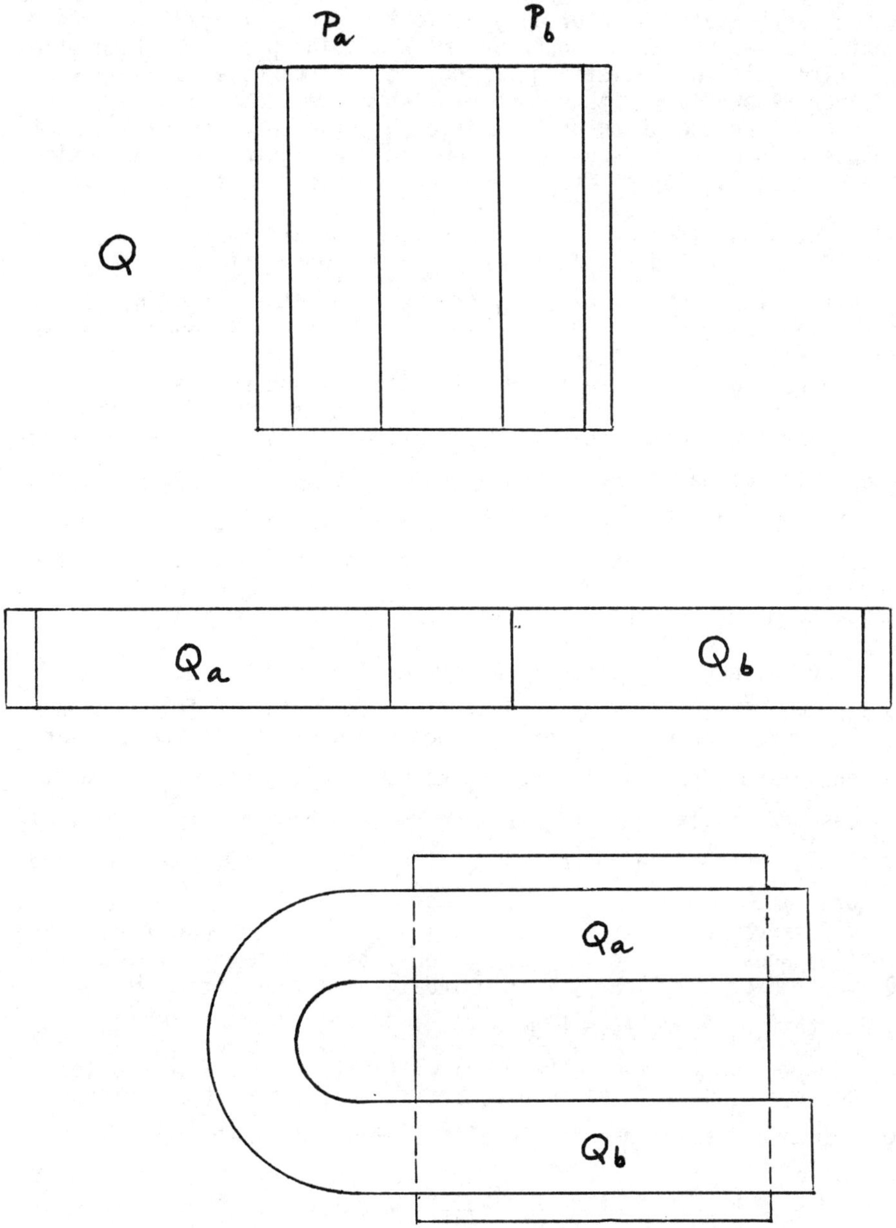

Figure 4.

Smale's horseshoe map is a mapping of a square Q into the plane which can be defined informally as follows (see Figure 4): imagine that Q is made of fresh dough and mark a point p on it. Then stretch Q horizontally while letting its height shrink (holes and tears are not allowed!). With enough stretching we can obtain a strip of dough long enough to be bent into the shape of a horseshoe that we position over the square as shown in the figure. By definition, the new location of p after these manipulations is the image $\Phi(p)$ of p under the mapping.

Note that $\Phi(p)$ may fall within Q or not and that, in the latter case, its image under Φ is not defined. However, there exists a subset Λ of Q with the property that all the iterates $\Phi^n(p)$ of any $p \varepsilon \Lambda$ belong to Q for $n = 0, \pm 1, \pm 2, \ldots$. The existence of Λ can be ascertained as follows.

First, consider the forward iterates of the square. In order for $p \varepsilon Q$ to belong to Λ it is necessary that its image $\Phi(p)$ fall in either of the rectangles Q_a and Q_b for, otherwise, its second iterate would not be defined. Hence $\Phi(p) \varepsilon Q$ only if p belongs to either of the preimages $P_a = \Phi^{-1}(Q_a)$ and $P_b = \Phi^{-1}(Q_b)$ of Q_a and Q_b, respectively. It is easy to see by reversing the above construction and thus bringing the horseshoe shaped piece of dough back into the shape of Q that P_a and P_b are vertical strips as shown in the figure.

Now assume that $p \varepsilon P_a$ or P_b so that $\Phi(p) \varepsilon Q_a$ or Q_b and apply Φ a second time. As we have just seen, for $\Phi^2(p)$ to fall within Q $\Phi(p)$ must also be in P_a or P_b. Hence p must lie in the preimage of one of the four little rectangles $Q_i \cap P_j$, i, j = a or b, and it is easy to verify that the preimage of each of these is a vertical strip P_{ij} which is a subset of P_i, i.e. $P_{ij} = \Phi^{-1}(Q_i \cap P_j) \subset P_i$ (see Figure 5).

Observe that the image of Q_i, i = a or b, is a horseshoe lying in $\Phi(Q)$ whose intersection with Q consists of two horizontal strips Q_{ij}, j = a or b. Clearly we can label these four strips in such a way that $Q_{ij} \subset Q_i \subset Q$ (see Figure 6).

If we apply Φ once more $\Phi^3(p)$ will fall in Q provided $\Phi^2(p)$ belongs to P_a or P_b. As before, it is readily seen that for this to happen p must belong to one of the eight vertical strips

$$P_{ijk} = \Phi^{-2}(Q_{ij} \cap P_k) \subset P_{ij} \subset P_i \quad , \; i,j,k = a \text{ or } b.$$

Finally, if we apply Φ an infinite number of times we obtain an infinite set Λ_1 of vertical strips of zero width since from the construction of Φ horizontal distances shrink by a factor smaller

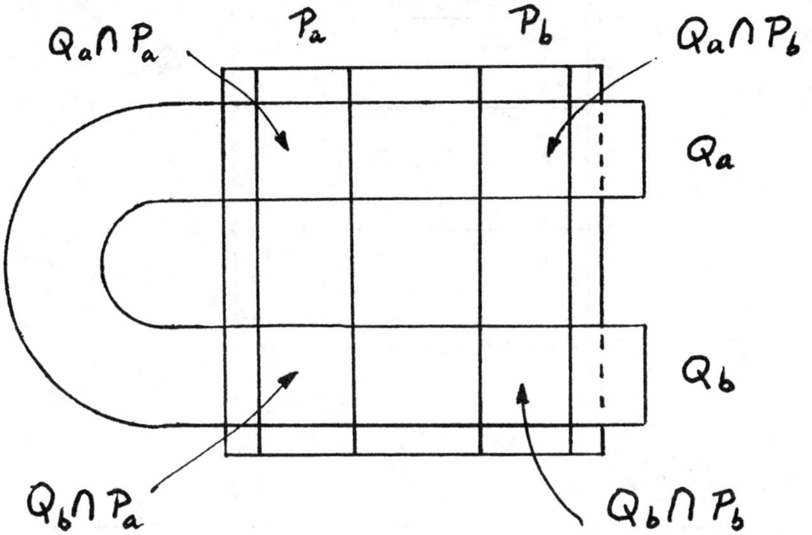

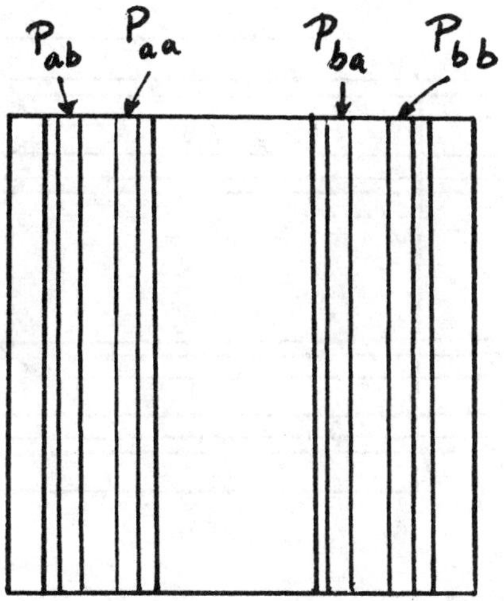

Figure 5.

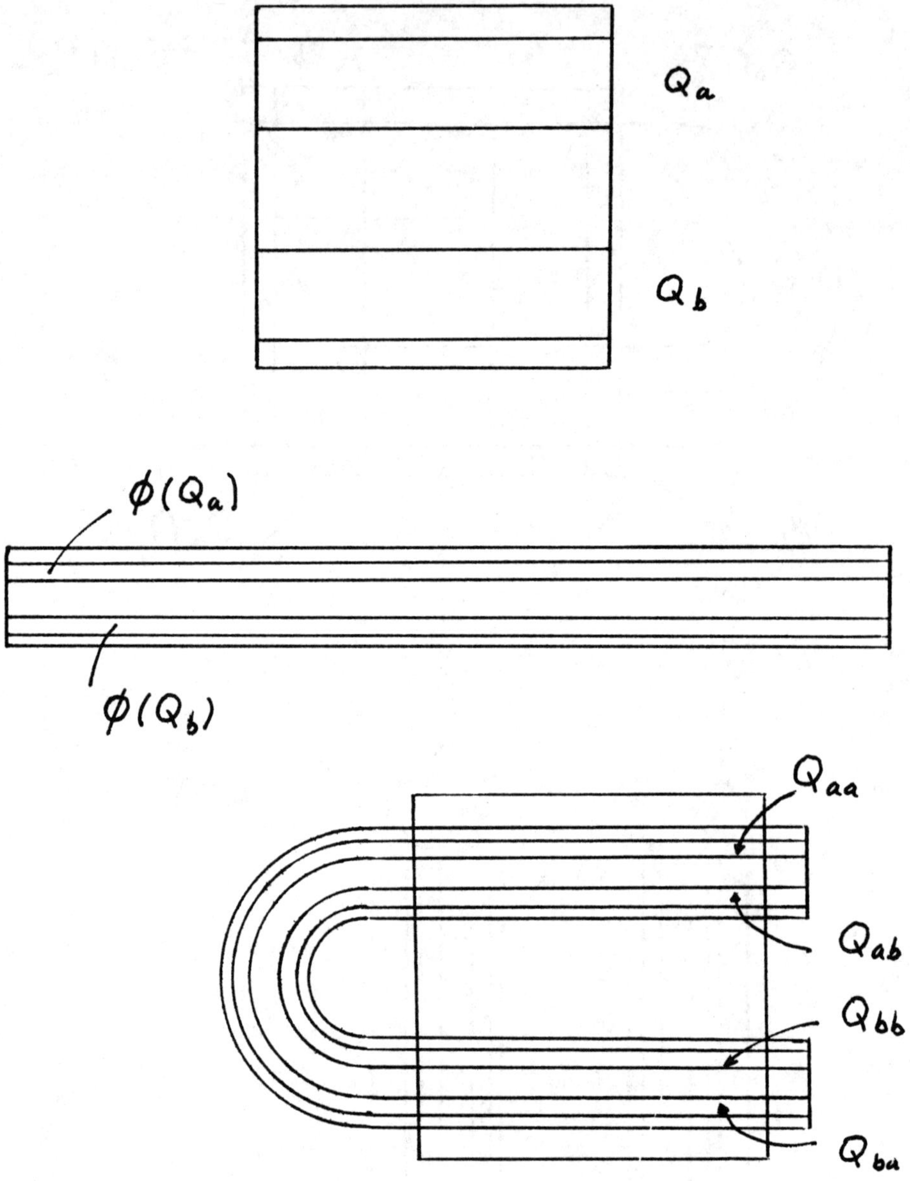

Figure 6.

than one under the application of Φ^{-1}. In other words we know at this stage that Λ is a subset of an infinite set Λ_1 of vertical segments $P_{ijk...}$ where i, j, k, ... = a or b.

From the way we have labeled the elements $P_{ijk...}$ of the set Λ_1 we see that they are in one-to-one correspondence with the elements of the set of infinite sequences of two symbols {i, j, k, ...}, where i, j, k, ... = a or b. Because the elements of the so-called ternary set of Cantor are also in one-to-one correspondence with this set of sequences, the set Λ_1 is equivalent to this Cantor set and, for this reason, is itself called a Cantor set (see Appendix D).

Next we consider the backward iterates $\Phi^{-n}(p)$, n = 1, 2, ... of a point $p \varepsilon Q$ under the inverse mapping (see Figure 6).

From the foregoing $\Phi^{-1}(p)$ is defined only if p belongs to Q_a or Q_b, $\Phi^{-2}(p)$ is defined only if p belongs to either of the four horizontal strips Q_{ij}, i, j = a or b, and, in the limit, $\Phi^{-\infty}(p)$ is defined only if p belongs to one of infinitely many horizontal segments $Q_{ijk...}$ whose set we denote by Λ_2. Just as Λ_1, Λ_2 is a Cantor set.

It is now obvious that the set Λ that we are looking for is the intersection of Λ_1 and Λ_2, that is the set of points of intersection of the vertical segments of Λ_1 with the horizontal segments of Λ_2. Λ is thus the product of two Cantor sets, i.e. a Cantor set in two dimensions, whose points are in one-to-one correspondence with the elements of the set S of doubly infinite sequences of two symbols. Indeed a point $p \varepsilon \Lambda$ is at the intersection of a vertical segment $P_{s_1 s_2 ...}$ with a horizontal segment $Q_{s_{-1}\, s_{-2}}$ so that, in obvious notation, the infinite sequence $s = (\ldots s_{-2}s_{-1}\ ,\ s_1 s_2 \ldots)$ is uniquely associated with p and vice-versa.

Each point of Λ can therefore be labeled by its corresponding sequence s and it is clear that the sequence corresponding to $\Phi(p)$ is obtained from s by simply shifting the "decimal" point by one symbol to the left.

Shifts over doubly infinite sequences of symbols are commonly used in probability theory since they constitute convenient representations of chance processes like the consecutive outcomes of spins of a roulette wheel. They are usually called Bernoulli shifts.

The set Λ has the following important properties:

1) it contains a non-denumerable set of periodic points of Φ with arbitrarily large period. These periodic points evidently correspond to those doubly infinite sequences of S which are periodic.

2) an arbitrarily small neighborhood of a point of Λ contains periodic points of Φ. In topological language this property is ex-

pressed by saying that the set of periodic points of Φ is dense in Λ.

3) the set Λ is hyperbolic by which is meant the following: if we consider the effect of Φ on a small neighborhood of a point $p\varepsilon\Lambda$, Φ contracts distances in the vertical direction and expands them in the horizontal direction. This is analgous to the behavior of a Poincaré map at a hyperbolic point, hence the terminology. Such a behavior is often referred to as a hyperbolic splitting. The conditions under which it occurs can be expressed in a precise manner for higher dimensional mappings in terms of the spectrum and eigenspace of the Jacobian matrix of the mapping: if we denote by $D\Phi$ the linear map generated by this matrix these conditions are that the eigenspace can be split into independent components E^s and E^u such that $D\Phi$ contracts on E^s and expands on E^u. The importance of this notion of hyperbolicity stems from the fact that mappings which possess an invariant hyperbolic set often generate random behavior under successive iteration and are ergodic.

We are now ready to discuss the crucial physical consequences of these properties: if we view the horseshoe map as the Poincaré map of a discrete dynamical system, the behavior of the system is random at any macroscopic level of description. The dynamical system that we have in mind is defined as follows: the possible states of the system are represented by the points of the plane with the law of evolution

$$p_{n+1} = \Phi(p_n),$$

i.e. if the system is in the state p_n at time n, n an integer, its state at time n+1 is the image of p_n under Φ.

Suppose now that we observe this system for a limited time and with limited precision. We can for instance fix our attention on the two vertical strips P_a and P_b and record in which of these p_n falls at times n, n + 1, ..., n + k, thereby obtaining a string of k letters $s_1 s_2 \ldots s_k$ (s_i = a or b). From the foregoing discussion the actual orbit either belongs to Λ or not but we have no way of telling which is the case from our limited observations. Worse, even if we knew that the actual orbit is in Λ we could not tell anything about its future history because there is a continuum of possibilities corresponding to the infinite subset of sequences which contain a string of k letters identical to the string that we have recorded. In other words, no matter how large k is, we cannot predict in which of P_a or P_b p_{n+k+1} is going to fall the probability of either events being 1/2 (provided of course that $p_n \varepsilon \Lambda$).

We now return to the problem of the behavior of the Poincaré map in the neighborhood of a homoclinic point.

We shall content ourselves to state the results obtained by Smale, Moser, and Conley (cf 18, p. 99).

Smale has shown that in any neighborhood of a homoclinic point there exists an invariant set I_2 for some fixed power, say Φ^{ℓ}, of Φ such that, under the mapping Φ^{ℓ}, the points of I_2 are in one-to-one correspondence with the doubly infinite sequences of two symbols. More generally, Conley has shown that the restriction to a fixed power of Φ can be relaxed in which case there exists an invariant set I_{∞} whose points are in one-to-one correspondence with the doubly infinite sequences of an infinity of symbols. The sets I_2 and I_{∞} have the same properties as the set Λ of the horseshoe map. Consequently, not only is there an infinity of periodic points (a result already obtained by Birkhoff, cf ref. 14, Vol. 2, p. 530) but, more importantly, the behavior of the Poincaré map is of the same unpredictable nature as that of the horseshoe map.

Now that we have given an idea of the complexity of the neighborhood of a hyperbolic fixed point when homoclinic points are present we can indicate what happens in the neighborhood of an elliptic point.

The KAM theorem says that under a small perturbation most of the invariant ellipses surrounding an elliptic fixed point of the unperturbed system will be slightly deformed into closed invariant curves corresponding to the uncommensurable frequencies in the set Ω(*). In the gaps left out by these invariant curves there necessarily are homoclinic points as shown by Zehnder (51). These homoclinic points belong to hyperbolic points which occur in pairs with elliptic points. Each of these elliptic points is in turn surrounded by a similar configuration of invariant curves separated by gaps containing homoclinic, hyperbolic, and elliptic points, etc. in an endless fashion. To give an idea of the complexity of such a configuration we reproduce in Figure (7) the picture given by Arnold (12). Note that this picture corresponds to a perturbation which

Figure 7

*i.e. the set of uncommensurable frequencies corresponding to the surviving tori (cf p. 17)

is large enough for the gaps to be visible but so small that on the scale of the figure a definite geometric structure is still discernable. As ε is increased however this structure eventually disappears completely in wide regions. The sequence of morphological changes which take place can be described as follows:

For $\varepsilon = 0$ the hyperbolic fixed points of the unperturbed integrable system are smoothly connected by the invariant curves W^s and W^u defined before which intersect each other only at these fixed points. These invariant curves are then called separatrices because they separate the regions of the plane of section containing the various families of stable periodic orbits (resonances) of the unperturbed system. When a small perturbation is turned on W^s and W^u still exist but, as we have seen, they intersect each other at an infinity of homoclinic points and meander in a complicated way. They no longer deserve the name of separatrices since in the regions now bordered by them there exist infinitely many elliptic and hyperbolic points corresponding to resonances between the unperturbed system and the perturbation. As long as ε is very small, the area of the region filled by the invariant curves W^s and W^u is also very small and can be viewed as a network of very narrow roads connecting the slightly displaced hyperbolic points of the unperturbed system. There is, in addition to these, an intricate network of new roads connecting the hyperbolic points created in the gaps by the perturbation but their width is so small that in most practical problems they can be ignored.

Physicists call "stochastic layers" the roads associated with the separatrices of the unperturbed system. For clarity we will call them "primary stochastic layers" while all the additional roads created by the perturbation will be called "secondary stochastic layers".

As the size of the perturbation is further increased the primary stochastic layers widen abruptly at a certain critical value ε_1 creating regions of appreciable size where the motions are of a random nature. In the complement of these so-called stochastic regions, the invariant curves corresponding to the preserved tori still exist but they are interspersed with some of the secondary layers whose width has considerably increased but is still very small since they were initially extremely thin.

If we increase ε past ε_1 nothing spectacular happens for a while until a new critical value is reached at which those secondary stochastic layers which, on some scale, were barely visible at $\varepsilon = \varepsilon_1$, in their turn suddenly widen at $\varepsilon = \varepsilon_2$, thereby creating new stochastic regions of appreciable size containing random motions.

Further increases of ε beyond ε_2 will similarly result in the observation of successive widenings of more stochastic layers which will merge with the already existing wide stochastic regions at critical values $\varepsilon_3, \varepsilon_4, \ldots$.

An additional complication must be expected in general, that is, depending upon the region of the surface of section one is looking at, the sequence of critical values will be different.

Finally, what happens for large values of ε is a matter of speculation at this time but it seems plausible that this will depend on whether the system associated with the Hamiltonian $\varepsilon H_1(q,p)$ describing the perturbation is integrable or not.

The determination of the critical values $\varepsilon_1, \varepsilon_2, \ldots$ and of the size of the corresponding stochastic regions is of great importance to plasma physics, for there are already three problems known to involve stochastic phenomena of the kind under discussion. We have mentioned one before, namely the motion of an ion in a uniform magnetic field under the influence of an electrostatic wave. The others are the motion of an ion in a trapped-ion mode (52), and the effect of magnetic field irregularities in plasma confining devices which may cause the destruction of some or all magnetic surfaces depending on the size of the irregularities (S-10). The methods of analysis under discussion are applicable to the latter problem because the divergence-free character of the magnetic field implies that, for instance in a tokamak, the field lines induce a mapping which has the property of conserving the integral $\int B.dS$ where dS is the area element of any meridian plane of section (54; cf also 13, n°305, Vol. III). Consequently, all the pathological phenomena occuring in area-preserving mappings must also happen when a perturbation is applied to a well behaved configuration of nested toroidal magnetic surfaces. Note that in general the mapping is not area-preserving in the usual sense contrary to a common belief. This problem could be analyzed using the techniques developed for two-degree of freedom Hamiltonian systems since Hamiltonians can be constructed whose orbits are the field lines of a given magnetic field (66, 67).

There is presently no satisfactory analytical method for the calculation of the sequence of critical values $\varepsilon_1, \varepsilon_2, \ldots$ for the simple reason that stochasticity still lacks a precise mathematical characterization. In view of the importance of the problem let us review briefly the state of our knowledge in this particular context.

As indicated before it is known that random motions are present in the neighborhood of any homoclinic point and that the Poincaré map possesses an invariant hyperbolic set there. Since homoclinic points are associated with hyperbolic fixed points it is believed that stochastic motions are due to the presence of a dense set of hyperbolic fixed points which are often viewed as scattering centers. According to Smale (68) this is no more than a conjecture at the present time (see also 69, p. 76). Yet this belief is supported by the following observation made in many numerical experiments: in a stochastic region orbits very close initially depart from each other at a rate which, on the average, is exponential while in the

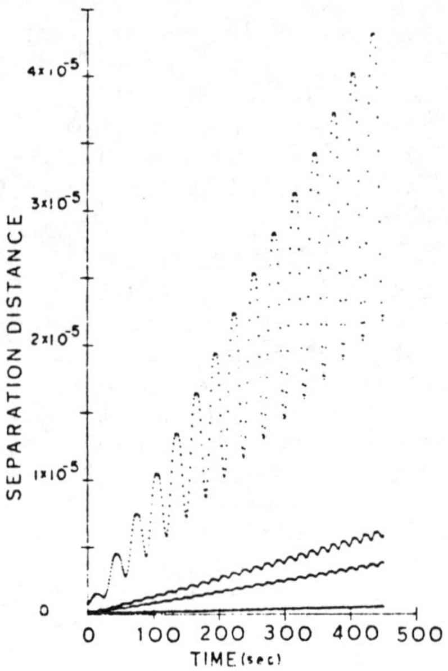

Figure 8. This figure shows the growth of separation distance between two trajectories initially started about 10^{-7} apart in the full phase space. Separation distance versus time is plotted for four distinct orbit-pairs. Each orbit-pair starts in a smooth level-curve region and the linear growth of separation distance with time is apparent.

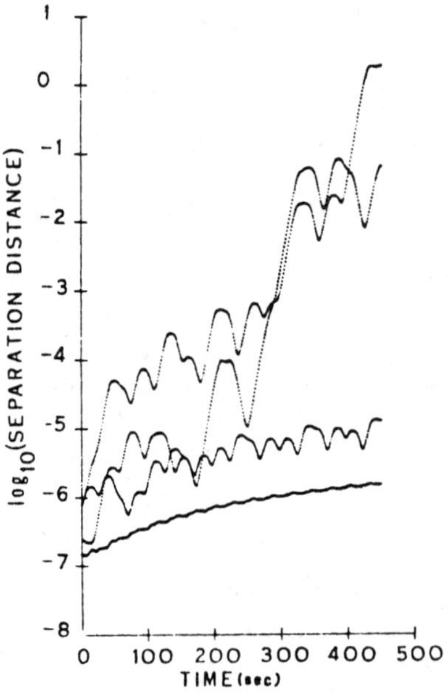

Figure 9. This figure plots $\log_{10}$ of the separation distance versus time for four distinct orbit-pairs started in a smooth level-curve region, and their generally negative curvature indicates the linear growth of separation distance with time. The generally upper lying curves are for orbit-pairs started in a stochastic region of the level-curve plane, and the rapid exponential-order growth of separation distance with time is relatively clear.

rest of the plane of section the rate of departure is only linear. To illustrate what we mean by "on the average" we reproduce in Figs. (8) and (9) graphs of the variation versus time of the distance between two adjacent orbits (70). It is evident that the rate of change of this distance varies widely over each oscillation and that according to the case it is only its average which displays an exponential or linear growth.

Motivated by this empirical evidence, several authors have attempted to determine the onset of stochasticity by associating it with a change in the behavior of the distance between adjacent orbits.

For instance Mo (71) considers the distance between the origin in phase-space and a point on an orbit as a random variable obeying a generalized Langevin equation that she derives using Zwanzig's projection operator technique. From this equation she computes certain frequencies which would be the basic real frequencies in the case of an integrable system. It turns out that, as the size of the perturbation increases, one of these frequencies becomes pure imaginary for a value of ε close to the one obtained via numerical integration of three different well studied systems*. The whole approach, however, lacks rigorous mathematical justification, and it is not clear how it can be applied to the case where the ε for which stochasticity sets in depends on the region of phase-space.

Also Brumer and Duff (72) have proposed that the onset of stochasticity can be determined from the behavior of the solutions of the so-called variational equation associated with the Hamiltonian system. Recall that the variational equation of a vector differential equation $dx/dt = f(x)$ essentially describes the behavior of the distance η between two solutions $x_1(t)$ and $x_2(t)$ which are infinitesimally close at $t = 0$. The vector form of the variational equation for η is $\dot{\eta} = A(t)\eta$ where $A(t)$ is the Jacobian matrix of f evaluated at $x = x_1(t)$. Brumer and Duff then argue that if the matrix A possesses eigenvalues with positive real parts in a region of phase-space then the orbits will depart exponentially and stochasticity will be present. They have applied this line of reasoning to the Hamiltonians studied by Hénon and Heiles (30) and by Barbanis (43) and obtained estimates which agree well with numerical results. However, although their criterion is probably necessary for the presence of stochasticity, it is certainly not sufficient since Rod has constructed integrable Hamiltonians for which the matrix A has positive eigenvalues (73). Note also that the positiveness of the real part of an eigenvalue implies only a local exponential departure of the orbits whereas, as indicated before, it is the average of the distance between orbits over relatively large times which shows an exponential increase in stochastic regions.

*Mo actually uses the energy as the parameter controlling the degree of stochasticity which, for the systems she considers, is equivalent to using ε.

Another avenue for determining the onset of stochasticity is based on the notion of system entropy introduced by Kolmogorov which seems somehow related to the degree of stochasticity and therefore can perhaps be used in this connection (74).

A third possibility is through the use of the notion of overlapping of resonances (75, 76). For systems with two degrees of freedom it corresponds to the observation that at $\varepsilon = \varepsilon_1$ the primary stochastic layers which are the first to widen as ε increases seem to merge together on their facing boundaries and eventually overlap. The notion has been used successfully in many instances but it requires a difficult determination of the location of the separatrices and does not seem to be applicable to higher dimensional systems.

To sum up, there is no satisfying method available for the determination of the onset of wide spread stochasticity in a given region of phase-space.

Among the various open questions related to this problem the following seem particularly important:

1) how does the number of periodic points per unit area of the plane of section vary as one of the frequencies tends to zero, i.e. as one approaches a separatrix?

2) are the fixed points in a given region elliptic or rather hyperbolic and, if so, with or without reflection?

3) is the conjecture true that as the perturbation increases in size elliptic points become hyperbolic with reflection as suggested by numerical experiments (68)?

To answer these questions when the perturbation is not small is certainly very difficult. But when the perturbation is small a powerful method is available for the determination of the periodic solutions which applies to a wide class of Hamiltonian systems. A modern version of a method invented by Poincaré (13, Chap. III, Vol. 1), the method that we have in mind was developed by a number of mathematicians notably Cesari, Hale, and Gambill, and will therefore be called the PCHG method. As Poincaré's method it reduces the determination of the periodic solutions to an algebraic problem, i.e. each periodic solution is associated to a particular root of a set of transcendental equations called the determining equations. But it differs from Poincaré's in the way the periodic solutions are calculated: Poincaré represents them by series while in the PCHG method they are obtained as the limit of an infinite sequence of functional iterations. Since the PCHG method does not seem to be known to many physicists we feel it worthwhile to present it heuristically in this report. It will be found in Appendix C together with indications for its application to slightly perturbed integrable Hamiltonian systems.

We now turn to the question of stability of the solutions of Hamiltonian systems which is intimately related to the preceding.

1.4 STABILITY IN THE LARGE OF HAMILTONIAN SYSTEMS

For slightly perturbed integrable systems the question of the stability in the large of their solutions is settled for the case of two degrees of freedom under the conditions of the KAM theorem. Indeed in the three-dimensional phase-space corresponding to a given value of the energy, each orbit remains enclosed between adjacent tori and is therefore bounded. But for more than two degrees of freedom this is very likely no longer true in general since Arnold has shown on a particular example that there are orbits connecting neighborhoods of distant tori (77, see also 11 and 78). A discussion of this phenomenon called "Arnold's diffusion" is presented in a forthcoming paper by Chirikov (76).

1.5 NON-CANONICAL PERTURBATIONS OF HAMILTONIAN SYSTEMS

Because real systems are never exactly conservative it is important to study their behavior under a small, non-canonical perturbation. The work of Murdock (79-82) on such systems is particularly interesting and does not seem too well known to physicists. One of his typical results is the following (79, 81).

Consider a sequence of pendulums each coupled to the next by a weak nonlinear spring, slightly damped, and with the first pendulum in the sequence subject to weak forcing. Then, except for solutions having amplitudes near zero, the steady state solutions are periodic with period a small integer multiple of the forcing period and with the frequencies of adjacent pendulums having small integer ratios.

These results may become important in wave-wave interactions, in weak turbulence, many mode coupling, and multiple resonance.

Another recent result which also seems interesting in this connection is that of Hale and Taboas (83). They consider the equation

$$\ddot{x} + g(x) = - \lambda\dot{x} + \mu\, f(t) \tag{17}$$

in a neighborhood of $x = \dot{x} = 0$ where $xg(x) > 0$, $g(0) = 0$, and f is 2π-periodic. The equation $\ddot{x} + g(x) = 0$ has a family of periodic solutions with frequency $\omega(a)$ which depends upon the "amplitude" a. Assume that there is $a = a_0$ such that $\omega(a_0) = 1$, $\omega'(a_0) \neq 0$, and denote by Γ the corresponding orbit. Then the 2π-periodic solutions of Eq. (17) which lie in a neighborhood of Γ for (λ,μ) in a small neighborhood of $(0, 0)$ are not continuous in general at $(\lambda,\mu) = 0$. Only when λ/μ tends to a constant as (λ,μ) tends to zero are these solutions continuous in λ, μ. It is found that a neighborhood of the origin in the λ, μ-plane is divided into a finite and even number of sectors bound by curves intersecting at the origin, such that the number of 2π-periodic solutions changes by two as the point (λ,μ) crosses any one of these curves. The number of sectors and the number of such solutions within each of them depend upon the number of roots of a certain transcendental equation.

An important consequence of the theorem for the applications is that an attempt at finding these 2π-periodic solutions by classical perturbation procedures will fail, unless one takes account of the peculiar properties just described, and limits oneself to values of λ, μ lying within one of the sectors with λ/μ tending to a constant as these parameters tend to zero simultaneously.

1.6 METHOD OF AVERAGING

The method of averaging is widely used for the study of differential systems which depend periodically on some of the variables. The aim is to simplify the system by averaging out its dependence on these variables when their rate of change is much larger than that of the others. The method is powerful and can yield valid and useful results in many instances. However, it does not always work even for relatively simple problems and it must therefore be used with some caution. We feel that the attention of the physicists has not been sufficiently drawn to the pitfalls of the method and want therefore to give examples borrowed from Arnold (84); see also 11, p. 102) which show that the solution obtained via the averaging method may have little to do with the actual solution.

The first equations considered by Arnold are

$$\dot{\Phi}_1 = I_1, \dot{\Phi}_2 = I_2, \dot{I}_1 = \varepsilon, \dot{I}_2 = \varepsilon \text{ a } \cos (\Phi_1 - \Phi_2) \qquad (18)$$

where a > 1 and its averaged version is

$$\dot{J}_1 = \varepsilon, \dot{J}_2 = 0 .$$

For initial conditions

$$I_1 = I_2 = J_1 = J_2 = 1, \Phi_1 = 0, \Phi_2 = \cos^{-1} (1/a)$$

the solution is

$$I_1(t) = I_2(t) = 1 + \varepsilon t,$$

$$J_1(t) = 1 + \varepsilon t, J_2(t) = 1.$$

One usually expects that the solution of the averaged system will differ from the actual solution by $\sim \varepsilon$ over an interval $O(1/\varepsilon)$. But in the present case

$$|I(1/\varepsilon) - J(1/\varepsilon)| = 1,$$

i.e. the averaged motion loses any relation to the actual motion.

The other example given by Arnold shows the effect of the passage through a resonance. The system is

$$\dot{\Phi}_1 = I_1 + I_2, \dot{\Phi}_2 = I_2, \dot{I}_1 = \varepsilon, \dot{I}_2 = \varepsilon \cos (\Phi_1 - \Phi_2) \qquad (19)$$

with averaged equations

$$\dot{J}_1 = \varepsilon, \ \dot{J}_2 = 0 \ .$$

For the initial conditions

$$\Phi_1(0) = \Phi_2(0) = I_1(0) = 0, \ I_2(0) = 1.$$

which correspond to resonance, it is easily found that

$$|I(t) - J(t)| = |I_2(t) - 1| = \sqrt{2\varepsilon} \int_0^\tau \cos\,(x^2)dx$$

where $\tau = t\sqrt{\varepsilon/2}$. For $t = 1/\varepsilon$

$$|I(1/\varepsilon) - J(1/\varepsilon)| \simeq C\,\sqrt{\varepsilon}$$

where the constant C does not depend on ε, which shows that the error is not of the order of ε over $t \sim 1/\varepsilon$. The effect of the resonance is to disperse the bundle of trajectories which initially differ only in their phases. The scattering of I_2 after going through the resonance is $O(\sqrt{\varepsilon})$.

Arnold has also considered the system (84)

$$\begin{aligned} \dot{\Phi} &= \omega(I) + \varepsilon f(I,\Phi) \\ \dot{I} &= \varepsilon\, F(I,\Phi) \end{aligned} \tag{20}$$

where Φ, ω, and f are k-dimensional and I, F are ℓ-dimensional. The averaged system is

$$\dot{J} = \varepsilon\, \overline{F}(J)$$

where

$$\overline{F}(J) = (2\pi)^{-k} \int F(J,\Phi)d\Phi \ .$$

When $k = 2$, i.e. $\Phi = (\Phi_1, \Phi_2)$, $\omega = (\omega_1, \omega_2)$, he has obtained the following result: if the quantity

$$A(I,\Phi) = (\frac{\partial\omega_1}{\partial I}\ , F)\ \omega_2 - (\frac{\partial\omega_2}{\partial I}\ , F)\omega_1$$

does not vanish then

$$|I(t) - J(t)| < C\,\sqrt{\varepsilon}\,(\log(1/\varepsilon))^2 \text{ for all } 0 < t < 1/\varepsilon \ .$$

Note that the condition $A \neq 0$ is violated by example (18) because then $A = I_2 - I_1$ a cos $(\Phi_1 - \Phi_2)$ changes sign at $I_1 = I_2$ if $a > 1$ as it was assumed; hence the difference in the estimates of the error.

Finally, it must be borne in mind that the effects of the passage through resonance have not been rigorously studied for more than two degrees of freedom and, consequently, that there is no guarantee that the error will still be of order $\sqrt{\varepsilon}$ over $t \sim 1/\varepsilon$.

For a Hamiltonian system $k = \ell = n$ (the number of degrees of freedom) in Eqs. (20) which now are of the form

$$\dot{\Phi} = \omega(I) + \varepsilon \frac{\partial H_1}{\partial I}$$

$$\dot{I} = - \varepsilon \frac{\partial H_1}{\partial \Phi}$$

where $H = H_o + \varepsilon H_1$, $\omega = \partial H_o / \partial I$.

The situation is now entirely different since $\overline{F}(J) \equiv 0$ so that J = constant. This is consistent with the conclusions of the KAM theorem, since the conservation of quasi-periodic motions under perturbation implies that $|I(t) - J(t)| < K$, a constant, for all times, for all initial conditions in the non-degenerate case (Δ_1 or $\Delta_2 \not\equiv 0$) of two degrees of freedom, and for most initial conditions for $n > 2$.

If H_o depends upon fewer actions than degrees of freedom, i.e. $H_o = H_o(I_1, \ldots, I_k)$, $k < n$, (the case of so-called proper degeneracy) and if the averaged system is either integrable or close to an integrable one than there exist quasi-periodic solutions corresponding to the unperturbed system (85; see also 11, 12). These quasi-periodic solutions have k "fast" frequencies $O(1)$ coming from the unperturbed system and (n-k) "slow" frequencies $O(\varepsilon)$ that arise from the averaged system.

But, in the general case, when the averaged system is neither integrable or close to an integrable one, the relation between the solutions of the perturbed and averaged systems is unknown even over a time interval $O(1/\varepsilon)$.

From the examples and results given above it should be clear that the method of averaging must be used with great care since the conditions under which the actual error is of the order suggested by intuitive arguments are not well known. The lack of information regarding this question is due to the fact that estimates of the error usually are extremely difficult to obtain. Good sources about these matters are Hale's books (86) and (87) where the averaging method is discussed at length.

2. CHAOTIC MOTIONS IN DISSIPATIVE SYSTEMS: THE CASE OF STRONG TURBULENCE

The study of dissipative dynamical systems has taken a new direction since the discovery about ten years ago of a new class of mathematical entities which have come to be known as strange attractors. Anticipating somewhat we can say that a strange attractor is a subset of asymptotically stable orbits in which the motions are highly sensitive to the initial conditions and display a chaotic behavior. It must be emphasized that the chaotic character of such motions is intrinsic to the model and does not require any additional assumption about the occurence of randomness. Strange attractors possess a property which is most appealing to physicists: they are structurally stable by which is meant, roughly, that their orbits are not qualitatively affected by small perturbations. The possibility of occurence of strange attractors in certain well known mathematical models is of great importance for it implies that the long term behavior of the solutions is of an apparently random nature, a feature quite unexpected in our conventional view of deterministic systems. Also there are indications that they provide better explanations for the observed random characteristics of the corresponding natural phenomena thereby validating, at least qualitatively, some mathematical models.

Strange attractors have already been found in a classic MHD problem, namely the origin of the reversals of the earth's magnetic field (88), and in a model for the continuous mode laser where they are believed to be responsible for the so-called spike instability (89). Such an attractor is also known to be present in a model for the convective instability of an atmospheric layer heated from below (90) which suggests that the turbulent phenomena that we observe in fluids are linked to the presence of strange attractors in the infinite-dimensional phase-space of the Navier-Stokes equations. This is particularly relevant to plasma physics since Dupree has shown (91) that the two-dimensional Navier-Stokes equation is identical to the drift approximation of the Vlasov equation for $k_{\parallel} = 0$ and $k_{\perp} a_i \ll 1$ ($k_{\parallel}$ and $k_{\perp}$ are wave numbers parallel and perpendicular to a constant magnetic field, a_i is the ion gyroradius). Also, because strange attractors do not exist in linear systems, any theory of strong plasma turbulence based on the linearized version of the MHD equations is bound to fail if, as seems by now very likely, strange attractors are present in the phase-space of the full equations. These new considerations strongly reinforce Montgomery's arguments against the common practice of throwing away the non-linear terms (92).

We shall therefore attempt in this section to acquaint the reader with the contemporary work of various mathematicians on the problem of turbulence. Our task is not easy because of the advanced and abstract character of the mathematical methods involved. We hope that without getting bogged down in technical details our review describes with sufficient clarity the gist of the approach to the problem of turbulence taken by these mathematicians.

To begin with, the Navier-Stokes equations are adopted as a model for the motion of an incompressible viscous fluid. This is certainly a valid choice, at least for Reynolds numbers not too high, since the hypothesis used in deriving these equations are well satisfied by the turbulent flows that we observe. In vector form these equations are

$$\frac{\partial u}{\partial t} + (u \cdot \nabla)u = -\frac{1}{\rho}\nabla p + \nu\,\nabla^2 u + f$$

$$\nabla \cdot u = 0$$

$$u = 0 \text{ or prescribed on } \partial\Omega,$$

where Ω is the region containing the fluid, $\partial\Omega$ its boundary, u the velocity field of the fluid, ρ the density, p the pressure, and f the external force. The parameter ν represents the kinematic viscosity which in dimensionless variables is simply equal to the reciprocal of the Reynolds number R, i.e. $\nu = 1/R$.

Let us recall in passing the major difficulties presented by the analysis of these equations in the neighborhood of $\nu = 0$, that is for large Reynolds numbers. (See for instance 93, p. 287).

First, the order of the partial differential equation for the momentum is two for $\nu \neq 0$ and one for $\nu = 0$. Thus standard perturbation theory cannot give us an approximation to the solution for ν small starting from the solution of the equations with $\nu = 0$ - the Euler equation.

Second, for $\nu \neq 0$ the velocity u is prescribed on the boundary $\partial\Omega$ because of viscosity while for the Euler case u must be parallel to $\partial\Omega$.

The usual way of attacking the problem is through the use of space averages or correlation functions that one tries to determine from the equations without solving them. This always results in a closure problem: one obtains an infinite sequence of differential or integro-differential equations which are formally satisfied by an infinite sequence of unknown functions; but the equation of rank n in the sequence of equations involve functions of rank (n+1) in the sequence of unknowns. Consequently, truncating the sequence at the n-th equation can yield an approximation of known accuracy to the unknown functions of rank up to n included only if some appropriate information on the behavior of the functions of rank (n + 1) can be obtained by some other means. One of the goals of the mathematicians is to extract just this kind of information from the Navier-Stokes equations through a qualitative analysis of the changes in the behavior of their solutions as the Reynolds number is varied.

Before dealing with the contemporary mathematical work on turbulence we must give an account of the conventional ideas about the problem as they have been proposed and developed by Landau (94; see also 95) and Hopf (96, 97).

Landau proposed that the sequence of events observed when the Reynolds number is increased corresponds to concomitant changes in the stability of the solution of the equations. In his picture, there exists a steady laminar solution which is stable with respect to small perturbations at low values of R. As R is increased there is a critical value R_1 at which this steady solution becomes unstable; this results in the appearance of a time dependent and stable flow which is periodic with frequency ω_1, say. For values of R slightly above R_1 this flow can be represented by superposing on the steady flow a periodic flow with an amplitude initially small but one that will increase with R. As R is further increased there is a second critical value R_2 at which the unsteady periodic flow itself becomes unstable; again, for R slightly above R_2 the flow can be looked upon as the result of the superposition over the previous flow of a periodic flow of relatively small amplitude and frequency ω_2 (in general ω_2 will be incommensurable with ω_1), and so on. For high Reynolds numbers the velocity field would then depend on many irrationally related frequencies and therefore would look chaotic to the observer.

The plausibility of Landau's picture was later reinforced when Hopf pointed out that the successive transitions envisaged by Landau could be explained if the solutions of the fluid partial differential equations undergo successive bifurcation phenomena similar to those which can happen in ordinary differential systems depending in a certain manner on a parameter (96).

However, as it has been realized since, the Hopf-Landau picture has at least three serious defects:

1) it can be shown that it leads to a Gaussian statistics which is not observed (98, 99);

2) the increase in the complexity of the flow would be gradual, whereas the turbulent regime sets in at a usually sharply defined value of the Reynolds number;

3) the ultimate turbulent state would not depend in a sensitive manner on the initial conditions while the opposite is observed.

This picture was therefore in great need of improvement but it took some time until progress in mathematics made it possible for Ruelle and Takens (S-11) in 1971 to offer a new picture which is more flexible than Landau's in its ability to model the various stages of unsteady flows observed in nature and is free of the defects mentioned above (99). To be sure, these mathematicians' picture, like Landau's, is still no more than a guess but it is a "better educated" one because it incorporates certain newly discovered properties of differential systems which were not available to Landau. This, plus the fact that recent and more accurate experiments seem to confirm its validity, has contributed to its gaining increasing interest since it was proposed.

The rest of this section is devoted to an exposé of the Ruelle-Takens picture of turbulence.

We shall begin with a presentation of two mathematical phenomena of the theory of differential equations which form the basis upon which this new picture is built.

One is the possibility for the solutions to become trapped forever in regions of phase-space where their behavior is chaotic. These regions have fewer dimensions than the phase-space itself and possess extremely complex geometric structures. They are the strange attractors mentioned earlier.

The other phenomenon, the so-called Hopf bifurcation consists in the appearance or disappearance of a periodic solution as a parameter crosses a certain critical value.

We will need some definitions and properties related to autonomous systems of differential equations of the form

$$x = X_\mu(x) \qquad (\ = d/dt). \tag{21}$$

We should consider the case where x is an element of a functional space and X_μ an operator acting on it, μ being a scalar parameter that we will later identify with the Reynolds number. But it will be sufficient for the purpose of our heuristic presentation to restrict ourselves to the much simpler case where x and X_μ are n-dimensional real vectors in R^n (S-12).

We assume existence and uniqueness of a solutions $x(t) = \Phi_t(x_o)$ satisfying the initial conditions $x(0) = x_o$, i.e. $\Phi_o(x_o) = x_o$. The function $\Phi_t(x)$ is then called the flow of equation (21). Note the group property

$$\Phi_{s+t}(x) = \Phi_s\,(\Phi_t(x)) \tag{22}$$

for all x, t, s, as long as both sides are defined.

A set $A \subset R^n$ is said to be invariant with respect to Eq. (21) if $\Phi_t(x) \in A$ for all $x \in A$ and all $t \in R$. Examples of such sets are a stationary solution a for which $\Phi_t(a) = a$ for all t, a closed orbit corresponding to a T-periodic solution ($\Phi_{t+T}(x) = \Phi_t(x)$ for all t if x is any point of the orbit), or, more generally, a k-dimensional manifold with $k \leq n$.

Of special importance among the various types of invariant sets are those which are called attractors. Although there is no generally accepted definition of what an attractor is, we can say loosely that an invariant set A is an attractor if it has a neighborhood U such that the distance between the point $\Phi_t(x)$ and A tends to zero as t tends to infinity for all $x \in U$. For example, let a be a stationary solution, i.e. $X_\mu(a) = 0$, such that the Jacobian matrix $\partial X_\mu/\partial x$ evaluated at a has all its eigenvalues in the left half plane. Then it is known that there is a neighborhood of a with the above property. Other examples of attractors are a closed orbit or a torus when they are asymptotically stable (Here and in the following we mean the surface of the torus, not the solid torus, unless otherwise specified).

All these examples of attractors have been known for a long time but they do not exhaust all the possibilities, for it has by now been established that there exist attractors whose structure is topologically much different from those of a point, a closed orbit, or a torus. The possible occurence of such attractors was first pointed out by Smale (50) who gave a recipe for constructing flows possessing such an attractor from a certain class of discrete dynamical systems. The attractors of the flows thus constructed by Smale possess very unusual properties which led Ruelle and Takens to give them the epithet "strange" (100). At the present time the expression "strange attractor" is used to designate any attractor in which the motions have a chaotic character.

As an example of a strange attractor whose structure is relatively easy to understand we will now describe one invented by Shub. We borrow from Smale (103) and Bowen (69).

We begin with a description of the mapping and of its properties.

Consider a solid torus T whose points are referred by the coordinates (θ, r, s) where θ is the azimuth of the meridian plane containing the point, and r and s are rectangular coordinates in the meridian plane as shown in Figure 10. Let $D(\theta)$ denote the disc intersection of T by the meridian plane.

The mapping h is then defined by the correspondence

$$(\theta, r, s) \overset{h}{\to} (2\theta, \varepsilon_1 \cos\theta + \varepsilon_2 r, \varepsilon_1 \sin\theta + \varepsilon_2 s) \qquad (23)$$

where $0 < \varepsilon_2 < \varepsilon_1 < 1/2$.

To see what the image of T is under h note first that the image of the particular disc $D(0)$ is smaller by a factor ε_2; indeed it is easily seen from (23) that this image is a disc centered at $r = \varepsilon_1$, $s = 0$, and of radius ε_2. But there are other points in $D(0)$ which land there under h: they come from the disc $D(\pi)$ and lie in the disc centered at $r = -\varepsilon_1$, $s = 0$, with same radius ε_2 as shown in Figure 11.

For an arbitrary angle $\theta \neq 0, \pi$, the points of $D(\theta)$ obtained through the application of h come from the disc $D(\theta/2)$ according to

$$(\theta/2, r, s) \overset{h}{\to} (\theta, \varepsilon_1 \cos\frac{\theta}{2} + \varepsilon_2 r, \varepsilon_1 \sin\frac{\theta}{2} + \varepsilon_2 s)$$

and from the disc $D(\theta/2 + \pi)$ according to

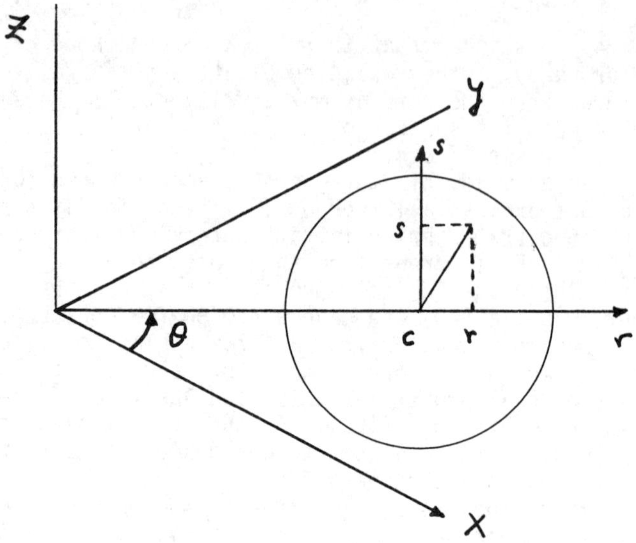

Figure 10.

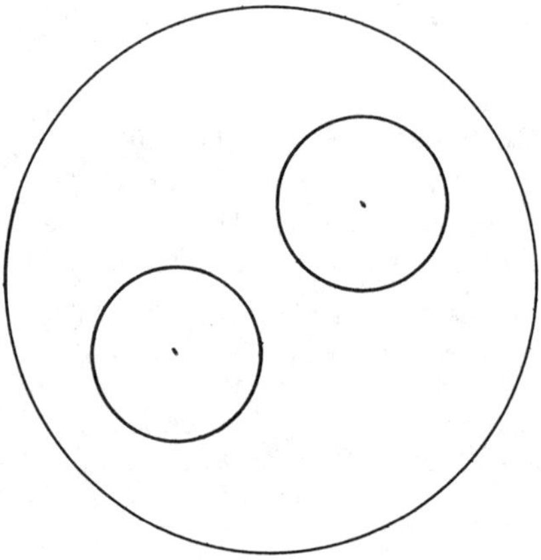

Figure 11.

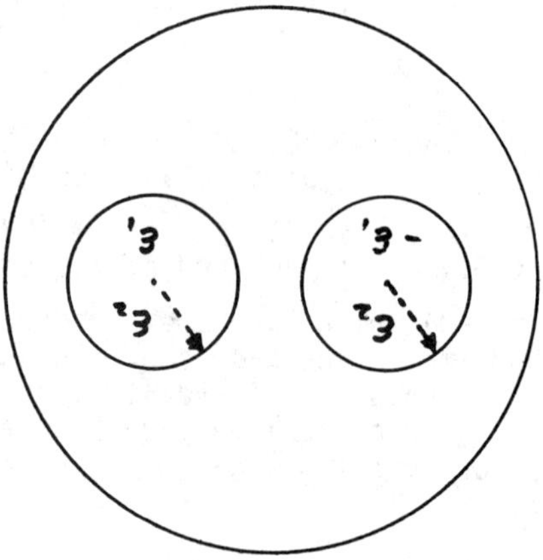

Figure 12.

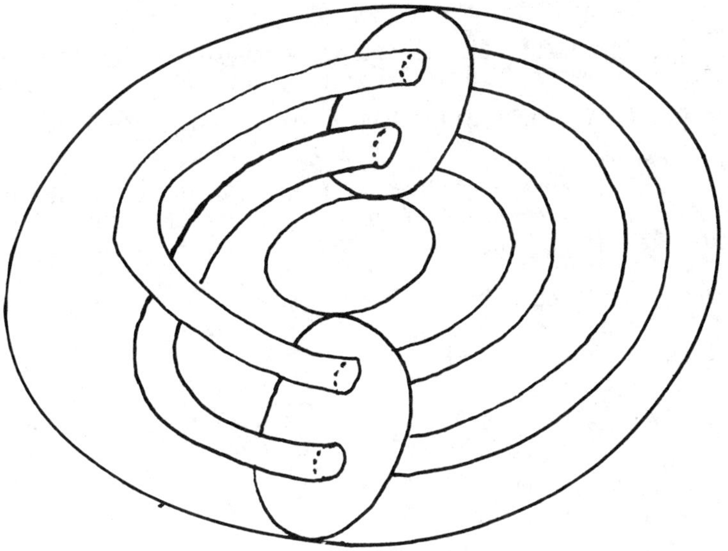

Figure 13.

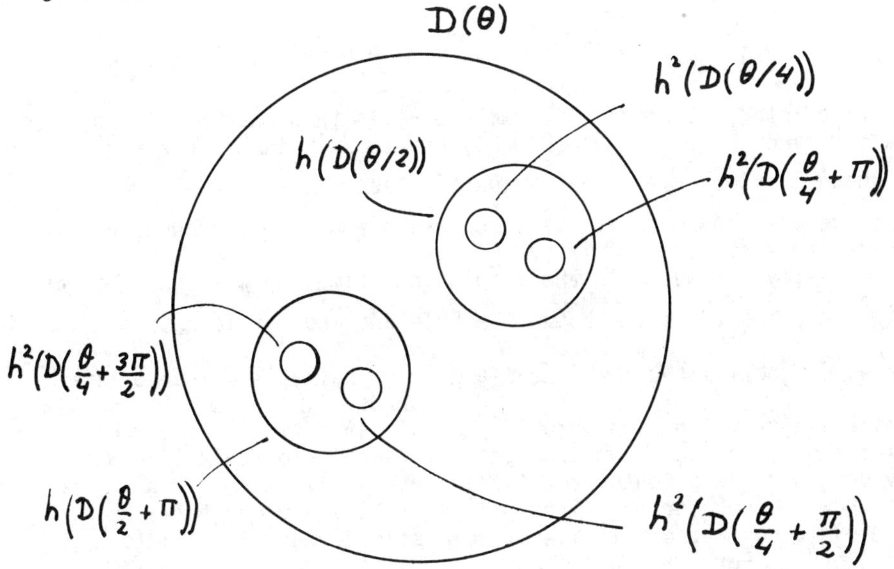

Figure 14.

$$(\frac{\theta}{2} + \pi, r, s) \overset{h}{\to} (\theta, - \varepsilon_1 \cos \frac{\theta}{2} + \varepsilon_2 r, - \varepsilon_1 \sin \frac{\Theta}{2} + \varepsilon_2 s).$$

Hence $D(\theta/2)$ and $D(\theta/2 + \pi)$ are mapped onto smaller discs centered at $(\varepsilon_1 \cos \theta/2, \varepsilon_1 \sin \theta/2)$ and $(- \varepsilon_1 \cos \theta/2, - \varepsilon_1 \sin \theta/2)$, respectively. Again, both discs have radius ε_2 (Figure 12).

It is then easily seen that the image $h(T)$ of the torus T under h is a solid torus contained in T which wraps twice around the center hole of T as sketched in Figure 13. Note that the effect of h is to make the new torus longer and thinner than the original one.

Now apply h a second time. The formula is

$$h^2(\theta, r, s) =$$

$$(4\theta, \varepsilon_1 \cos 2\theta + \varepsilon_2 (\varepsilon_1 \cos \theta + \varepsilon_2 r), \varepsilon_1 \sin 2\theta + \varepsilon_2 (\varepsilon_1 \sin \theta + \varepsilon_2 s)).$$

Proceeding as before it is found that

$$h^2(D(\theta/4)) \text{ and } h^2(D(\frac{\theta}{4} + \pi)) \subset h(D(\frac{\theta}{2})) \subset D(\theta)$$

and

$$h^2(D(\frac{\theta}{4} + \frac{\pi}{2})) \text{and } h^2(D(\frac{\theta}{4} + 3 \frac{\pi}{2})) \subset h(D(\frac{\theta}{2} + \pi)) \subset D(\theta)) .$$

The intersection of $h^2(T)$ with $D(\theta)$ is thus made up of four smaller discs, two within $h(D(\theta/2))$ and two within $h(D(\theta/2 + \pi))$ as shown in Figure 14. The radius of these discs is ε_2^2. The map h^2 therefore produces a torus which wraps four times around the central hole of T.

Clearly, if we apply the mapping h n times we will get a torus T_n which wraps 2^n times around the hole and whose thickness $\sim \varepsilon_2^n$. Hence, in the limit $n \to \infty$, the set $\Lambda = \bigcap_{n>0}^{\infty} h^n(T)$ is a line of infinite length which winds around the hole an infinite number of times and it can be shown that the intersection of this line with any disc $D(\theta)$ is a Cantor set (see Appendix C).

Now we look for the fixed points of h. In order for a point of $D(\theta)$ to be mapped into itself we must have $2\theta = \theta + 2k\pi$ for $k = 0$ or 1. For $k = 0$ we get $\theta = 0$ and then the coordinates of the fixed points which are in $D(0)$ are solutions of the equations

$$\varepsilon_1 + \varepsilon_2 r = r \text{ and } \varepsilon_2 s = s$$

which have the unique solution $r = \varepsilon_1/(1-\varepsilon_2)$, $s = 0$. For $k = 1$ we find another fixed point in $D(\pi)$ located at $r = -\varepsilon_1/(1-\varepsilon_2)$, $s = 0$.

In general, the n-th power of the mapping has one fixed point in each of the disc $D(\theta)$ such that

$$2^n\theta = \theta + 2k\pi, \text{ i.e. } \theta = 2k\pi/(2^n - 1)$$

where $k = 0, 1, \ldots, 2^n - 2$. Hence h^n has $2^n - 1$ fixed points of period n and in the limit of $n \to \infty$ we see that there is an infinity of fixed points of arbitrarily large period.

Among the various properties of h the following are essential for our purpose:

1) Λ attracts all points of T under repeated application of h;

2) at each point of Λ there is a so-called "hyperbolic splitting" in the sense that h expands in one direction (it makes T always longer) and contracts in the other two dimensions (always shrinks the cross-section by a factor ε_2).

3) for the discrete dynamical system defined by $(\theta_{n+1}, r_{n+1}, s_{n+1}) = h(\theta_n, r_n, s_n)$ the set Λ is structurally stable.

We shall now show how a discrete dynamical system can be turned into a continuous one. The ability to effect this conversion is important because it leads to models, with known solutions, for systems such as the Navier-Stokes equations whose solutions are not known in general.

Recall from section 1 that given a flow defined by some autonomous ordinary differential equations we can associate to it a discrete mapping (the Poincaré map) provided that a surface of section exists (*). Smale's prescription for constructing a smooth flow associated to a given discrete mapping can be viewed as the reverse of a Poincaré map. He has called the flow thus obtained the suspension of the given mapping. We cannot give here a description of its construction as Smale did it in abstract topological terms. So, instead, we use a geometrical approach to show first how to construct the suspension of an arbitrary mapping of the real line $f: R \to R$. This real line will be the x-axis and the flow Φ_t that we want to construct will be defined in the strip B: $0 \leq y \leq 1$ of the plane of x and y (see Figure 15).

Clearly it is sufficient in order to define a flow to give the prescription for constructing the unique orbit going through an arbitrary point together with the time parametrization of the motion on the orbit. This parametrization we define by the differential equations

(*) Note that the notion of surface of section can be extended to n-dimensional differential systems, $n > 3$, the "surface" then having dimension $n - 1$.

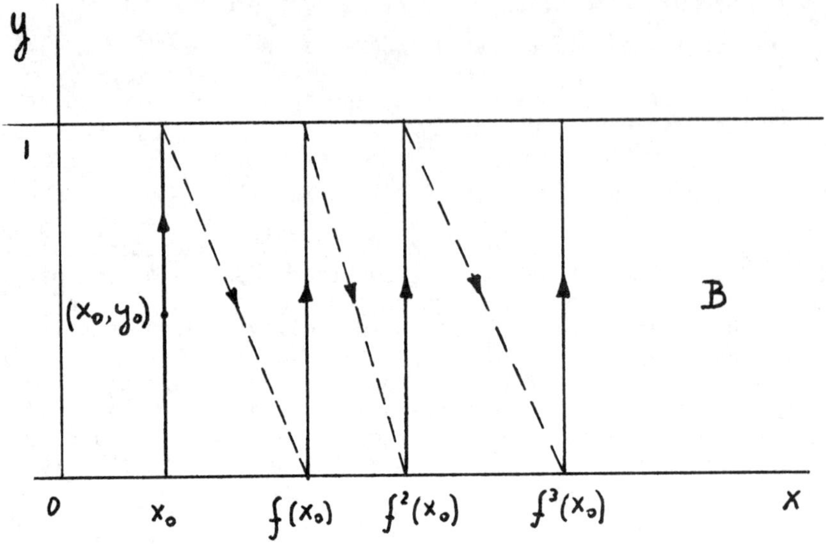

Figure 15.

$$\frac{dx}{dt} = 0, \frac{dy}{dt} = 1.$$

The autonomous character of these equations guarantees that the flow will have the group property (22).

Now, since x must be constant according to the first equation we define the orbit through a point $(x_0, y_0) \in B$ with $y_0 < 1$ as the line $x = x_0$. For $t > 0$, starting from (x_0, y_0) the point representing the motion moves with unit speed until $y = 1$. We then make it jump in zero time to the point $(f(x_0), 0)$; from there the orbit is the line $x = f(x_0)$ until $y = 1$ again. The point jumps in zero time to $(f^2(x_0), 0)$, moves up along $x = f^2(x_0)$, and so on.

Finally, for $t < 0$ we proceed in the same fashion using the iterates $f^{-n}(x_0)$, $n = 1, 2, \ldots$.

We have thus defined the suspension flow of f whose Poincaré map on the line of section $y = 0$ is nothing but f as required. Note that the fixed points of f^n give rise to closed orbits on which the motion is n-periodic.

The generalization of the above construction to a mapping $f: R^n \to R^n$, $n > 1$, is straightforward. Let B denote the subset of R^{n+1} defined by the inequality $0 \leq x_{n+1} \leq 1$. The suspension is then defined on B by the differential equations

$$\dot{x}_i = 0,\ 1 \leq i \leq n,\ \dot{x}_{n+1} = 1 .$$

The orbit through the initial point $x^0 \varepsilon B$ is the line $x_i = x_i^0$, $1 \leq i \leq n$, and the solution point moves on it with unit speed until $x_{n+1} = 1$. It then jumps in zero time to the point $x_i = f_i(x^0)$, $1 \leq i \leq n$, $x_{n+1} = 0$, moves uniformly on the line $x_i = f_i(x^0)$, $1 \leq i \leq n$, until $x_{n+1} = 1$, jumps to $f_i^2(x^0)$, $1 \leq i \leq n$, $x_{n+1} = 0$, and so on.

We can now apply the foregoing prescription, starting from Shub's mapping of a torus described before, to construct a flow with continuous vector field defined on R^4 for which the set Λ is a globally stable, strange attractor as $t \to \infty$.

The continuous dynamical system built this way has an extremely complicated behavior because of the hyperbolic splitting mentioned before. Indeed, consider solutions starting at $t = 0$ from two points P_o and P_o' a distance ε apart. However small ε is the distance between $\Phi_t(P_o)$ and $\Phi_t(P_o')$ will increase exponentially with time (excluding the case where P_o and P_o' are in the same disc $D(\theta)$) and the transverse motion in the r and s coordinates will be completely different for the two orbits.

Much work has been devoted to the study of strange attractors since their relatively recent discovery and more is known about them than we can possibly cover in this report. We only mention that the ergodic properties of some of them have been investigated by several authors (see for instance 69).

We now turn to the other essential element of the Ruelle-Takens picture of turbulence, namely the Hopf bifurcation phenomenon.

2.1 HOPF BIFURCATION PHENOMENON

The Hopf bifurcation deals with the qualitative changes in behavior of the solutions of a differential equation

$$\dot{x} = X_\mu(x)$$

as the parameter μ goes through certain critical values.

We assume as before that x and X_μ are finite, n-dimensional vectors. We also assume that $x = 0$ is a stationary solution, i.e. $X_\mu(0) = 0$.

Now let $A = \partial X_\mu/\partial x$ denote the Jacobian matrix of X_μ evaluated at 0 and suppose there exists an interval (a, b) such that for $a < \mu < b$ A has a pair of simple complex eigenvalues $\lambda(\mu) = \alpha(\mu) + i\,\omega(\mu)$ and $\overline{\lambda}(\mu)$ whose imaginary part does not vanish in (a, b) while the real part α is strictly increasing with μ and vanishes for

$\mu = \mu_1 \in (a, b)$. The other eigenvalues are supposed to stay in the left half plane for all μ in that interval.

It then turns out that, under mild smootheness conditions on X_μ, the following necessarily happens (93): as μ crosses the value μ_1 a single periodic solution appears in the neighborhood of the origin $x = 0$. More precisely: if a certain function U of the first, second, and third partial derivatives of $X_{\mu_1}(0)$ is positive the periodic solution appears for $\mu > \mu_1$ and is stable; as $\mu \to \mu_1$ from above, the corresponding orbit shrinks into the origin. If, instead, U is negative the periodic solution exists for $\mu < \mu_1$, is unstable, and its orbit shrinks into 0 as $\mu \to \mu_1$ from below. Moreover, in both cases, as $\mu \to \mu_1$ the frequency tends to $|\omega(\mu_1)|$ and the radius of the periodic orbit approaches zero as $\sqrt{|\mu - \mu_1|}$.

This is Hopf bifurcation theorem (96; see also 93 for a modern account and generalizations). When U is positive one says that the stationary solution undergoes a normal Hopf bifurcation whereas for $U < 0$ the phenomenon is called an inverted Hopf bifurcation. Note that a more complicated behavior can occur if $U = 0$ or/and if more eigenvalues cross the imaginary axis at $\mu = \mu_1$.

A remarkable feature of the Hopf theorem is that when its conditions are met no matter how large the (finite) dimension of the differential system is, we can predict the appearance of a periodic solution and determine its stability with great ease*.

An important extension of Hopf's theorem was provided by Ruelle and Takens (100) who showed that, under suitable technical conditions, a stable invariant two-dimensional torus appears in the vicinity of a closed orbit when the latter becomes unstable as the parameter is increased (see also 93).

Finally, very recently, Chenciner and Iooss have given sufficient conditions which guarantee that, as an invariant two-dimensional torus becomes unstable, a stable invariant three-dimensional torus will appear (104).

We are now in the position to present the Ruelle-Takens picture.

2.2 THE RUELLE-TAKENS PICTURE OF TURBULENCE

These authors view the Navier-Stokes equations as an evolution vector differential equation

$$\frac{du}{dt} = X_R(u) \tag{24}$$

*The evaluation of U may be cumbersome but is straightforward for it involves only differentiations and algebraic calculations.

where u is an element of the functional space U of vector fields defined over the region Ω containing the fluid and satisfying the boundary conditions, X_R is the vector field obtained through the application of the Navier-Stokes operator to u, and R is the Reynolds number.

The following assumptions are made:
1) for $1 \leq R < \bar{R}$ ($<\infty$) Eq. (24) has a stationary solution u(r, R) where r is the position vector (x, y, z) in Ω;
2) the eigenvalues of the Jacobian matrix $A = \partial X_R/\partial u$ evaluated at u(r, 1) have all strictly negative real part which implies that u(r, 1) is asymptotically stable;
3) as R increases from 1, successive pairs of complex eigenvalues cross the imaginary axis for $R = R_1, R_2, \ldots$ (with $1 < R_i < R_{i+1}$ for all i).

It follows from 3) that the stationary solution u(r,R) becomes unstable and a stable closed orbit T^1 appears for R slightly greater than R_1 assuming a normal Hopf bifurcation takes place. To this closed orbit corresponds a periodic flow with frequency ω_1, say.

As r increases further and reaches R_2 a second normal bifurcation occurs: the closed orbit T^1 becomes unstable and there appears a stable invariant torus T^2 on which the motion is quasi-periodic with frequencies ω_1, ω_2.

More generally it is assumed that as R crosses R_k from below a stable invariant torus T^{k-1} becomes unstable and that there appears a stable invariant torus T^k with one more dimension.

Now, although it is possible that the motion on the torus T^k, $k > 2$, be quasi-periodic, Ruelle and Takens conjecture that this is very unlikely, the most probable situation being the existence of strange attractors. Their conjecture is based upon their proof that the situation is indeed such on the particular torus T^4.

The novel and essential physical implication of these mathematical arguments is that the onset of turbulence can be associated with the sudden appearance of strange attractors in which the motion is intrinsically chaotic. Moreover, since the appearance of strange attractors is likely as soon as $R > R_3$, for $R_1 < R < R_3$ the unsteady flow should display only one or two well defined frequencies, its spectrum becoming continuous in the turbulent regime associated with a strange attractor.

Finally, the possible appearance of several unconnected strange attractors with their associated basins of attraction implies that the ultimate turbulent regime may depend critically on the initial conditions.

To sum up, strange attractors constitute new mathematical objects which should be added to the panoply of tools that can be utilized to model phenomena encountered in plasma physics.

As emphasized earlier, this picture of turbulence is no more than a guess at the present time because it relies upon many unproven assumptions. One of these is that the solutions of the Navier-Stokes equations in three dimensions are globally regular, i.e. that they can be continued for all times and remain smooth. This is true in two dimensions but still a conjecture in three dimensions. Note that if strange attractors do exist the proof for three dimensions will probably show that solutions are globally regular provided the initial conditions lie in one of their basins of attraction (93, p. 303).

Another assumption made by Ruelle and Takens is that the stability of the solutions can be deduced from an associated finite dimensional problem (93).

Finally, Kaplan and Yorke (105) have pointed out that there is no reason to believe at present that perturbations which change the system into one with a strange attractor are more common than those which change the system into one with a stable attracting periodic orbit. This remark is important because in the Ruelle-Takens argument tne non-linearities are assumed small and treated as perturbations. On the other hand, Kaplan and Yorke consider it plausible that the likelihood of strange attractors increases as the size of the non-linearities increases.

We now indicate some experiments and analytical work which seem to confirm the validity of the Ruelle-Takens picture.

Gollub and Swinney (106) have studied the radial velocity in a fluid (water) rotating between concentric cylinders using light-scattering measurements. They have observed three distinct transitions as the Reynolds number is increased and each of these adds a new frequency to the velocity spectrum. At a higher, sharply defined Reynolds number all discrete spectral peaks suddenly disappear leaving a broad hump. In a similar experiment, Swinney, Fenstermacher, and Gollub (107) have obtained the same sequence of transitions; the turbulent state however was characterized by a series of broad humps in the power spectrum.

These observations are clearly in conflict with the Landau picture but consistent with that of Ruelle and Takens.

On the analytical side, Sherman and McLaughlin (108) have studied the solutions of a model for these experiments based on the Orr-Sommerfeld equation. They have found that it is possible to adjust the free parameters of the model so as to reproduce the observed dependence of the spectrum. This model also is consistent with the Ruelle-Takens picture.

Earlier attempts at solving the fluid equations had produced a chaotic attractor after a sequence of events which agree more or less with this picture (109, 110). In particular, Lorenz' work (90) deserves special consideration in view of the rather extensive investigation to which his model has been subjected by various mathematicians (S-13).

From the point of view of plasma physics this model is instructive because it illustrates the complexity arising from the competition between convective instabilities and heat transport.

2.3 THE LORENZ MODEL

The problem considered by Lorenz is essentially the Bénard problem of the convective motion of a fluid layer of uniform depth heated from below in a two-dimensional container. Using a method that he had proposed earlier (113), this author expanded the stream function and the temperature in Fourier series of space variables with time dependent coefficients, assuming free boundary conditions on the velocity field and no heat flow through the wall of the container. Substitution of these series in the partial differential equations of the problem yields an infinite set of ordinary differential equations among which Lorenz retained only three. Although no mathematical justification has been given for this severe mutilation, it is believed that the main features of the solutions to the fluid equations are qualitatively preserved, at least within some range of values of the parameters. He thus obtained the dimensionless system

$$\dot{x} = -\sigma x + \sigma y$$

$$\dot{y} = rx - y - xz$$

$$\dot{z} = xy - bz \ ,$$

where x is the intensity of the convective motion, y is the temperature difference between the ascending and descending currents, and z is the distortion of the vertical temperature profile from linearity; σ is the Prandtl number, b is a constant related to the basic spatial frequency in the x direction, and r is the Rayleigh number.

Lorenz found through numerical integration that, in spite of its relative simplicity, this system displays a remarkable variety in its behavior the most interesting phenomenon being the appearance, for r large enough, of an attractor which is neither a point nor a closed orbit.

Among the properties of the Lorenz system the following are elementary and easy to verify:

1) the system is invariant under the transformation $x \to -x$, $y \to -y$, $z \to z$. This is a consequence of the invariance of the equations of motion under a reflection through a vertical line at the center of the container.

2) the divergence of the vector whose components are the right-hand sides of the equations has a constant value $-(\sigma + b + 1)$. Consequently, the corresponding flow shrinks volumes in the phase-space at a uniform rate which, for parameter values of interest ($\sigma = 10$, $b = 8/3$) is -13.66 and therefore is quite large; a unit volume is transformed into the volume $\exp(-13\ 2/3) \simeq 10^{-6}$ over a unit interval of time.

3) all solutions are bounded as $t \to \infty$. In fact one can show that there exists a ball B in phase-space such that all solution curves

eventually get trapped inside it forever. The shrinking properties of the flow then implies that the volume of the image of this ball tends to zero at $t \to \infty$ and all solutions tend to a set which is closed and has measure zero.

Since the ultimate behavior of all solutions depends upon the nature of this set we shall review how its structure changes as r is increased from zero, keeping the other parameters fixed as before ($\sigma = 10$, $b = 8/3$).

1) $0 \leq r \leq 1$: the origin $x = y = z = 0$ is then the only singular point and is globally attracting; physically, it corresponds to conduction: the fluid layer is at rest and the temperature varies linearly between the two surfaces.

2) $r > 1$: there are now three singular points: the origin which has become unstable and the two points

$$C = (\sqrt{b(r-1)}, \sqrt{b(r-1)}, r-1), \ C' = (-\sqrt{b(r-1)}, -\sqrt{b(r-1)}, r-1)$$

which correspond to steady convection.

For all $r > 1$ there is a two-dimensional manifold (a surface) containing the origin which is invariant under the flow and on which the solution curves approach the origin as $t \to \infty$. There is also a one-dimensional invariant manifold (a curve) containing 0 along which the point is reached as $t \to -\infty$.

The points C and C' are locally stable for $1 < r \leq r_2 = 470/19 \simeq 24.74$. For $r > r_2$ they become unstable because two complex conjugate eigenvalues move from the left half plane to the right one. The corresponding Hopf bifurcation is of the inverted type, i.e. in the neighborhood of each of these points there exists an unstable periodic orbit for r slightly smaller than r_2, whose size shrinks to zero as $r \to r_2$ from below.

There are two important transition values for r which have been found numerically:

At $r = r_0 \simeq 13.926$ the unstable trajectories of the origin return to 0 as $t \to \infty$ (they are homoclinic orbits in the terminology of section 1). For $1 < r < r_0$ all trajectories tend to one of the rest points 0, C, or C', while for $r > r_0$ there are infinitely many periodic orbits and infinitely many unstable "turbulent orbits" (105) which do not tend to any point or periodic orbit. The set of these orbits has zero measure.

At $r = r_1 \simeq 24.06$ each of the unstable trajectories from 0 tend to an unstable periodic orbit as $t \to \infty$. These two periodic orbits are of the saddle type regarding stability, i.e. a Poincaré map of their neighborhood gives rise to a hyperbolic point whose stable manifold contains the corresponding trajectory coming from the origin.

For $r > 24.06$ there is a chaotic stable attractor which probably disappears at some value of r between 28 and 50 since it was not observed in Lorenz' numerical integrations for $r = 50$.

We now describe in more detail what happens in the intervals (r_0, r_1) and (r_1, r_2) successively (105).

1) $r_0 < r < r_1$: the stationary points C and C' have neighborhoods N_C and $N_{C'}$ such that the trajectory starting from any point in N_C (resp. $N_{C'}$) tends to C (resp. C') as $t \to \infty$. Because two eigenvalues are complex at C and C', this trajectory approaches its limit point like a spiral so that s, y, z, display damped harmonic oscillations about their limiting values.

If the initial point is far from C and C', the trajectory first oscillates a number of times between two regions located one in $x > 0$ the other in $x < 0$ until it becomes trapped either in N_C or in $N_{C'}$. On the average, the number of such oscillations is larger the closer r is to r_1 and seems to tend to infinity as $r \to r_1$. Also this number of wide oscillations critically depends upon the initial point so that it is impossible to predict how many have already taken place. The corresponding solutions look chaotic for some time which may be very long if r is only slightly smaller than r_1 and then rapidly settle down at C or C' after few damped harmonic oscillations about it. Kaplan and Yorke call such behavior metastable chaos and suggest that it may be what Creveling et al. (114) have observed in the flow of a fluid through pipes: more than one hundred chaotic oscillations took place before the oscillations become regular and damp out.

2) $r_1 < r < r_2$: C and C' are still locally stable but there is a chaotic stable attractor which attracts for ever trajectories coming from a set of positive measure.

We will now give an idea of the behavior of the orbits in the strange attractor for $r = 28 > r_2$. The points C and C' are then unstable like the origin. At each point there is a manifold (a surface) $W^u_{C,C'}$ made up of orbits which spiral away from the point and, transverse to this manifold, there is a stable manifold $W^s_{C,C'}$ (a curve) along which the point is approached as $t \to \infty$. The situation is sketched in Figure 16.

The overall behavior of an orbit in the strange attractor is then as follows: supposing that we start from a point close to the origin the orbit first gets close to C, say, and then, moving very close to W^u_C spirals away from C. In the course of its motion around the line W^s_C its minimum distance from the origin decreases at each revolution until, suddenly, it goes towards the other singular point C'. It then executes a certain number of revolutions around the line $W^s_{C'}$, spiralling away from C', then goes back to the vicinity of C, spirals away from it, etc. A remarkable feature of the motion is that the number of consecutive revolutions about C or C' varies widely in a manner that is unpredictable in practice. This can be explained with the help of Figure 17 in which, for clarity,

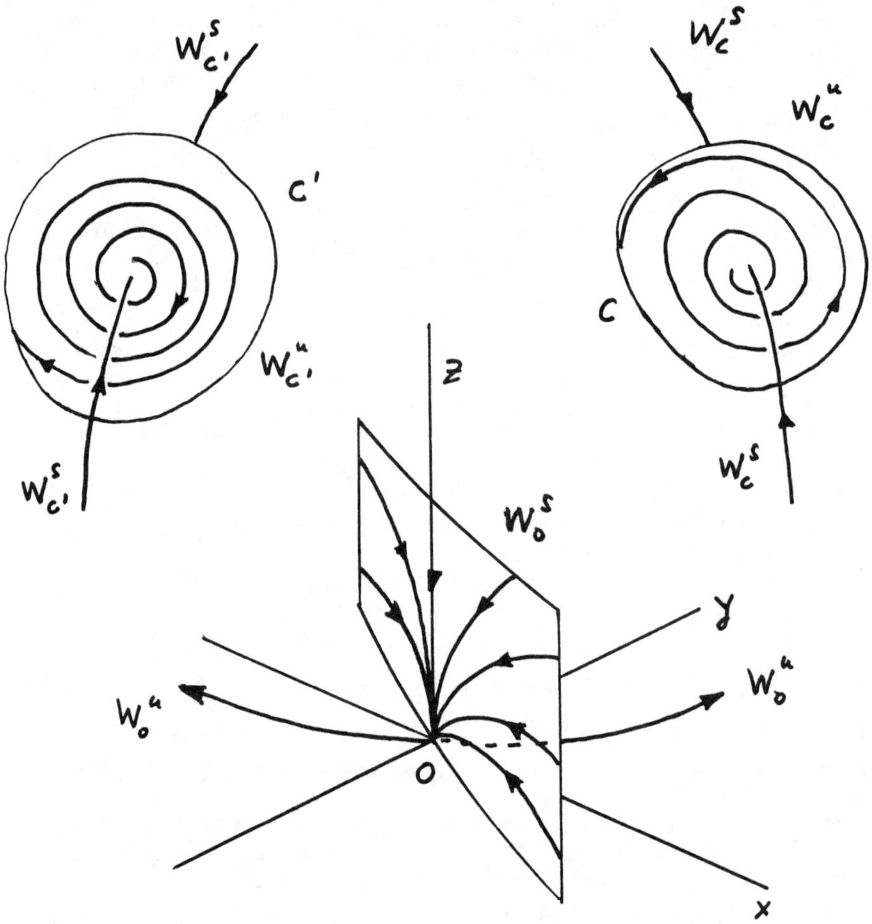

Figure 16.

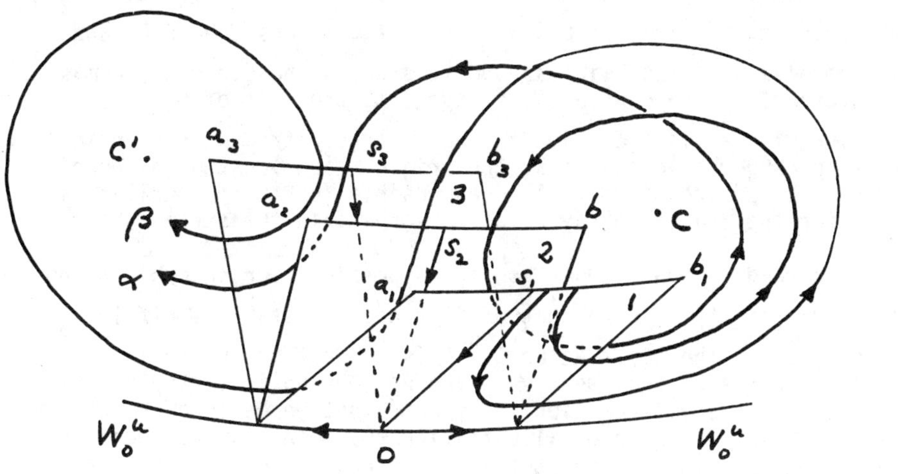

Figure 17.

we have not drawn the manifolds $W^u_{C,C'}$ and $W^s_{C,C'}$ of C and C'. In a neighborhood of the origin the strange attractor (which is a surface) is made up of uncoutably many sheets, all containing the unstable manifold W^u_0 (a line) of the origin, which intersect transverse arcs in Cantor sets. This geometrical object looks locally like a book of infinitely many pages which are bound along the line W^u_0. We show three of these pages labeled 1, 2, and 3. On each page the configuration of the orbits is hyperbolic; on page 1, for instance, there is only one orbit $S_1 0$ going to 0 which is the intersection of this page with the stable manifold W^s_0 (a surface).

We are now going to trace two different orbits labeled α and β respectively, which start from points of the segment $S_1 b_1$ on page 1. First α leaves the vicinity of 0 and is attracted by the stable manifold W^s_0 of C. It gets close to the surface W^u_C and after one revolution returns close to the origin where it moves on page 2. Because we have chosen to make it cross the segment $a_2 b_2$ to the right of S_2 its next revolution again takes place around C. However when it comes back towards 0 it lands on the segment $a_3 b_3$ of page 3 at a point to the left of S_3; as a result its next revolution will be around C'.

We now look at the orbit β which also leaves the vicinity of 0 on page 1 but from a point of S_1b_1 closer to S_1. It winds around C like α initially but after a single revolution it intersects a_2b_2 at a point to the left of S_2; as a result, it goes around C' and will revolve at least twice around it since we have chosen to make it intersect the segment a_3b_3 of page 3 to the left of S_3.

The picture is then clear: the strange attractor has an infinity of hyperbolic points at the origin, each located in one of its sheets and these points are responsible, by their peculiar nature, for the chaotic behavior of the orbits in the strange attractor.

The reader might wonder how it is possible for an orbit to go from the vicinity of C to that of C' since the stable manifold W_o^s seems to be in the way. Actually it is not but its shape is so complicated to sketch that we have given up showing it in the figure.

Finally, we must caution the reader that we designed Figure 17 for the sole purpose of illustrating the mechanism of chaos in the strange attractor: the topological inconsistencies that it may contain are of no importance in this connection.

The preceding outline of the properties of the Lorenz system covers most of what is known at the present time. For more details we refer the reader to the original reports.

It is to be expected that much further research will be devoted to the study of the Lorenz system in spite of its questionable value as a valid approximation to the fluid equations. The reason for this is that the system is probably one of the simplest dissipative systems that can display the kind of chaotic behavior believed to be at the roots of an explanation of turbulence. Another reason for a continuing interest in this system is that it is also encountered in laser physics where it is believed to explain the onset of self-pulsing (89).

Among the various questions being investigated one is whether the motion in the Lorenz attractor is ergodic in a sense consistent with the dissipative structure of the system. This is of central importance to physics in general and to plasma physics in particular since the notion of "anomalous" transport coefficients like that of thermal conductivity of the Bénard convective layer has a meaning only if it can be shown that for all motions in the attractor the asymptotic value of time averages does not depend upon initial conditions. An outline of a possible approach to this problem has been given by Lanford (102).

APPENDIX A

SUPERCONVERGENT PERTURBATION METHOD

In the perturbation method described in the text the size of the perturbation decreases by a factor ε at each stage of approximation. But this rate of decrease is too small to overcome the difficulties due to the presence of the small denominators. Fortunately, the perturbation scheme can be modified so that the n-th approximation is of order ε^{2^n}, the fast convergence of the resulting series then compensating for the divergence related to the small denominators. This scheme due to Kolmogorov (15) is essentially equivalent to Newton's method of successive approximations to the root of a scalar equation f(x) = 0. More precisely, since the objects to be approximated are functions rather than numbers, it is Kantorovitch's generalization (115) of Newton's method to equations in function space that was proposed by Kolmogorov and used by Arnold in his proof of the KAM thorem.

We shall outline the principal features of the method for the case of a Hamiltonian of the form

$$H(p,q) = H_o(p) + \varepsilon H_1(p,q) + \varepsilon^2 H_2(p,q) + \dots$$

where $H_k(p,q)$ is 2π-periodic in each of the q's for all k's.

Rewrite this as

$$H(p,q) = H_o(p) + \varepsilon \overline{H}_1(p) + \varepsilon \tilde{H}_1(p,q) + \varepsilon^2 H_2(p,q) + \dots$$

where for an arbitrary function f(p,q) 2π-periodic in the q's

$$\overline{f}(p) = \frac{1}{(2\pi)^n} \int f(p,q)dq$$

and $\tilde{f}(p,q) = f(p,q) - \overline{f}(p)$. Hence f(p,q) has mean value zero.

Apply the canonical transformation generated by the function $p'q + \varepsilon S_1(p',q)$ where S_1 is to be determined so that in the new Hamiltonian the perturbation is $O(\varepsilon^2)$. From the Hamilton-Jacobi theory the old momentum p is given by

$$p = p' + \varepsilon \frac{\partial S_1}{\partial q} . \qquad \text{(A1)}$$

Replace p by this expression in H and expand about p = p' obtaining

$$\begin{aligned} H(p',q) = H_o(p') &+ \varepsilon \frac{\partial H_o}{\partial p_i'} \frac{\partial S_1}{\partial q_i} + \frac{\varepsilon^2}{2!} \frac{\partial^2 H_o}{\partial p_i' \partial p_j'} \frac{\partial S_1}{\partial q_i} \frac{\partial S_1}{\partial q_j} + \dots \\ &+ \varepsilon \overline{H}_1(p') + \varepsilon^2 \frac{\partial \overline{H}_1}{\partial p_i'} \frac{\partial S_1}{\partial q_i} + \dots \end{aligned} \qquad \text{(A2)}$$

$$+ \varepsilon \tilde{H}_1(p',q) + \varepsilon^2 \frac{\partial \tilde{H}_1}{\partial p_i'} \frac{\partial S_1}{\partial q_i} + \ldots$$

$$+ \varepsilon^2 H_2 (p',q) + \ldots \qquad \text{(A2 - cont.)}$$

where summation over repeated indices is implied and the omitted terms are of order ε^3.

In order to make the q-dependent terms in ε disappear, choose S_1 such that

$$\omega_i^{(1)}(p') \frac{\partial S_1}{\partial q_i} + \tilde{H}_1 (p',q) = 0$$

where $\omega_i^{(1)}(p') = \partial H_o/\partial p_i'$.

Then solve Eq. (A1) together with the equation

$$q' = q + \varepsilon \frac{\partial S_1}{\partial p'}$$

for q in terms of p', q' and substitute the resulting expressions into Eq. (A2). This can be done using the generalization of Lagrange's formula to vector equations (115).

Drop the superscripts on p' and q' and write the new Hamiltonian as

$$H(p,q) = H_o(p) + \varepsilon \overline{H}_1^{(1)}(p) + \varepsilon^2 \overline{H}_3^{(2)}(p) + \varepsilon^3 H_3^{(2)}(p)$$

$$+ \varepsilon^2 \tilde{H}_2^{(2)}(p,q) + \varepsilon^3 \tilde{H}_3^{(2)}(p,q) + O(\varepsilon^4)$$

Now apply a canonical transformation generated by p'q + $\varepsilon^2 S_2(p',q)$. Substitution of

$$p = p' + \varepsilon^2 \frac{\partial S_2}{\partial q}$$

into H and expansion about p = p' yields

$$H(p',q) = H_o(p') + \varepsilon^2 \frac{\partial H_o}{\partial p_i'} \frac{\partial S_2}{\partial q_i} + \varepsilon^4 \ldots$$

$$+ \varepsilon \overline{H}_1^{(1)}(p') + \varepsilon^3 \frac{\partial \overline{H}_1^{(1)}}{\partial p_i'} \frac{\partial S_2}{\partial q_i} + \varepsilon^5 \ldots$$

$$+ \varepsilon^2 \overline{H}_2^{(2)}(p') + \varepsilon^4 \ldots$$

$$+ \varepsilon^3 \overline{H}_3^{(2)}(p') + \varepsilon^5 \ldots$$

$$+ \varepsilon^2 \tilde{H}_2^{(2)}(p',q) + \varepsilon^4 \ldots$$

$$+ \varepsilon^3 \tilde{H}_3^{(2)}(p',q) + \varepsilon^5 \ldots$$

$$+ \ldots$$

It is readily seen that not only the terms in ε^2 but also those in ε^3 can be eliminated provided S_2 satisfies the equation

$$\left(\frac{\partial H_o}{\partial p_i'} + \varepsilon \frac{\partial \overline{H}_1}{\partial p_i'}\right) \frac{\partial S_2}{\partial q_i} + \tilde{H}_2^{(2)}(p',q) + \varepsilon \tilde{H}_3^{(2)}(p',q) = 0 .$$

It can be shown by induction that the procedure may be continued indefinitely, i.e. the n-th Hamiltonian will contain a perturbation of order ε^ν with $\nu = 2^n$.

Notes: 1) Howland has very recently discussed this method and indicated how it can be implemented without having to use Lagrange's inversion formula through the use of Lie transforms (117).

2) Bogoliubov (118) and Bogoliubov et al. (119) have extended the method to non-Hamiltonian systems.

Remarks: As illustrated by its success in the proof of the KAM theorem the superconvergent method is certainly a powerful tool in pure analysis. However, as pointed out by Cary (120), it is not evident at this time that the computational labor that it requires for a given order of approximation is less than when using traditional methods. On the other hand, a symbol manipulation program such as MACSYMA (121) may reduce the labor sufficiently to make this approach more useful (122).

APPENDIX B

NOTE ON THE KAM THEOREM

In the main text, for the simplicity of the presentation, we have considered Hamiltonians in which the smallness of the perturbation is due to the presence of a small factor ε in the perturbing part. Actually, in his proof of the KAM theorem, Arnold only assumes that

$$H(p,q) = H_o(p) + H_1(p,q)$$

with H_1 bounded in a compact region of the phase-space. Denoting the bound by M, he proves the theorem for M sufficiently small.

Regarding the conditions (10) of the text, i.e.

$$\Delta_1 = \det \left(\frac{\partial^2 H_o}{\partial p_i \partial p_j} \right) \not\equiv 0 \text{ or } \Delta_2 = \det \begin{pmatrix} \frac{\partial^2 H_o}{\partial p_i \partial p_j} & \frac{\partial H_o}{\partial p_i} \\ \frac{\partial H_o}{\partial p_j} & 0 \end{pmatrix} \not\equiv 0$$

note that they are not equivalent each corresponding to a different version of the theorem; but each of them is sufficient for the validity of the conclusions. (See for instance 11, p. 269.)

APPENDIX C

PERIODIC SOLUTIONS OF A SLIGHTLY PERTURBED INTEGRABLE HAMILTONIAN SYSTEM BY THE PCHG METHOD

1. THE PCHG METHOD

Designed for the determination of the periodic solutions of differential equations, the method takes various forms depending on the nature of the equations of interest. We will concentrate on the case where the equations can be written in vector form as

$$\dot{x} = \mu\, q\, (t,\, x,\, \mu) \qquad \text{(C1)}$$

for, as we shall show, the equations of a slightly perturbed integrable system can be brought to this form by an appropriate change of variables. In (C1) x and q are n-vectors, q is T-periodic in t and μ is a small parameter. For $\mu = 0$ the solution is $x(t) = a_o$, a constant vector, and the goal is to find for μ small the T-periodic solutions, if any, which reduce to constant vectors as $\mu \to 0$.

The method rests upon four basic theorems (cf ref. 86, theorems 6-1 to 6-4, Chap. 6) whose contents, leaving aside technical details, are as follows:

Given an n-dimensional constant vector a there exists for μ sufficiently small a unique vector function $x^*(t, a, \mu)$ with the following properties:

a) x^* is continuous and T-periodic in t, and its mean value is precisely equal to the given vector a, i.e.

$$\frac{1}{T} \int_0^T x^*(t,\, a,\, \mu)\, dt = a\ .$$

b) it satisfies the differential equation

$$\dot{x}^* = \mu\, q(t, x^*, \mu) - \mu \frac{1}{T}\int_0^T q(t, x^*(t,a,\mu), \mu)\, dt \qquad (C2)$$

c) $x^*(t, a, 0) = a$, i.e. for $\mu = 0$ x^* reduces to the given constant vector.

Moreover the function x^* is the limit as k tends to infinity of the sequence of functions defined recursively by

$$x^{(o)}(t) = a$$

$$x^{(k+1)}(t) = a + \mu \int^t d\xi\{q(\xi, x^{(k)}(\xi, a, \mu),\mu) - \frac{1}{T}\int_0^T q(t, x^{(k)}(t, a, \mu),\mu)dt\} \qquad (C3)$$

which converges for μ sufficiently small and where the integral $\int^t$ denotes the (unique) primitive (i.e. indefinite integral) whose mean value is zero. Note that all $x^{(k)}$ are free from secular terms (i.e. of terms in t^m, m = 1, 2, ...) since the integrand in (C3) also has zero mean value.

As a consequence of this theorem if there exists a vector $a(\mu)$ depending continuously on μ and such that

$$\frac{1}{T}\int_0^T q(t,x^*(t,a(\mu),\mu),\mu)dt = 0 .$$

then $x^*(t,a(\mu),\mu)$ is a T-periodic solution of Eq. (C1) as is readily seen from Eq. (C2).

In the limit $\mu \to 0$ we have: if there exists an n-vector $a(0) = a_o$ such that

$$\text{(a) } q_o(a_o,0) = 0 \text{ and (b) } \det\left(\frac{\partial q_o(a_o,0)}{\partial x}\right) \neq 0 \qquad (C4)$$

where

$$q_o(x,\mu) = \frac{1}{T}\int_0^T q(t,x,\mu)\, dt \qquad (C5)$$

then there exists for μ sufficiently small a T-periodic solution $x(t,\mu)$ such that $x(t,0) = a_o$.

Finally, the method is exhaustive, i.e. it produces all T-periodic solutions which reduce to a constant vector as $\mu \to 0$.

The equations $q_o(a_o,0) = 0$ and more generally

$$q_o(x^{(k)}(t,a(\mu),\mu),\mu) = 0$$

are called the determining equations.

The following comments are in order at this point:
a) the periodic solutions of Eq. (C1) necessarily have period T or an integer multiple of T in which case they are called subharmonics. The PCHG method can be implemented so as to determine all of these solutions (123; cf also ref. 124, p. 399).
b) in practice it may happen that the equation $q_o(a_o,0) = 0$ is identically satisfied for all a_o; one must then look for the smallest value of k such that the determining equation $q_o(x^{(k)},\mu) = 0$ is not an identity and thus determines the vector $a(\mu)$ (for details see ref. 123, p. 46).

2. SLIGHTLY PERTURBED INTEGRABLE HAMILTONIAN SYSTEMS

We assume that the Hamiltonian is of the form

$$H(J, w) \equiv H_o(J) + \varepsilon H_1(J, w, \varepsilon) \qquad \text{(C6)}$$

where J-w are action-angle variables and H_1 is 2π-periodic in w. The differential system is

$$\begin{aligned} \dot{\omega} &= \omega(J) + \varepsilon \frac{\partial H_1}{\partial J} \qquad \left(\omega(J) = \frac{\partial H_o}{\partial J}\right) \\ \dot{J} &= - \varepsilon \frac{\partial H_1}{\partial w} . \end{aligned} \qquad \text{(C7)}$$

For $\varepsilon = 0$ the solution is

$$\begin{aligned} w(t) &= w(J^o)t + w^o \pmod{2\pi} \\ J(t) &= J^o \end{aligned} \qquad \text{(C8)}$$

where w^o, J^o are constants of integration.

Following Poincaré (13, n° 42, Vol. 1) we choose a vector J^o such that $\omega_i(J^o) = m_i\nu$ where ν is the greatest common frequency and the m's are integers. We are going to look for periodic solutions with period $T = 2\pi/\nu$ which reduce for $\varepsilon = 0$ to functions w(t) and J(t) of the form (C8). The values of w^o corresponding to these solutions will be determined by the determining equations of an auxiliary system.

If we define new variables and parameter I, ϕ, μ by

$$w = \omega(J^o)t + \phi$$

$$J = J^o + \mu\, I$$

$$\mu = \sqrt{\varepsilon}$$

the variables ϕ and I obey the differential equations

$$\dot{\phi} = \mu\, \Omega(J^o)\, I + \mu^2 \frac{\partial}{\partial J} H_1(J^o + \mu I,\, \omega(J^o)t + \phi,\, \mu^2) + \mu^2 F(I,\mu)$$

$$\dot{I} = -\, \mu \frac{\partial}{\partial w} H_1(J^o + \mu I,\, \omega(J^o)\, t + \phi,\, \mu^2)$$

where

$$\Omega(J^o) = \frac{\partial\omega(J^o)}{\partial J} = \left(\frac{\partial^2 H_o}{\partial J_i \partial J_j}\right)$$

is the Hessian matrix of H_o evaluated at $J = J^o$ and the function F is defined by

$$\omega(J^o + \mu I) - \omega(J^o) = \mu\, \Omega(J^o)\, I + \mu^2 F(I,\mu)\ .$$

These equations are in the required form and the determining equations for the zero-th approximation ϕ^o, I^o (the vector a_o of parag. 1 above) are

$$\frac{1}{T}\int_0^T \Omega(J^o)\, I^o\, dt = \Omega(J^o)I^o = 0 \qquad \text{(C9-a)}$$

$$\frac{1}{T}\int_0^T \frac{\partial}{\partial w} H_1(J^o,\, \omega(J^o)\, t + \phi^o,\, 0)\, dt = 0\ . \qquad \text{(C9-b)}$$

The vector I^o is uniquely determined by the linear homogenous equation (C9-a) provided that $\det \Omega(J^o) \neq 0$ in which case $I^o = 0$. The existence of periodic solutions then depends upon whether Eq. (C9-b) have real solutions in ϕ^o. To each root ϕ^o of these equations corresponds a T-periodic solution of system (C7) which for $\mu = 0$ reduces to the functions given by (C8) where $w^o = \phi^o$. Note that from the form of Eq. (C8) if ϕ^o is a root so is $\phi^o + \omega(J^o)\tau$ where τ is an arbitrary constant.

The case where det Ω $(J^o) \doteq 0$ must be handled differently:

If all the components of J appear in $H_o(J)$ it is often possible to find an equivalent Hamiltonian for which this determinant does not vanish (13). This possibility is based on the fact that if U(x) is any scalar function the Hamiltonian system corresponding to the new Hamiltonian

$$H'(J, w, E, \varepsilon) \equiv U(H(J, w, \varepsilon))/U'(E)$$

is easily seen to be identical to system (C7) provided the value of the energy E be appropriately chosen. The difficulty will be removed since the Hessian of H'(J, w, E, 0) in general will be different from that of H_o. Hamiltonian (12) provides an example of such a situation for which the choice $H'(J, w, E, \varepsilon) \equiv H^2(J, w, \varepsilon)/(2E)$ resolves the difficulty as long as $\beta \neq 0$. When $\beta = 0$ the Hamiltonian $H_o = J_1 + \nu J_2$ represents two uncoupled harmonic oscillators and it is not necessary to change parameter. Indeed for an n-degree of freedom Hamiltonian

$$H \equiv \sum_{i=1}^{n} \omega_i J_i + \varepsilon H_1(J, w, \varepsilon)$$

the variables ϕ and I defined by the equations

$$w = \omega t + \phi, \; J = I$$

obey the differential system

$$\dot{\phi} = \varepsilon \frac{\partial}{\partial J} H_1 \; (I, \omega t + \phi, \varepsilon)$$

$$\dot{I} = - \varepsilon \frac{\partial}{\partial w} H_1(I, \omega t + \phi, \varepsilon)$$

which is in the required form (C1).

APPENDIX D

CANTOR SETS

As an example of a Cantor set we describe Cantor's ternary set:

To construct it mark the points 1/3 and 2/3 on the closed interval I = [0, 1] and then delete the points of the open interval (1/3, 2/3) called the "middle third". Call T_1 the remainder of the points in I, i.e.

$$T_1 = [0, \tfrac{1}{3}] \cup [\tfrac{2}{3}, 1] \; .$$

Now trisect the segment [0, 1/3] at 1/9 and 2/9, and the segment [2/3, 1] at 7/9 and 8/9. Then delete the middle third of each segment, i.e. the open sets (1/9, 2/9) and (7/9, 8/9). Call T_2 the remainder of the points in T_1, i.e.

$$T_2 = [0, \frac{1}{9}] \cup [\frac{2}{9}, \frac{1}{3}] \cup [\frac{2}{3}, \frac{7}{9}] \cup [\frac{8}{9}, 1] .$$

We can continue this process indefinitely obtaining a sequence of sets such that $T_1 \subset T_2 \subset T_3 \subset \ldots$ as shown:

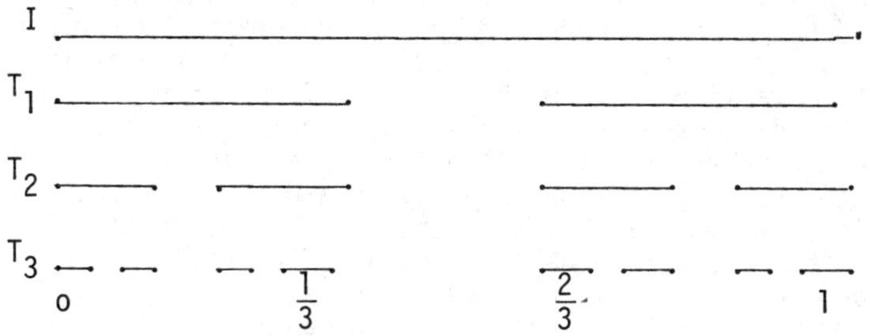

Cantor's ternary set T is then defined as the intersection of all T_n's i.e. $T = \bigcap_{n=0}^{\infty} T_n$ with $T_o = I$. It has the following properties:

Clearly T_n consists of 2^n disjoint closed intervals that we can number consecutively from left to right. We say that one of these intervals is odd or even according to the parity of its number.

Now let $x \in T$ be a point in T and define the infinite sequence of symbols $a = (a_1, a_2, \ldots)$ by the prescription

$$a_n = \begin{cases} 0 \text{ if } x \text{ belongs to an odd interval in } T_n \\ 2 \text{ if } x \text{ belongs to an even interval in } T_n \end{cases}$$

The sequence a corresponds to the representation of the number x in the system of base 3:

$$x = a_1 (1/3) + a_2 (1/9) + \ldots + a_n (1/3^n) + \ldots$$

Intuitively we see therefore that all points in T are in one-to-one correspondence with the infinite sequences of two symbols (here 0 and 2).

Using the above representation it can be shown that T has the power of the continuum (i.e. is non-denumerable) and yet has measure zero. Note that there are Cantor sets with positive measure (125).

SOURCES

S-1 The theory of Hamiltonian systems can be found in many books on classical mechanics notably Whittaker (1), Goldstein (2), Corben and Stehle (3), Ter Haar (4), and Meirovitch (5). A minimum background in traditional calculus, linear algebra, elementary analysis, and a little topology is sufficient for a thorough understanding of the material presented in these books. None of these authors makes any reference to the possibility of existence of random motions. The only book written in the traditional vein that we have found where these motions are discussed at an elementary level is Bartlett's (6).

The modern theory of Hamiltonian systems is constructed within the framework of modern analysis and can be found in Abraham's text book (7). Its understanding requires familiarity with "basic undergraduate calculus and algebra, and a limited amount of classical analysis, point set topology, and elementary mechanics". A background in modern advanced calculus on manifolds, exterior algebra, and general topology is given which is barely sufficient for the understanding of the new formulation of the theory. Although the book contains a translation of Kolmogorov's lecture in which he announced his theorem on slightly perturbed Hamiltonian systems, random motions are not covered. Another modern presentation of the theory at an advanced level is given in Sternberg's book on celestial mechanics which deals with random motions (8).

Finally, a very extensive coverage of Hamiltonian systems theory both in the traditional and in the new formulation including random motions can be found in Hagihara's monumental treatise on celestial mechanics (9).

Note added in proof: We have just received a copy of a French translation of a text book by Arnold published in Russian in 1974 and entitled "Mathematical Methods of Classical Mechanics" where the author gives a modern, comprehensive, and self-contained presentation of the subject. For the interested reader the reference is "Méthodes mathématiques de la mécanique classique" par V. Arnold, traduit du russe par Djilali Embarek, Traduction fransaise Editions Mir, Moscou 1976.

The following new books should also be useful:

- Differential topology with a view to applications, by D. R. J. Chillingworth, Pitman Publishing, London 1977.
- Analysis, Manifolds, and Physics, by Y. Choquet-Bruhat, C. Dewit-Morett, and M. Dillard-Bleick, Elsevier North-Holland Publishing Co., New York, 1977.

Nonlinearity and Functional Analysis, by Melvyn S. Berger, Academic Press, New York, 1977.

S-2 Arnold's theorem can be found in (7) where an outline of the proof is given. For a full proof see his original paper (10), Arnold and Avez (11), or (12).

S-3 For a modern exposition of action-angle variables see (5) or (12).

S-4 Besides the proof of Kolmogorov's theorem Arnold's paper (5) contains a clear discussion of the stability problems of classical mechanics. See also (11) and (12).

S-5 Actually Moser (17) gave an analogous result corresponding to systems of two degrees of freedom in the differentiable case. See also (18).

S-6 A comprehensive exposé of the available perturbation methods can be found in Giacaglia (19) and Nayfeh (20). See also Born (21) and Meirovitch (5).

S-7 Many papers have appeared which deal with the problem of ion motion in a uniform magnetic field under the influence of an electrostatic wave; lists of references can be found in (22-25).

S-8 In connection with this question Hénon makes the following remarks in (28): "Recall that the transformation associated with a conservative dynamical system possesses an integral invariant, i.e. it conserves the quantity

$$\iint_D I(x,y)dxdy$$

evaluated over any domain D. x and y are the coordinates of the surface of section and I is a function which depends on the surface of section chosen. It is only for certain particular choices that I is identically equal to 1 and consequently that the transformation conserves areas in the strict sense".

S-9 There is a considerable literature on the properties of area-preserving mappings; the basic references are (9, 11, 13, 17, 18, 27, 29). For specific area-preserving mappings, see (6, 30-43).

S-10 The random behavior of magnetic field lines was apparently first observed in numerical calculations by Gelfand et al. (53) in 1961. Discussions of the problem can be found in Morozov and Soloveev (54) and Grad (55). The destruction of magnetic surfaces caused by various types of perturbations has been studied notably by Rosenbluth et al. (56), Filonenko et al. (57), Finn (58), and Finn and Kaw (59), Stix (60, 61), and Rechester and Stix (62) where references to other works of physicists can be found. Mathematical studies of the problem have been carried out by Melnikov (63) who pointed out that the widely used method of averaging of Bogoliubov and Mitropolsky (64) may not be applicable to this problem (see his reference #7 in 63), and by Moser (65) who restricts himself to the region immediately surrounding the central, circular field line of a toroidal configuration in the case where curl B = 0.

S-11 The Ruelle-Takens picture of turbulence was presented in (100). See also (93) and (99).

S-12 For an introduction to ordinary differential equations with emphasis on their dynamical aspects see the book by Hirsch and Smale (101). See also Lanford's lecture notes (102).

S-13 Mathematical investigations of the Lorenz system have been done by Williams (111), Guckenheimer in (93), Ruelle (99), Kaplan and Yorke (105). We have not had access to the contents of a recent paper by Afraimovitch, Bikov, and Shilnikov (112).

REFERENCES

1. Whittaker, E. T., "Analytical Dynamics of Particles and Rigid Bodies", 4th edition, Cambridge Univ. Press, 1959.
2. Goldstein, H., "Classical Mechanics", Addison-Wesley, Reading 1950.
3. Corben, C. and P. Stehle, "Classical Mechanics", Wiley, New York 1950.
4. Ter Haar, D., "Elements of Hamiltonian Mechanics", Pergamon Press, New York 1971.
5. Meirovitch, L., "Methods of Analytical Dynamics", McGraw-Hill, New York 1970.
6. Bartlett, J. H., "Classical and Modern Mechanics", Univ. of Alabama Press, University of Alabama 1975.
7. Abraham, R. (with assistance of J. E. Marsden), "Foundations of Mechanics", Benjamin, New York 1967.
8. Sternberg, S., "Celestial Mechanics", Benjamin, New York 1969.
9. Hagihara, Y., "Celestial Mechanics", MIT Press, Cambridge 1970.
10. Arnold, V. I., A Theorem of Liouville Concerning Integrable Problems of Dynamics, Sibirsk. Mat. Z., 4 (1963) 471-474; Amer. Math. Soc. Translation, Series 2, 61 (1967) 292-296.
11. Arnold, V. I. and A. Avez, "Ergodic Problems of Classical Mechanics", Benjamin, New York 1968.
12. Arnold, V. I., Small Denominators and Problems of Stability of Motion in Classical and Celestial Mechanics, Russian Math. Surveys, 18, 6 (1963) 85-191.
13. Poincaré, H., "Les méthodes nouvelles de la mécanique céleste", 3 vol., Gauthier-Villars, Paris 1892-99; reprinted by Dover, New York 1957. English translation: "New Methods of Celestial Mechanics", NASA TF-450, April 1967.
14. Birkhoff, G. D., Collected mathematical papers, Amer. Math. Soc., Providence 1950; reprinted by Dover, New York 1968.
15. Kolmogorov, A. N., Preservation of Conditionally Periodic Movements with Small Change in the Hamilton Function, Dokl. Akad. Nauk SSSR, 98 (1954) 527-530; English translation: Los Alamos Scientific Laboratory, Los Alamos, New Mexico, LA-TR-71-67.

16. Arnold, V. I., Proof of a Theorem of A. N. Kolmogorov on the Preservation of Conditionally Periodic Motions under a Small Perturbation of the Hamiltonian, Russian Math. Surveys, 18, 5 (1963) 9-36.
17. Moser, J., On Invariant Curves of Area Preserving Mappings of an Annulus, Nachr. Akad. Wiss. Gottingen, Math. Phys. Klasse, Nr 1 (1962) 1-20.
18. Moser, J., "Stable and Random Motions in Dynamical Systems", Princeton Univ. Press, Princeton 1973.
19. Giacaglia, G. E. O., "Perturbation Methods in Non-Linear Systems", Springer-Verlag, New York 1972.
20. Nayfeh, A., "Perturbation Methods", Wiley, New York 1973.
21. Born, M., "The Mechanics of the Atom", translation of the original 1925 edition by J. W. Fisher and D. R. Hartree, Ungar Publ. Co., New York 1960.
22. Karney, C. F. F. and A. Bers, Stochastic Ion Heating by a Perpendicularly Propagating Electrostatic Wave, PRR 76/35-2, Res. Lab. of Electronics, MIT, March 1977a.
23. Karney, C. F. F., Stochastic Heating of Ions in a Tokamak by RF Power, Ph.D. Thesis, MIT, Dept. of EE & CS, May 1977.
24. Smith, G. R. and A. N. Kaufman, Stochastic Acceleration by a Single Wave in a Magnetic Field, Phys. Rev. Lett., 35 (1975) 1613-1616.
25. Smith, G. R. Stochastic Acceleration by a Single Wave in a Magnetized Plasma, Ph.D. Thesis, Univ. of California, Berkeley 1977.
26. Niven, I., Irrational Numbers, Math. Assoc. of America 1965.
27. Birkhoff, G. D., Surface Transformations and Their Dynamical Applications, Acta Math., 43 (1920) 1-119; or in Vol. 2 of ref. 14.
28. Hénon, M., Problèmes Numériques Liés à la Recherche des Solutions des Transformations Ponctuelles Conservatives, Colloques Internationaux du CNRS N° 229, "Transformations Ponctuelles et leurs Applications", Toulouse 10-14 Sept. 1973, CNRS Paris 1976, pp. 387-398.
29. Greene, J. M., Two-Dimensional Measure-Preserving Mappings, J. Math. Phys., 9 (1968) 760-768.
30. Hénon, M., and C. Heiles, The Applicability of the Third Integral of Motion: Some Numerical Experiments, Astron. J., 69 (1964) 73-79.
31. Hénon, M., Numerical Study of Quadratic Area-Preserving Mappings, Quart. Appl. Math., 27 (1969) 291-312.
32. Hitzl, D. L., The Swinging Spring - Invariant Curves Formed by Quasi-Periodic Solutions II, Astron. & Astrophys., 41 (1975) 197-198.
33. Helleman, R. G., On the Iterative Solutions of a Stochastic Mapping, in "Statistical Mechanics and Statistical Methods, Theory and Applications", U. Landman ed., Plenum, New York 1977.
34. Danby, J. M. A., Wild Dynamical Systems and the Role of Two or More Small Divisors, in "Periodic Orbits, Stability and Resonances", G. E. O. Giacaglia ed., D. Reidel Publ. Co., Dordrecht 1970.

35. " " " ", The Evolution of Periodic Orbits Close to Homoclinic Points, Celest. Mech., 8 (1973) 273-280.
36. Dragt, A. J. and J. M. Finn, Insolubility of Trapped Particle Motion in a Magnetic Dipole Field, J. Geophys. Res., 81 (1976) 2327-2340.
37. Brahic, A., Numerical Study of a Simple Dynamical System, I The Associated Plane Area-Preserving Mapping, Astron. & Astrophys., 12, (1971) 98-110.
38. Rannou, F., Numerical Study of Discrete Plane Area-Preserving Mappings, ibid., 31 (1974) 289-301.
39. Braun, M., On the Applicability of the Third Integral of Motion, J. Diff. Eqs., 13 (1973) 300-318.
40. Braun, M., Numerical Studies of an Area-Preserving Mapping, in "Dynamical Systems, An International Symposium", L. Cesari, J. K. Hale, and J. LaSalle eds, Vol. 2, Academic Press, New York 1976,
41. Aubry, S., On the Dynamics of Structural Phase-Transitions - Lattice Locking and Ergodic Theory, Preprint (Brookhaven National Lab., Associated Universities Inc., Upton, Long Island, NY 11973).
42. Froeschlé, C., A Numerical Study of the Stochasticity of Dynamical Systems with Two Degrees of Freedom, Astron. & Astrophys., 9 (1970) 15-23.
43. Barbanis, B., Trapping of Particles in a Time-Dependent Hamiltonian, Celest. Mech., 14 (1976) 201-208.
44. Moser, J., The Analytic Invariants of an Area-Preserving Mapping Near a Hyperbolic Fixed Point, Comm. Pure Appl. Math., 9 (1956) 673-692.
45. Hadamard, J., Les surfaces à courbures opposées at leurs lignes géodésiques, J. de Math., 4 (1898) 27-73.
46. Morse, M. and G. A. Hedlund, Symbolic Dynamics, Amer. J. Math., 60 (1938) 815-866.
47. Smale, S., Diffeomorphisms with Many Periodic Points, Differential and Combinatorial Topology, (A Symposium in Honor of Marston Morse), pp. 63-80, Princeton University Press, Princeton 1965.
48. Alekseev, V. M., Quasi-Random Oscillations of a One-Dimensional Oscillator, Soviet Math. Dokl., 8, 6 (1967) 1421-1424.
49. Alekseev, V. M., Quasi-Random Oscillations and Qualitative Aspects of Celestial Mechanics, 9th Summer Mathematics School, 2nd ed., revised, Yu. A. Mitropolski and A. N. Sharovsky eds, Acad. of Sci. of the Ukrainian S.S.R., Naukova Dumka Publ. House, Kiev 1976 (Russian); English translation: ERDA-tr-302, available from Technical Information Services, Oak Ridge, Tennessee.
50. Smale, S., Differentiable Dynamical Systems, Bull. Amer. Math. Soc., 73 (1967) 747-817.
51. Zehnder, E., Homoclinic Points Near Elliptic Fixed Points, Comm. Pure Appl. Math., 26 (1973) 131-182.
52. Smith, G. R., Overlap of Bounce Resonances and the motion of Ions in a Trapped-Ion Mode, Phys. Rev. Lett., 38 (1977) 970-973.

53. Gelfand, I. M., N. M. Zueva, A. I. Morozov and L. S. Soloveev, Magnetic Surfaces of the Three-Path Helical Magnetic Field Excited by a Crimped Field, Soviet Physics-Tech. Phys., 6 (1962) 852-855.
54. Morozov, A. I. and L. S. Soloveev, The Structure of Magnetic Fields, Rev. Plasma Phys., 2 (1966) 1-101.
55. Grad, H., Toroidal Containment of a Plasma, Phys. of Fluids, 10 (1967) 137-154.
56. Rosenbluth, M. N., R. Z. Sagdeev, J. B. Taylor and G. M. Zaslavsky, Destruction of Magnetic Surfaces by Magnetic Field Irregularities, Nuclear Fusion, 6 (1966) 297-300.
57. Filonenko, N. N., R. Z. Sagdeev and G. M. Zaslavsky, Destruction of Magnetic Surfaces by Magnetic Field Irregularities: Part II, Nuclear Fusion, 7 (1967) 253-265.
58. Finn, J. M., The Destruction of Magnetic Surfaces in Tokamaks by Current Perturbations, ibid, 15 (1975) 845-854.
59. Finn, J. M. and P. K. Kaw, Coalescence Instability of Magnetic Islands, Phys. of Fluids, 20 (1977) 72-78.
60. Stix, T. H., Current Penetration and Plasma Disruption, Phys. Rev. Lett., 36 (1976) 521-524.
61. Stix, T. H., Aspects of Stochastic Heating, Proc. Conf. on Plasma Heating, Varenna 1976.
62. Rechester, A. B. and T. H. Stix, Magnetic Braiding Due to Weak Asymmetry, Phys. Rev. Lett., 36 (1976) 587-590.
63. Melnikov, V. K., Lines of Force of a Magnetic Field, Soviet Phys. Dokl., 7 (1962) 502-504.
64. Bogoliubov, N. N. and Y. A. Mitropolski, Asymptotic Methods in the Theory of Non-Linear Oscillations, Hindustan Publ. Corp., Delhi 1961.
65. Moser, J., Lectures on Hamiltonian Systems, Memoirs Amer. Math. Soc., 81 (1968) 1-60.
66. Kerst, D. W., The Influence of Errors on Plasma-Confining Magnetic Fields, Plasma Phys. (J. Nucl. Energy Part C), 4 (1962) 253-263.
67. Whiteman, K. J., Invariants and Stability in Classical Mechanics, Rep. Prog. Phys., 40 (1977) 1033-1069.
68. Smale, S., private communication.
69. Bowen, R., On Axiom-A Diffeomorphisms, Lectures Given at the NSF Regional Conference held at North Dakota State Univ. (Fargo), June 1977.
70. Ford, J. and G. H. Lunsford, Stochastic Behavior of Resonant Nearly Linear Oscillator Systems in the Limit of Zero Non-Linear Coupling, Phys. Rev. A, 1 (1970) 59-70.
71. Mo, K. C., Theoretical Prediction for the Onset of Widespread Instability in Conservative Nonlinear Oscillator Systems, Physica, 57 (1972) 445-454.
72. Brumer, P. and J. W. Duff, A Variational Equation Approach to the Onset of Statistical Intramolecular Energy Transfer, J. of Chem. Phys., 65 (1976) 3566-3574.
73. Ford, J., private communication.

74. Benettin, G., L. Galgani, and J. M. Strelcyn, Kolmogorov Entropy and Numerical Experiments, Phys. Rev. A, 14, 6 (1976) 2338-2345.
75. Zaslavskii, G. M., and B. V. Chirikov, Stochastic Instability of Non-linear Oscillations, Soviet Phys. Uspekhi, 14, 5 (1972) 545-672.
76. Chirikov, B. V., A Universal Instability of Many-Dimensional Oscillator Systems, Proc. of the 1977 Como Conference on Stochastic Behavior in Classical and Quantum Hamiltonian Systems, G. Casati and J. Ford eds, to appear in Phys. Reports.
77. Arnold, V. I., Instability of Dynamical Systems with Several Degrees of Freedom, Soviet Math. Dokl. , 5 (1964) 581-585.
78. Arnold, V. I., The Stability Problem and Ergodic Properties for Classical Systems, Proc. Internat. Cong. Math. (Moscow 1966) 387-392; Izdat. "Mir", Moscow 1968 (Russian); Amer. Math. Soc. Transl. (2) 70 (1968) 5-11.
79. Murdock, J., Resonance Capture in Certain Nearly Hamiltonian Systems, J. Diff. Eqs., 17 (1975) 361-374.
80. Murdock, J., Nearly Hamiltonian Systems in Two Degrees of Freedom, Int. J. Non-Linear Mech., 10 (1975) 259-270.
81. Murdock, J., Nearly Hamiltonian Systems in Non Linear Mechanics: Averaging and Energy Methods, Indiana Univ. Math. J., 25 (1976) 499-523.
82. Murdock, J., Global Results by Local Averaging for Nearly Hamiltonian Systems, in "Dynamical Systems: An International Symposium", L. Cesari, J. K. Hale, and J. LaSalle eds, Vol 2, pp. 24-27, Academic Press, New York 1976.
83. Hale, J. K. and P. Z. Taboas, Interaction of Damping and Forcing in a Second Order Equation, Preprint.
84. Arnold, V. I., Conditions for the Applicability, and Estimate of the Error, of an Averaging Method for Systems Which Pass Through States of Resonance in the Course of Their Evolution, Soviet Math. Dokl., 6 (1965) 331-334.
85. Arnold, V. I., On the Classical Theory of Perturbations and the Problem of Stability of Planetary Systems, ibid., 3 (1962) 1008-1011.
86. Hale, J. K., "Oscillations in Non-Linear Systems", McGraw-Hill, New York 1963.
87. Hale, J. K., "Ordinary Differential Equations", Wiley, New York 1969.
88. Robbins, K. A., A New Approach to Subcritical Instability and Turbulent Transitions in a Simple Dynamo, Math. Proc. Cambridge Soc., 82 (1977) 309-325.
89. Haken, H., Analogy Between Higher Instabilities in Fluids and Lasers, Phys. Lett., 53A (1975) 77-78.
90. Lorenz, E. N., Deterministic Non Periodic Flow, J. of the Atmos. Sci., 20 (1963) 130-141.
91. Dupree, T. H., Theory of Two-Dimensional Turbulence, Phys. of Fluids, 17 (1974) 100-109.
92. Montgomery, D., Implications of Navier-Stokes Trubulence Theory for Plasma Turbulence, Proc. Indian Acad. Sci., Sect. A, Vol. 86, 1977, 87-110.

93. Marsden, J. E. and M. McCracken, "The Hopf Bifurcation and Its Applications", Springer-Verlag, New York 1976.
94. Landau, L. D., C. R. Acad. Sci. URSS, 44 (1944) 311.
95. Landau, L. D. and E. M. Lifschitz, "Fluid Mechanics", Pergamon Press, Oxford 1959.
96. Hopf, E., Abzweigung einer periodischen Lösung von einer stationären Lösung eines Differentialsystems, Berichten der Math.-Phys. Klasse der Sächs. Akad. Wiss. Leipzig. XCIV Band Sitzung vom 19. Jan. 1942. (An English translation can be found in 93).
97. Hopf, E., A Mathematical Example Displaying Features of Turbulence, Comm. Pure Appl. Math., 1 (1948) 303-322.
98. Marsden, J., Attempts to Relate the Navier-Stokes Equations to Turbulence, in Lect. Notes in Math #615, Springer-Verlag, New York 1977.
99. Ruelle, D., The Lorenz Attractor and the Problem of Turbulence, in "Turbulence and Navier-Stokes Equations", Orsay 1975, Lect. Notes in Math. #565, Springer-Verlag, New York 1976, p. 146.
100. Ruelle, D. and F. Takens, On the Nature of Turbulence, Comm. Math. Phys., 20 (1971) 167-192; Note concerning our paper "On the Nature of Turbulence", ibid., 23 (1971) 343-344.
101. Hirsch, M. W. and S. Smale, "Differential Equations, Dynamical Systems, and Linear Algebra", Academic Press, New York 1974.
102. Lanford III, O. E., Qualitative and Statistical Theory of Dissipative Systems, revised text of a series of lectures delivered at the 1976 CIME School of Statistical Mechanics (preprint).
103. Smale, S., Dynamical Systems and Turbulence, in "Turbulence Theory", P. Bernard and T. Ratiu eds, Lect. Notes in Math. #615, Springer-Verlag, New York 1977.
104. Chenciner, A. and G. Iooss, Bifurcation d'un tore T^2 en un tore T^3, C. R. Acad. Sci. Paris, série A, 284 (1977) 1207-1210.
105. Kaplan, J. L. and J. A. Yorke, Preturbulence: a Regime Observed in a Fluid Flow Model of Lorenz, preprint March 1977.
106. Gollub, J. P. and H. L. Swinney, Onset of Turbulence in a Rotating Fluid, Phys. Rev. Lett., 35 (1975) 927-930.
107. Swinney, H. L., P. R. Fenstermacher and J. P. Gollub, paper presented at the Symposium on Turbulent Shear Flows, April 18-20, 1977.
108. Sherman, J. and J. McLaughlin, Power Spectra of Nonlinearly Coupled Waves, preprint.
109. McLaughlin, J. B. and P. C. Martin, Transition to Turbulence in a Statically Stressed Fluid System, Phys. Rev. A, 12 (1975) 186-203.
110. McLaughlin, J. B., Successive Bifurcations Leading to Stochastic Behavior, J. Stat. Phys., 15 (1975) 307-326.
111. Williams, R. F., The Structure of the Lorenz Attractors, in "Turbulence Theory", P. Bernard and T. Ratiu eds., Lect. Notes in Math. #615, Springer-Verlag, New York 1977.
112. Afraimovitch, V. S., V. V. Bikov, and L. P. Shil'nikov, The Origin and Structure of the Lorenz Attractor, Dokl. Akad. Nauk USSR, 234, n°2 (1977) 336-339; English Trans.: Sov. Phys. Dokl.,

112. (cont.) 22 (1977) 253-255.
113. Lorenz, E. N., Maximum Simplification of the Dynamic Equations, Tellus, 12 (1960) 234-254.
114. Creveling, H. F., J. F. DePaz, J. V. Balladi, and R. J. Shoenhals, Stability Characteristics of a Single Phase Free Convection Loop, J. Fluid Mech., 67 (1975) 65-84.
115. Kantorovitch, L. V., Functional Analysis and Applied Mahtematics, Uspekhi Mat. Nauk, 3, n° 6 (1948) 89-185; English translation: NBS Report 1509, Washington, DC 1952.
116. Feagin, T. and R. G. Gottlieb, Generalization of Lagrange's Implicit Function Theorem to n Dimensions, Celest. Mech., 3 (1971) 227-231.
117. Howland, Jr., R. A., An Accelerated Elimination Technique for the Solution of Perburbed Hamiltonian Systems, Celest. Mech., 15 (1977) 327-352.
118. Bogoliubov, N. N., On Quasi-Periodic Solutions in Nonlinear Problems of Mechanics, First Math. Summer School, Kanev, 1963, Naukova Dumka, Kiev 1964, Vol. 1, p. 11 (in Russian).
119. Bogoliubov, N. N., Uy. A. Mitropolsky, and A. M. Samoilenko, English translation: Methods of Accelerated Convergence in Nonlinear Mechanics Hindustan Pub. Corp., Dehli 1976 (distributed by Springer-Verlag). Naukova Dumka, Kiev 1969 (in Russian).
120. Cary, J. R., Hamiltonian Perturbation Theory Using Lie Transforms, LBL-635.
121. The Mathlab Group, MACSYMA Reference Manual, Massachusetts Institute of Technology, 1977.
122. McNamara, B., Super-convergent Adiabatic Invariants with Resonant Denominators by Lie Transforms, preprint UCRL-79843 Rev. 1, December 2, 1977, to be submitted to J. Math. Phys..
123. Hale, J. K., private communication.
124. Hahn, W., Stability of Motion, Springer-Verlag, New York 1967.
125. Bowen, R., A Horseshoe with a Positive Measure, Inventiones Math., 29 (1975) 203-204.

SOME ILLUSTRATIONS OF STOCHASTICITY

L. Jackson Laslett
University of California, Lawrence Berkeley Laboratory
Berkeley, California*

A complex, and apparently stochastic, character frequently can be seen to occur in the solutions to simple Hamiltonian problems. Such behavior is of interest, and potentially of importance, to designers of particle accelerators -- as well as to workers in other fields of physics and related disciplines. Even a slow development of disorder in the motion of particles in a circular accelerator or storage ring could be troublesome, because a practical design requires the beam particles to remain confined in an orderly manner within a narrow beam tube for literally tens of billions of revolutions. The material I shall present is primarily the result of computer calculations I and others have made to investigate the occurrence of "stochasticity," and is organized in a manner similar to that adopted for presentation at a 1974 accelerator conference.[1]

As an introductory example, one can consider the longitudinal motion of a particle subjected to the radio-frequency electric fields employed to bunch, and sometimes accelerate, a beam within a synchrotron type of accelerator. If the electric field is regarded as equivalent to a simple travelling wave, having the speed of a reference particle in a "coasting beam," the motion is characterized by the pair of differential equations.

$$\frac{dy}{dn} = -K \sin \pi x \tag{1a}$$

$$\frac{dx}{dn} = \lambda' y \tag{1b}$$

wherein

y = fractional departure of energy from the reference value,
πx = electrical phase angle of field <u>vs</u>. particle,
$K \propto$ applied voltage, and
$\lambda' \propto$ derivative of revolution period with respect to energy.

K and λ' will be regarded as specified constants. The differential equations will be recognized as derivable from a Hamiltonian function

$$H = \frac{1}{2} \lambda' y^2 - \frac{K}{\pi} \cos \pi x, \tag{2}$$

* Work supported by the U. S. Dept. of Energy, Office of Energy Research.

ISSN: 0094-243X/78/221/$1.50

in which the independent variable (n) is the revolution number and does not appear explicitly in the Hamiltonian. Because n does not appear explicitly, the Hamiltonian of course is a constant of the motion. One accordingly obtains simple phase trajectories (in x,y space) -- of the familiar type characteristic of a physical (non-linear) pendulum (as was recognized by McMillan in connection with discovery of the principle of phase stability[4]).

In practice, however, the radio-frequency fields in fact are provided by localized cavities, so that the travelling-wave description constitutes an idealization and the motion is more appropriately represented by difference equations:

$$y_{n+1} = y_n - K \sin \pi x_n \tag{3a}$$

$$x_{n+1} = x_n + \lambda' y_{n+1}, \tag{3b}$$

with y_n measuring energy at the entrance to the nth cavity. These transformation equations are readily shown to be area preserving [$\partial(x_{n+1},y_{n+1})/\partial(x_n,y_n) = 1$] -- the motion in fact could be described through use of a Hamiltonian function, but one that would contain a periodic δ-function of the independent variable as a factor multiplying the term $- K/\pi \cos \pi x$. There thus is no evident simple constant of the motion, and the non-linearity of the equations precludes application of Floquet theory to this problem. (The use of a Hamiltonian formulation nonetheless can be helpful in analytic work, but difference equations of course are convenient for computational investigations).

It is of interest to take a quick look at some computational results obtained through use of a transformation equivalent to (3a,b) but written in terms of working variables $Y = y - (K/2) \sin \pi x$, $X = x$, so that the transformation assumes the form

$$\left.\begin{aligned} X_{n+1} &= X_n + \lambda'[Y_n - (K/2) \sin \pi X_n] \\ Y_{n+1} &= Y_n - (K/2)[\sin \pi X_n + \sin \pi X_{n+1}] \end{aligned}\right\} , \quad \begin{aligned} &(3a') \\ &(3b') \end{aligned}$$

with the result that the resulting phase diagrams will necessarily have a desirable symmetry about both the X- and Y-axes. With $K/\pi = 0.1$ and $\lambda' = 0.1$ we find what appear to be conventional bucket diagrams with buckets separated in Y by $2/\lambda'$ for successive harmonic modes, although we may wish to return to the question of whether the bucket boundaries are as simple and definite as appears on Fig. 1.

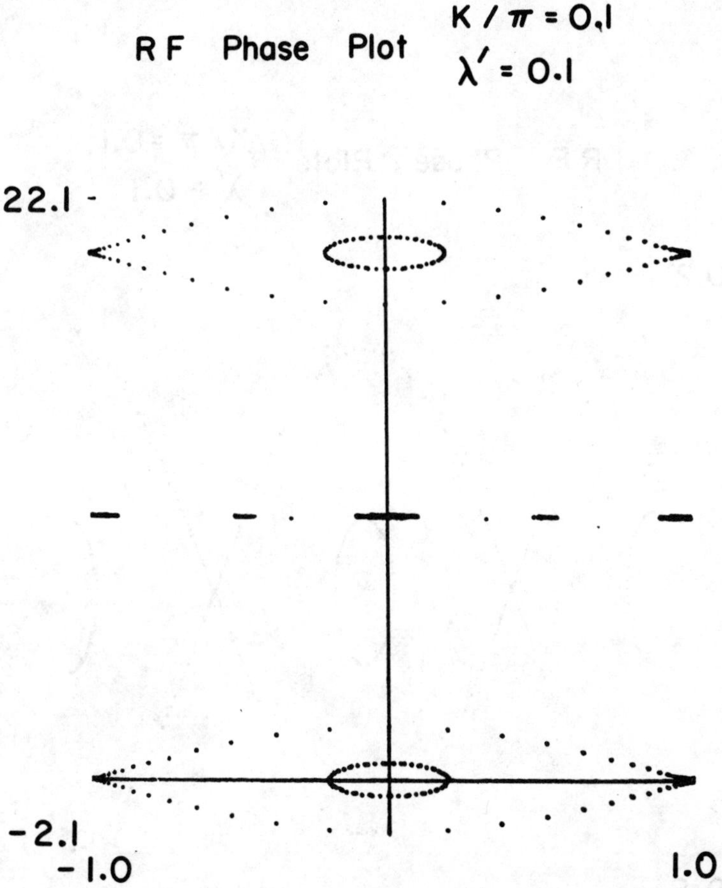

Fig. 1. X, Y phase plot for a coasting beam under the influence of an R.F. cavity with $K/\pi = 0.1$, $\lambda' = 0.1$ -- as computed by Eqs. (3a',b'). X is plotted mod. 2.

We also find evidence of some "sub-harmonic" structure (with higher order fixed points) that, if enlarged some 60X, has the appearance shown in Fig. 2.[5]

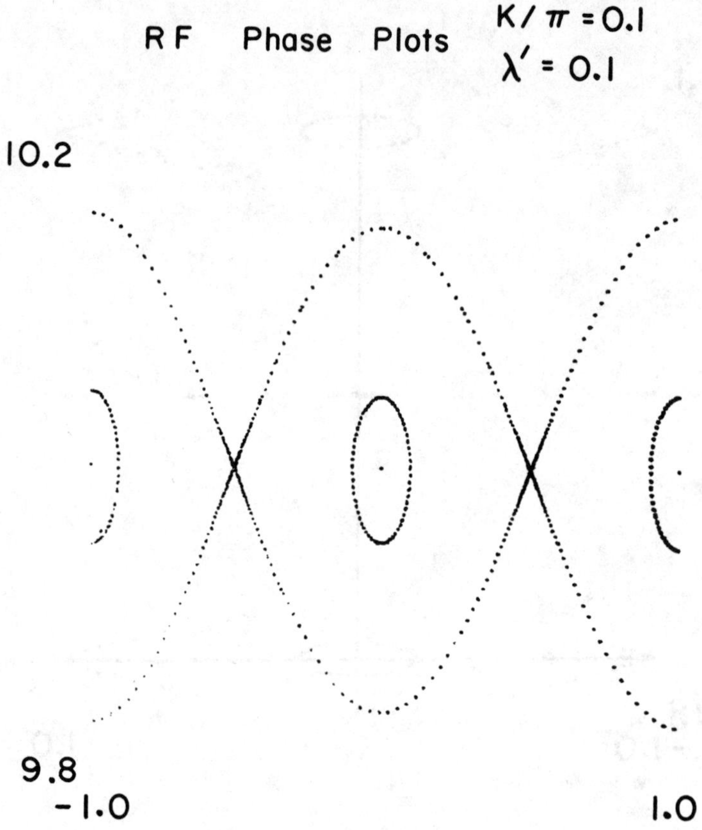

Fig. 2. Circa 60-fold vertical enlargement of central portion of Fig. 1, near Y = 10.0, showing sub-harmonic structure.

If the cavity voltage is increased eight-fold (so $K/\pi = 0.8$), the bucket areas are expected to become larger, and we indeed find this to be the case (Fig. 3), with an accompanying very marked increase of complexity that is immediately apparent in the phase

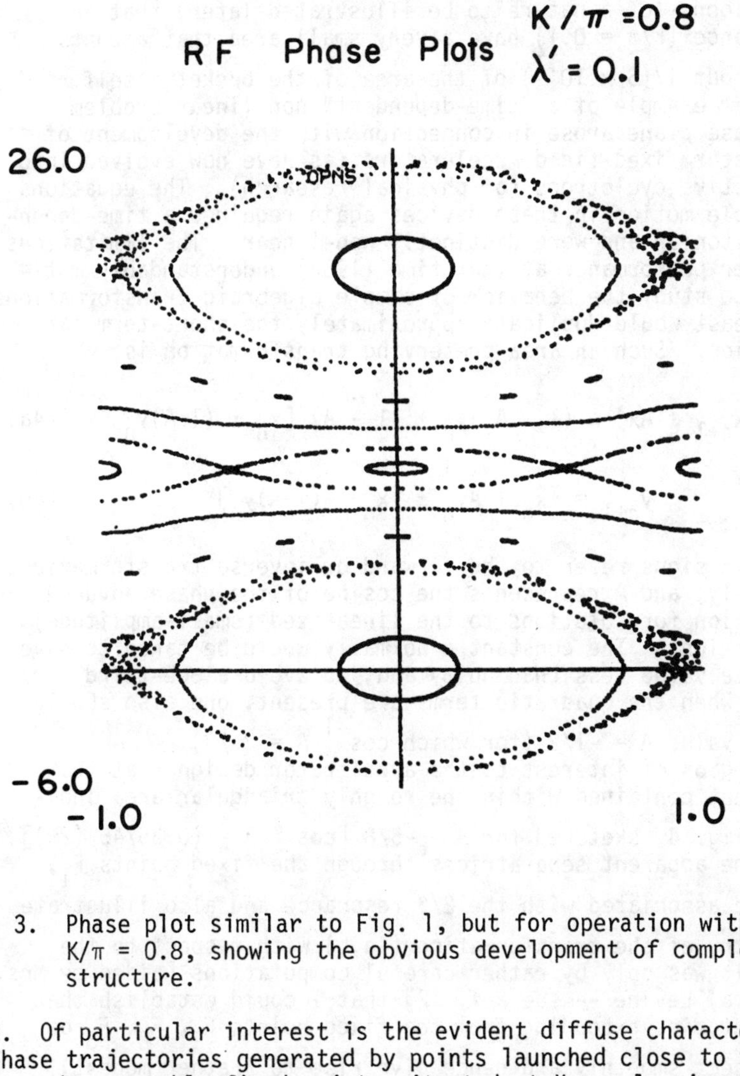

Fig. 3. Phase plot similar to Fig. 1, but for operation with K/π = 0.8, showing the obvious development of complex structure.

plot. Of particular interest is the evident diffuse character of phase trajectories generated by points launched close to the first-order unstable fixed points situated at X = ± 1, since the bucket boundary in consequence no longer appears clearly defined.

In the first example (K/π = 0.1), on the other hand, where the bucket width is some two and one-half times smaller in relation to the bucket separation, the presence of structure in the separatrix can be revealed computationally only with considerable care[6]. To do this, one can extend from the unstable fixed points the eigenvector directions of the transformation linearized about these fixed points, and examine whether such curves intersect smoothly. One finds in fact that they do not quite do so, but generate loops (of a nature to be illustrated later) that in this instance ($K/\pi = 0.1$) have a very small area that amounts to only about $1/(5 \times 10^{11})$ of the area of the bucket itself.

Another example of a "time-dependent" non-linear problem in the phase plane arose in connection with the development of spiral-sector fixed-field accelerators (as have now evolved into very effective cyclotrons for physical research). The equations for particle motion in these devices again required a time-dependent Hamiltonian and were distinctly non-linear. The limitations of computer performance at that time (1956) understandably motivated us to study the behavior of simple algebraic transformations that at least would duplicate approximately the short-term particle motion. Such an area-preserving transformation is

$$x_{n+1} = Ax_n \pm (1 - A^2)y_n + (1 - A)\,[x_n \pm (1-A)y_n]^2 \qquad (4a)$$

$$y_{n+1} = \mp x_n + Ay_n \pm [x_n \pm (1-A)y_n]^2 \;, \qquad (4b)$$

where the ± signs refer to the forward or inverse transformation, respectively, and A represents the cosine of the phase advance per iteration for solutions to the linearized (small-amplitude) transformations. The constant A normally would be taken to have an absolute value less than unity and, to avoid a one-third resonance when the quadratic terms are present, one also should avoid the value $A = -1/2$ (for which $\cos^{-1} A = 2\pi/3$).

The region of interest to the accelerator designer at that time is that contained within the roughly triangular area indicated on Fig. 4, sketched for $A = -5/8$ [$\cos^{-1} A \cong (0.35745)(2\pi)$], wherein the apparent separatrices through the fixed points F_1, F_2, F_3 are associated with the 2/3 resonance and also illustrate the symmetry of the transformation (4a,b) with respect to the x-axis. It was only by rather careful computations [aided by Mrs. H. (Barbara) Levine -- see Ref. 12] that I could establish that the trajectories extending from the fixed points F_1, F_2, F_3 do not intersect smoothly and hence give rise to (rather modest) regions of erratic behavior similar to those seen in phase diagrams for the earlier example. Outside the area F_1, F_2, F_3 indicated in Fig. 4, however, the transformation (4a,b) develops gross loops in phase trajectories extending from an order-1 fixed point at (1,0), and in this respect exhibits a behavior similar to that shown by a transformation of deVogelaere which will be mentioned later.

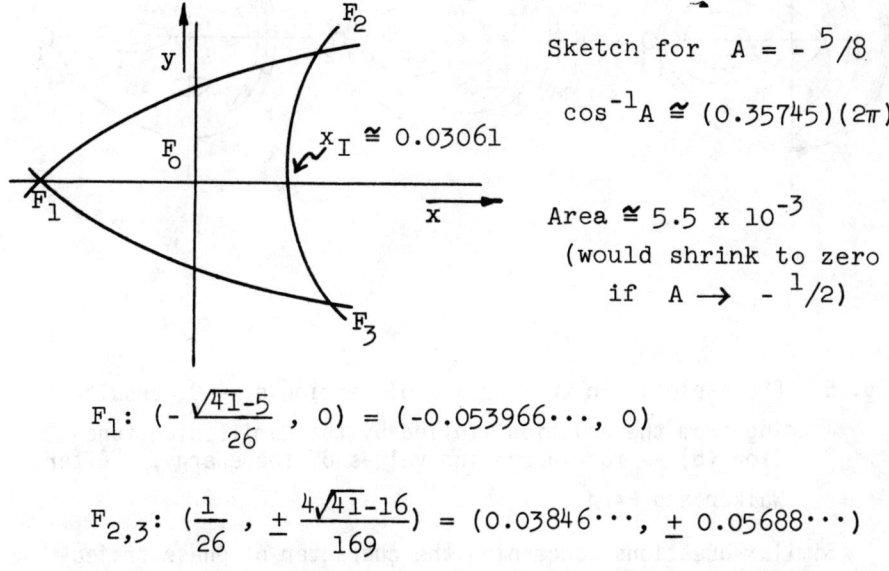

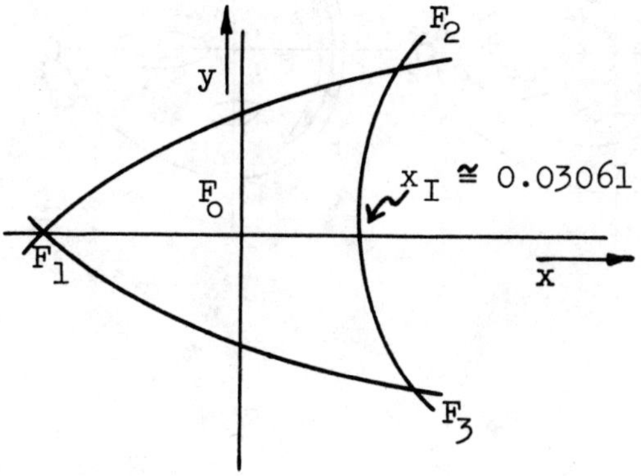

Sketch for $A = -5/8$

$\cos^{-1}A \cong (0.35745)(2\pi)$

Area $\cong 5.5 \times 10^{-3}$
(would shrink to zero if $A \to -1/2$)

$$F_1: \left(-\frac{\sqrt{41}-5}{26}, 0\right) = (-0.053966\cdots, 0)$$

$$F_{2,3}: \left(\frac{1}{26}, \pm\frac{4\sqrt{41}-16}{169}\right) = (0.03846\cdots, \pm 0.05688\cdots)$$

Fig. 4. Apparent separatrices through the third-order unstable fixed points of the transformation (4a,b), with A = -5/8.

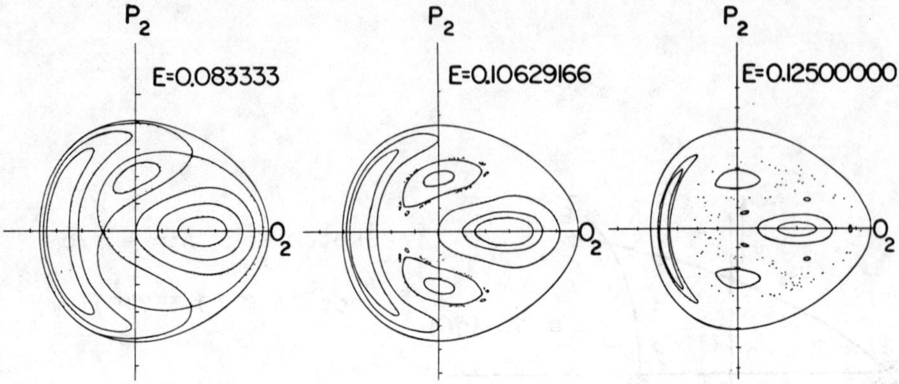

Fig. 5. Phase plots, in the surface of section $q_1 = 0$, resulting from the equation implied by the Hamiltonian function (5) -- for increasing values of the energy. [After Walker and Ford[8].]

Similar questions concerning the character of phase trajectories and the possible erratic or stochastic behavior of canonical mappings can arise in problems with more than one degree of freedom. As an example, Hénon and Hiles[7] and subsequently Walker and Ford[8] studied a model of an astronomical system, for which the Hamiltonian function was taken to be

$$H = \frac{1}{2}(p_1^2 + p_2^2 + q_1^2 + q_2^2) + q_1^2 q_2 - \frac{1}{3} q_2^3. \qquad (5)$$

The cubic terms appearing here as coupling terms become increasingly significant for increasingly large values of H -- which is itself a constant of the motion. With the coupling terms present, however, and in the absence of any simple constant of the motion other than H, a given phase trajectory might be expected to wander (ergodically) over virtually all of a three-dimensional surface specified by H = Constant (and that will be a closed surface for values of H below the dissociation energy). If, on the other hand, some additional integral of the motion were in fact also acting, the phase points of a given trajectory then would be constrained to lie on a two-dimensional surface, and graphs of the intersection of such surfaces with some selected plane or other surface (a "surface-of-section") would lead to simple curves in this plane rather than to a scattering of points. Computations of this nature indicated that for sufficiently small values of energy (e.g., $H \lesssim 1/12$) only curves that to computer accuracy were smooth (and relatively simple) were formed by intersection with the plane $q_1 = 0$ (and $p_1 \geqslant 0$). Examples in which the energy of the particles was successively raised, however, resulted in the development of ragged island structures or of apparent stochastic behavior over increasingly large portions of this surface-of-section (Fig. 5).

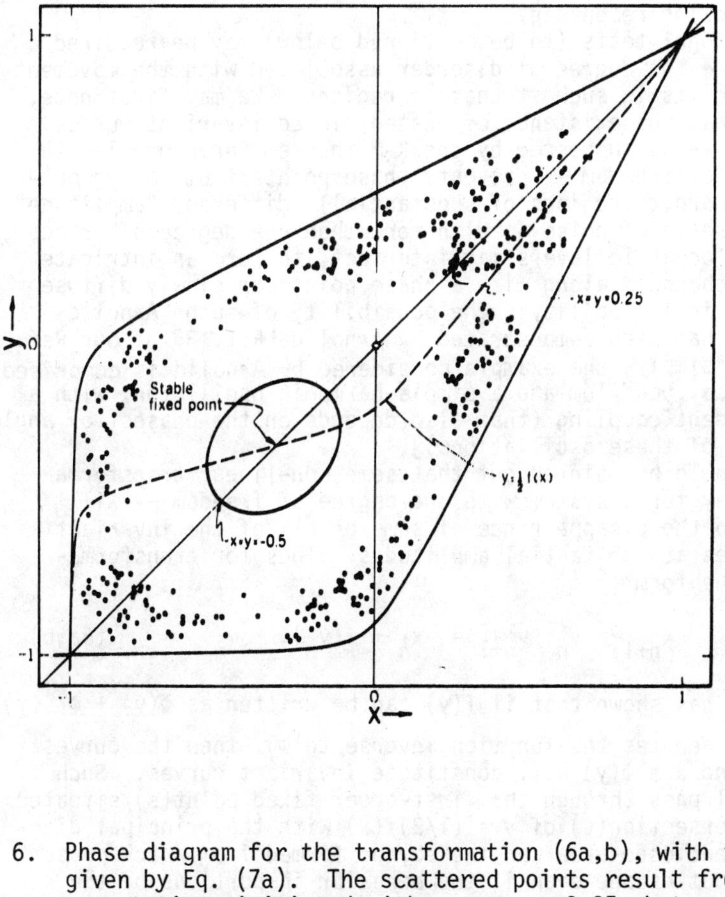

Fig. 6. Phase diagram for the transformation (6a,b), with f(y) given by Eq. (7a). The scattered points result from computations initiated with $x_0 = y_0 = 0.25$, but must remain within the separatrix defined by the function Φ [Eq. (7b)]. $k = 0.1$.

Such behavior appears concordant with the "KAM" (Kolmogorov-Arnol'd-Moser) theory (see Ref. 58, 59, & 60 of our Ref. 2c), which suggests that many of the invariant curves or surfaces present in the absence of the perturbation will persist, with only minor distortion, in the presence of a sufficiently small perturbation (see, however, Note 9). It is of interest, of course, to determine or to estimate the circumstances (e.g., perturbation strength) at which the KAM theory becomes inapplicable and extended regions of erratic (or stochastic) behavior develop. As we suggested by our first examples, and has been expounded more extensively by Zaslavskij and Chirikov,[2c,10] one means for obtaining such estimates may be by determining the ratio of resonance width [$\delta\omega = (d\omega/dI)_r \delta I$] to the distance ($\Delta\omega$) to the nearest neighboring resonance.

Additional tests (to be mentioned below) may be required to determine the degree of disorder associated with the movement of phase points in such stochastic regions. We may first note, however, that the existence of nested closed invariant curves in a plane -- as suggested by the KAM theorem for a problem in one degree of freedom -- prevents phase points from moving outward or inward to regions of substantially different "amplitude" (in the absence of noise). With more than one degree of freedom, however, stochastic layers may intersect, to form an intricate system of channels along with a phase point can slowly diffuse and result in instability. The possibility of such "Arnol'd diffusion" has been demonstrated by Arnol'd [Ref. 35 of our Ref. 2c; stated simply, the example considered by Arnol'd is comprised of a physical pendulum and a simple-harmonic oscillator, with a time-dependent coupling (that also depends on the phases, or angle variables, of these oscillations)].

It should be pointed out that some non-linear transformations -- say for a system with one degree of freedom -- will not lead to the disappearance of some or all of the invariant phase curves at substantial amplitudes. Thus for transformations of the form

$$x_{n+1} = y_n;\ y_{n+1} = -x_n + f(y_n), \qquad (6a,b)$$

McMillan[11] has shown that if $f(y)$ can be written as $\Phi(y) + \Phi^{-1}(y)$ (where Φ^{-1} denotes the function inverse to Φ), then the curves $y = \Phi(x)$ and $x = \Phi(y)$ will constitute invariant curves. Such curves will pass through the first-order fixed point(s) situated at the intersection(s) of $y = (1/2)f(x)$ with the principal diagonal. An enclosed area can thereby be formed from which phase points cannot escape even if the behavior in portions of the interior becomes highly stochastic. This is illustrated by an example (Fig. 6) in which

$$f(y) = \frac{1}{2}(3y-1) - \frac{1}{2}\frac{k^2}{y+1} + \sqrt{y^2 + k^2} \qquad (7a)$$

and

$$\Phi(x) = x - 1 + \sqrt{x^2 + k^2} \ . \qquad (7b)$$

Such a situation also can develop when f(y) is a stepwise linear function of y with discontinuities of slope, as has been noted by Drs. Judd and McMillan [see, for example, Figs. 13 and 14 (pp. 27-28) of Ref. 12]. If f(y) is of the form

$$f(y) = -(By^2 + Dy)/(Ay^2 + By + C), \qquad (8)$$

moreover, the <u>entire</u> phase plane will be covered by a family of simple invariant curves -- see, for example, the cases[11] $f(y) = 2ky/(1+y^2)$, with the invariants $x^2y^2 + x^2 + y^2 - 2kxy = \text{Constant}$, and $f(y) = 2ky/(1-y^2)$, with the invariants $x^2y^2-x^2-y^2+2kxy = \text{Constant}$, illustrated by Figs. 7- 9.

It is of interest to examine the mechanism whereby irregular behavior can develop in the neighborhood of unstable fixed points, taking as an illustration an example suggested by Professor deVogelaere that [when generalized and rewritten in variables leading to the form (6a,b) advocated by McMillan] employs

$$f(y) = 2[Ty + (1 - T)y^2]. \qquad (9)$$

First-order fixed points appear at (0,0) and at (1,1). For T = 0, this transformation, when linearized about the unstable fixed point at (1,1), can be represented by the matrix $\begin{bmatrix} 0 & 0 \\ -1 & 4 \end{bmatrix}$, with eigenvalues and eigenvector slopes

$$\lambda = 2 \pm \sqrt{3}, \quad dy/dx = \lambda.$$

A line segment extending downward from the fixed point (1,1) with the slope $2 + \sqrt{3}$, if subjected to repeated applications of the transformation, generates the loops shown in Fig. 10; similarly a line segment of slope $2 - \sqrt{3}$, if extended by the inverse transformation, generates the mirror-image curve (mirrored about the principal diagonal). Points such as A, B, C ... progress toward the fixed point in smaller and smaller steps and, since the transformation is area-preserving, the associated loops clearly must become increasingly elongated as they become increasingly narrow from repeated applications of the forward transformation.

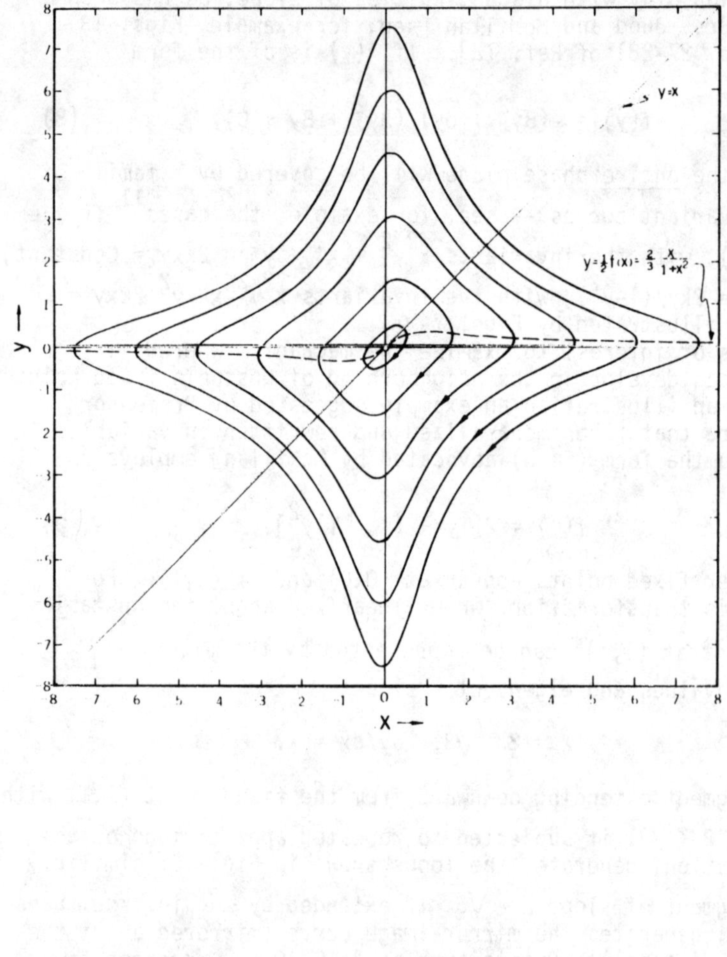

Fig. 7. Invariant curves for the transformation (6a,b) with $f(y) = 2ky/(1+y^2)$ and $k = 2/3$. [Figs. 7 - 11 after McMillan.[11]]

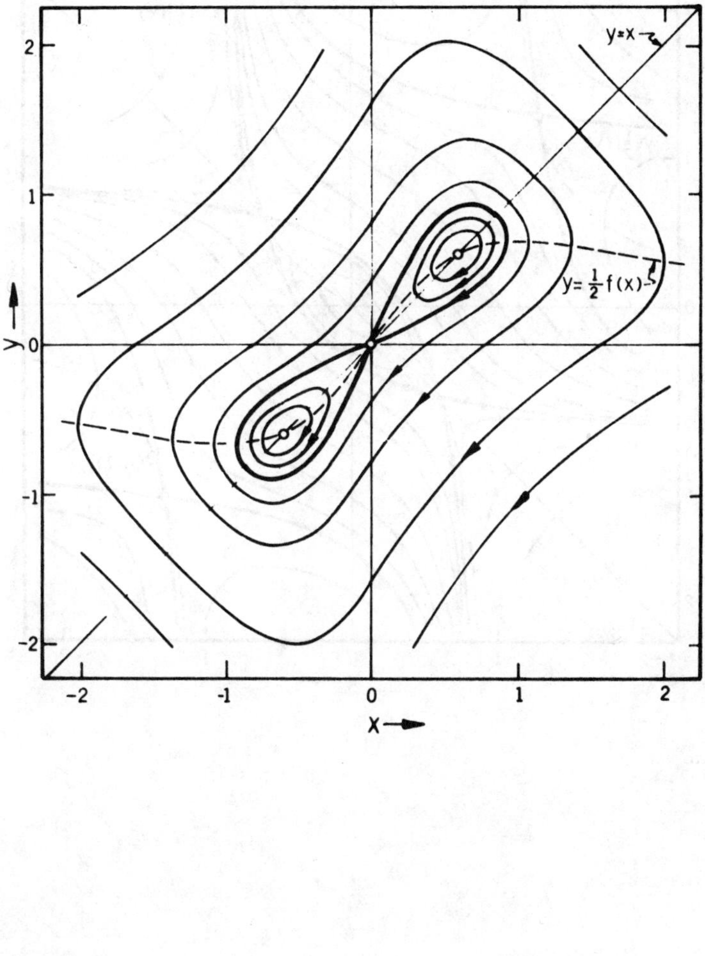

Fig. 8. Invariant curves for the same transformation as in Fig. 7, but with k = 1.36.

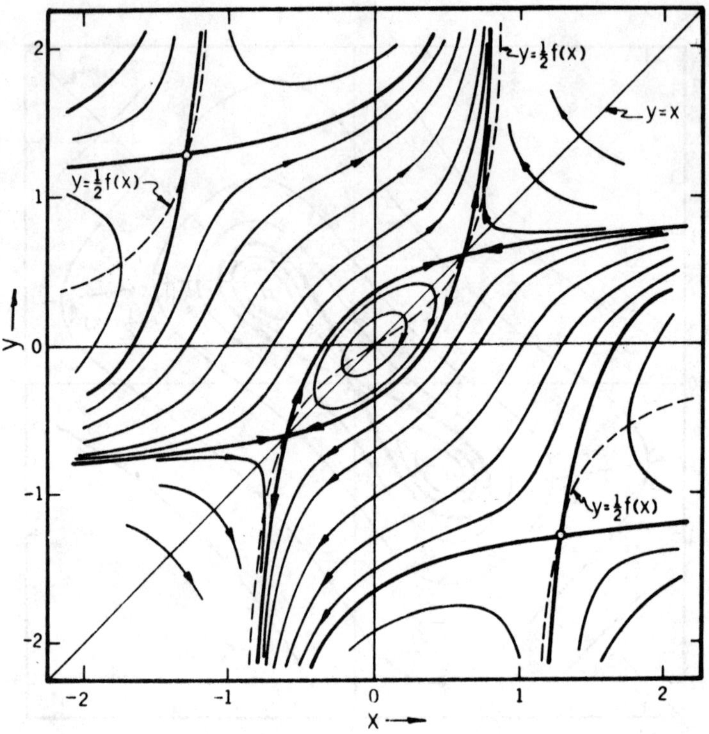

Fig. 9. Invariant curves for the transformation (6a,b) with $f(y) = 2ky/(1-y^2)$.

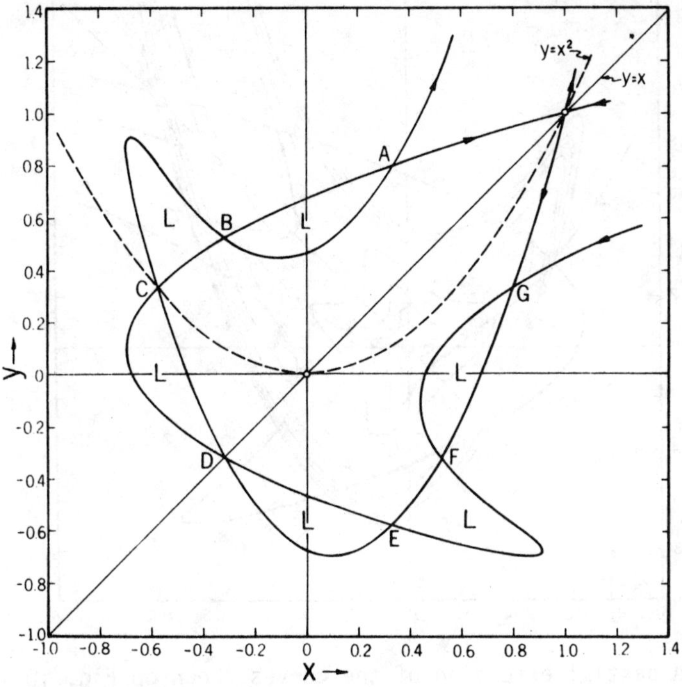

Fig. 10. Plot of the extensions of the eigenvector directions from the unstable fixed point at (1,1), for the deVogelaere transformation expressed in McMillan's variables [Eqs. (6a,b) and (9), with T = 0]. The areas of the loops marked L are all equal, by virtue of the area-preserving character of the transformation and the inherent symmetry about the principal diagonal.

The evolution of such loops clearly will become quite intricate (Fig. 11), but the loops apparently need not permeate the entire "interior". Portions of an inward loop can, in fact, enter, on a later iteration, into the interior of an outward-lying loop, as indicated on Fig. 11. A wealth of island structure, of course, can develop throughout the area of such phase diagrams.

In some instances a family of unstable fixed points for which the eigenvalues are negative may arise (in place of a stable family, for which λ is purely imaginary), and the appearance of phase trajectories can thereby be drastically affected. Phase trajectories in the neighborhood of two such eighth-order ("tune" = 2/8) fixed points are shown on Fig. 12 for the transformation of deVogelaere written to exhibit symmetry about the x-axis

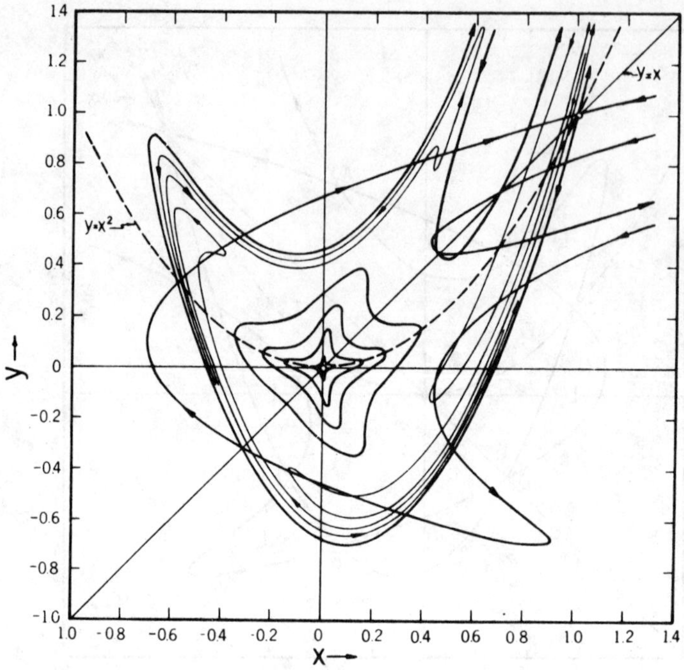

Fig. 11. A partial extension of the curves shown on Fig. 10.

$$x_{n+1} = y_n + Tx_n + (1 - T)x_n^2 \quad (10a)$$

$$y_{n+1} = -x_n + Tx_{n+1} + (1-T)x_{n+1}^2, \quad (10b)$$

with T = -1/8. In any case it is clear, however, that the development of a loop system such as that shown on Fig. 11, can readily give rise to an apparent stochastic motion of phase points in portions of the phase diagram -- most particularly near an unstable fixed point.

The existence of a firm separatrix, or of an extensive family of invariant curves generally, can be extremely sensitive to the exact form of the transformation.[14] A case of some physical interest arises in computational studies relating to the Toda Lattice.[15] This one-dimensional lattice consists of particles interacting through exponential pair potentials and can propagate certain non-linear wave forms ("solitons") without change of shape. One computational investigation[16] of stability for a three-particle lattice (with periodic boundary conditions) has commenced with a Hamiltonian function

$$H = \frac{1}{2}(P_1^2 + P_2^2 + P_3^2) + e^{-(Q_1-Q_3)} + e^{-(Q_2-Q_1)} + e^{-(Q_3-Q_2)}. \quad (11)$$

By a canonical transformation of variables, in recognition of the invariance of this system to translation -- so that $I_1 = P_1 + P_2 + P_3$ constitutes a constant of the motion -- the Hamiltonian (11) becomes expressible as a function of two pair of conjugate variables in the form

$$H = \frac{1}{2}(p_1^2+p_2^2)+ \frac{1}{24}[e^{(2q_2+2\sqrt{3}q_1)}+e^{(2q_2-2\sqrt{3}q_1)}+e^{-4q_2}], \quad (12)$$

which is identical to the Hénon-Heiles Hamiltonian function (5) through terms of third order. It is of interest to examine whether in the present case constants of the motion other than H act to restrict the motion. Computationally it was found -- again using the surface-of-section $q_1 = 0 (p_1 > 0)$ -- that in this case simple invariant curves apparently continue to exist in the q_2p_2, plane, even for very large values of H. Stimulated by this result Hénon[17] has directed attention to an additional integral of the motion that is valid in this case; the constants of the motion for the three-particle lattice then can be written in a form that we may express as[18]

$$H = \text{Constant} \quad (13a)$$

$$P_1 + P_2 + P_3 = \text{Constant, and} \quad (13b)$$

$$P_1P_2P_3 - P_1e^{-(Q_3-Q_2)} - P_2e^{-(Q_1-Q_3)} - P_3e^{-(Q_2-Q_1)} = \text{Constant.} \quad (13c)$$

Evidently[17], further analytic work in fact has now established that the n-particle Toda lattice with periodic boundary conditions (or with fixed ends) is a "completely integrable" system.

It is of some interest to seek means for anticipating whether stochastic behavior will occur in various portions of a phase diagram and to examine the character of such stochastic behavior as does occur. What we here have loosely termed stochastic behavior can be catalogued with respect to a hierarchy of properties (ergodicity, mixing, ...), indicative of increasing disorder, that are fundamentally significant for statistical mechanics.[2a,e] Of particular interest to the accelerator designer, of course,

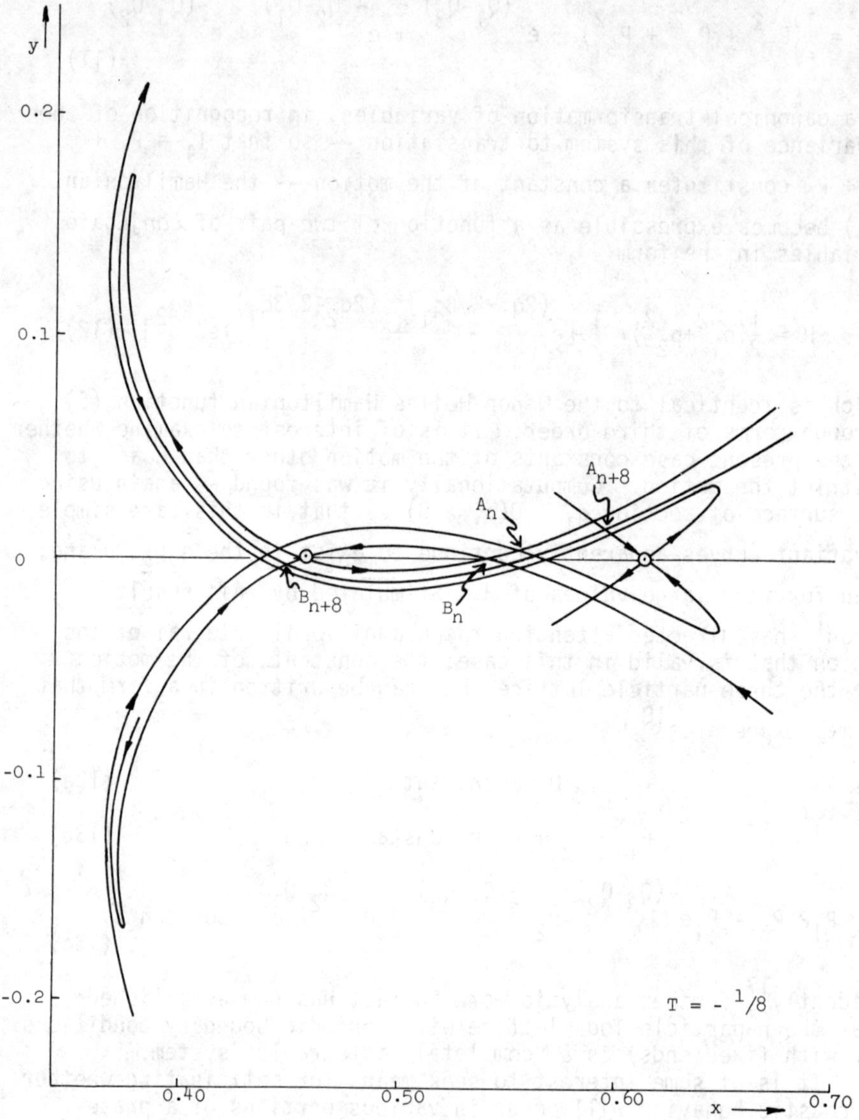

Fig. 12. Phase trajectories for the transformation (10a,b) with T = -1/8, in the neighborhood of two fixed points for which the eigenvalue is negative.

is the determination of a threshold beyond which stochastic behavior will set in and may act to carry a phase point to unacceptably large amplitudes. As noted earlier, stochastic behavior appears to be associated with overlapping resonances[2c], and this concept has served as the basis for some analytic estimates of stochasticity limits.[2c,19] It has been noted by René deVogelaere and confirmed in subsequent computations[20] that for a particular class of fixed-point families -- say those with rotation of the form m/(4m+1) -- there is a closely linear relationship between the order of the resonance (4m+1) and $\ell n|1 - 1/2\ \mathrm{Trace}|$ through many decades ("Trace" denoting the trace of the tangential-mapping or differential matrix associated with the 4m+1 iterations required to map a given fixed point onto itself). Such regularities, and others relating to the apparent size of the stable areas about high-order fixed points (*e.g.*, as estimated from the intersection angle of eigenvectors), have been considered useful indicators of the change in character of a mapping at certain amplitudes.[21,10,22]

A computational procedure of considerable interest for recognizing stochasticity is that in which one follows the evolution of the distance between two initially very close points in phase space. In practice it can prove desirable to reduce the separation from time to time by a recorded factor whenever the separation becomes excessive during the computations, or, perhaps preferably, to evaluate the growth of an *infinitesimal* vector through use of the cumulative tangential-mapping matrix. A high degree of stochasticity can be ascribed to the behavior of the transformation if there are such vectors whose length generally grows beyond the first iteration by a *factor* greater than unity (while others may similarly contract). (Ref. 2a, p. 55; for examples, see Ref. 23.) An analogous procedure -- that can be more attractive, although possibly of a less direct basic significance -- is an investigation of the growth of the eigenvalue(s) of the cumulative tangential mapping. Such eigenvalues can change sign repeatedly during the course of many interations, and hence will be seen to decrease from time to time, but an exponentially increasing trend in eigenvalue magnitude is likely to be associated with a similar type of increase for the lengths of the vectors mentioned previously. The nature of eigenvalue growth has been illustrated by Froeschlé[24] for the transformation[25]

$$x_{n+1} = x_n \cos\alpha - (y_n - x_n^2) \sin\alpha \qquad (14a)$$

$$y_{n+1} = x_n \sin\alpha + (y_n - x_n^2) \cos\alpha \ . \qquad (14b)$$

The general characteristics of this transformation, expressed in variables such that the transformation has the symmetry of McMillan's form, is seen on Fig. 13. On an expanded scale (X10), we see (Fig. 14) the sudden onset of erratic behavior as the starting values for the transformation are successively increased (in steps $\Delta x_o = 0.0025$, for $y_o = 0$), and on a scale expanded by a further factor 100/6 we see (Fig. 15) the presence of a great deal of additional structure within a portion of this "stochastic" region. Associated with the transition to the stochastic region there appears to be a marked change in the manner of growth of $\psi_n = \log|\lambda_n|$ (linear, vs. n, in the stochastic case -- indicative of an exponential trend for $|\lambda_n|$) or of the "Cesaro mean" $\mu_n = \frac{1}{n}\sum_{m=1}^{n}\frac{1}{m}\psi_m$ (constancy in the stochastic case, monotonically decreasing otherwise -- Fig. 16).[26] Such methods indeed may prove useful in investigating computationally the possible development of stochastic motion in storage-ring devices. Extended computations of this nature can present challenging problems with respect to computer accuracy.[27]

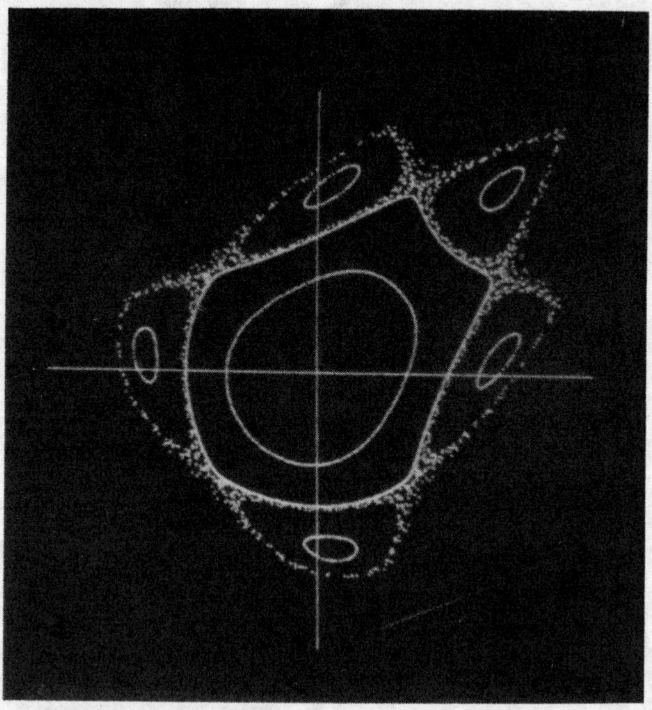

Fig. 13. Apparently smooth phase curves and a scattering of points resulting from iteration of the transformation (14a,b), with cos $\alpha = 0.22$ and coordinates X,Y appropriate to expressing the transformation in the form (6a,b).[25] Five islands of stability (containing stable fixed points of order 5) are seen surrounding the area associated with the order-1 fixed point at the origin. The outermost smooth curve, shown as bounding this inner area, resulted from the starting values $x_o = 0.5350$, $y_o = 0$ (Froschlé notation), and the scattered points result from $x_o = 0.5375$, $y_o = 0$. <u>Scale</u> (as indicated by the coordinate axes): -1.0 to 1.0.

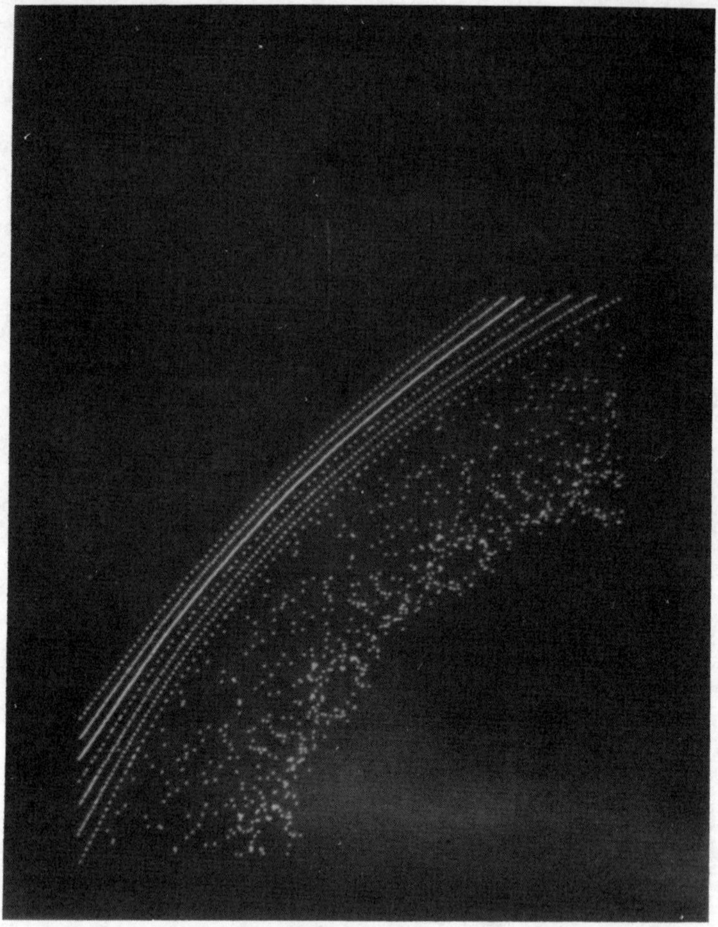

Fig. 14. Enlarged portion (10X) of Fig. 13, showing seven smooth phase trajectories resulting from starting values $x_o = 0.5200, 0.5225, \ldots 0.5350$ (and $y_o = 0$) and a scattering of points resulting from $x_o = 0.5375$, $y_o = 0$. Note the occurrence of open areas within the region covered by the scattered points -- for example the area surrounding an (unplotted) stable fixed point of order 65 at $X \cong 0.476$, $Y \cong 0.521$. <u>Scale</u>: 0.38 to 0.58.

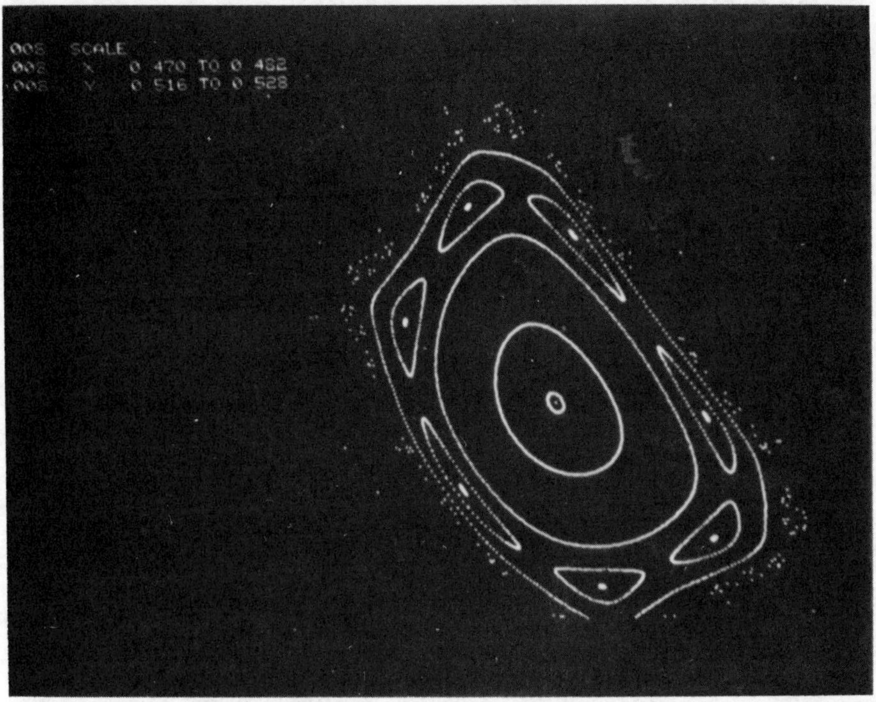

Fig. 15. Detailed multiple-island structure in the immediate neighborhood of an order-65 stable fixed point (shown here just below the center of the diagram) of which mention has been made in the caption to Fig. 14. <u>Scales</u>: 0.470 to 0.482 for X, 0.516 to 0.528 for Y.

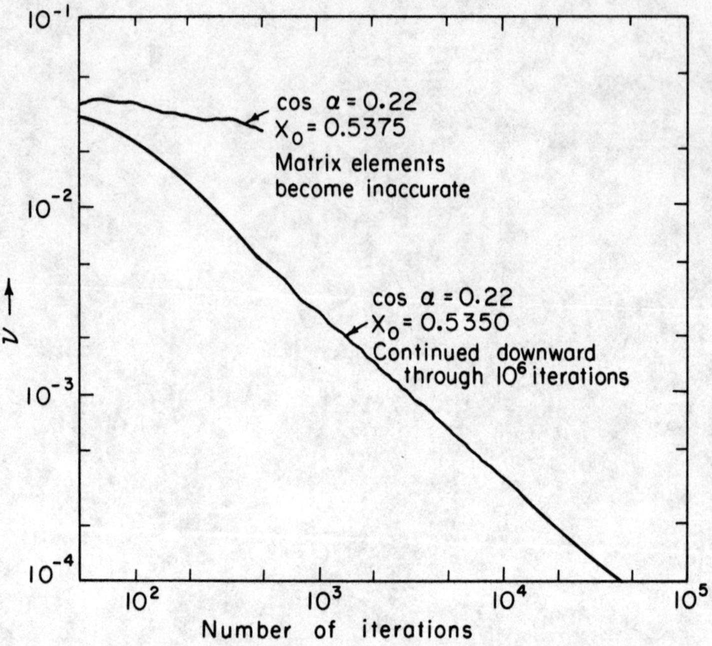

Fig. 16. Plots of the "sliding mean", ν_n (Note 26), vs. n, obtained from computations begun (i) with initial conditions leading to the last smooth curve of Fig. 14 (x_o = 0.5350) and (ii) with initial conditions leading to the scattered points on that Figure (x_o = 0.5375), of which only the results for the latter case indicate a general exponential upward trend of $|\lambda_n|$.

REFERENCES AND NOTES

1. L. Jackson Laslett, "Stochasticity", Proc. IX International Conf. on High Energy Accelerators, Stanford Linear Accelerator Center, CONF 740522 (S.L.A.C., Stanford, Calif.; 1974), pp. 394 - 400.
2. (a) An authoritative treatment of the mathematical aspects of the problems discussed here is given by V. I. Arnol'd and A. Avez, "Ergodic Problems of Statistical Mechanics" (Benjamin, New York, N.Y.; 1968) and (b) by Jürgen Moser, "Stable and Random Motions in Dynamical Systems" (Princeton Univ. Press, Princeton, N.J.; 1973); (c) an extended discussion of which portions relate more immediately to those of interest to an accelerator designer is presented, with many references, by G. M. Zaslavskij and B. V. Chirikov, Uspekhi Fizicheskikh, Nauk 105 and Engl. Transl. Sov. Phys. Usp. 14, 549-568 (1972), with further discussion and examples (d) by B. V. Chirikov in Nucl. Phys. Inst. Report 267 (Novosibirsk, USSR) with Engl. Transl. by A. T. Sanders, CERN Trans. 71-40 (CERN, Geneva, Switzerland; October 1971); and (e) related questions of ergodic theory in statistical mechanics are summarized by J. L. Lebowitz and O. Penrose, Physics Today, pp. 23-29 (February 1973).
3. CNRS Internat. Conf. on Point Transformations and Their Applications, Laboratoire d'Automatique et d'Analyse des Systèmes, Toulouse, France (10-14 September 1973). This conference was concerned with both Hamiltonian and non-Hamiltonian systems.
4. E. M. McMillan, Phys. Rev. 68, 143-144 (1945).
5. K. R. Symon and A. M. Sessler, Proc. CERN Symposium on High Energy Accelerators and Pion Physics 1, 44-58 (CERN, Geneva, Switzerland; 1956).
6. L. Jackson Laslett, ERAN-57 (Lawrence Berkeley Laboratory; 1970).
7. M. Henón and C. Heiles, Astron. J. 69, 73-79 (1964).
8. G. H. Walker and J. Ford, Phys. Rev. 188, 416-432 (1969).
9. The present example in fact is exceptional in that the unperturbed frequencies, being equal, are "rationally connected" and the analysis requires special treatment -- see F. G. Gustavson, Astron. J. 71, 670-686 (1966); J. Moser, "Lectures on Hamiltonian Systems", Memoirs Amer. Math. Soc., No. 81, 1-60 (1968).
10. See also V. K. Mel'nikov, Soviet Math. 4, 266-270 (1963).
11. Edwin M. McMillan, "A Problem in the Stability of Periodic Systems", in "Topics in Modern Physics -- A Tribute to Edward U. Condon", pp. 219-244 (Colorado Asso. University Press, Boulder, Colorado; 1971). A transformation written in the form (6a,b) is convenient for the study of area-preserving transformations in the plane because of the "double symmetry" pointed out by McMillan (p. 225). The transformation can

be interpreted as describing the effect of a simple linear focusing system supplemented by a periodic sequence of thin non-linear lenses that introduce at such points a Δy specified by f(x).

12. Laslett, McMillan, and Moser, Courant-Institute Report NYO-1480-101 (New York University, N.Y.; 1 July 1968).
13. Fig. 12 is taken from Fig. 2 (p.35) of the Appendix to Ref. 12. The fixed points shown are situated on the x-axis at $x \cong 0.4562733$ and 0.6130793, with the negative eigenvalue $\lambda \cong -8.369$ (in addition to its reciprocal, $\lambda \cong -0.1195$) for this family, and the point A_{n+8} is the eighth iterate of point A_n. For discussion of the occurrence and consequences of loop systems, see S. Smale, "Diffeomorphisms with Many Periodic Points ...", (Princeton University Press, Princeton, N.J.; 1965); E. Zehnder, Comm. Pure Appl. Math. 26, 131-182 (1973); Ref. 2c, Sect. 6.1; and Ref. 2d, Sect. 2.6.
14. The loss of a firm separatrix can be illustrated computationally for the transformation (6a,b) by modifying the function f(y) of (7a) so as to introduce the quantity 1 - b as a factor multiplying the second term on the right and setting $b \neq 0$ (for example, b = 0.05) -- L. Jackson Laslett, ERAN-239 (1974).
15. See, for example, M. Toda, Prog. Theoret. Phys. (Kyoto) Suppl. 45, 174-200 (1970); references cited therein; and related papers in this issue of the Supplement.
16. J. Ford, S. D. Stoddard, and J. S. Turner, Prog. Theoret. Phys. (Kyoto) 50, 1547-1560 (1973).
17. Cited in Ref. 16, p. 1558.
18. The validity of these (time-independent) expressions as constants of the motion of course can be confirmed directly by forming their Poisson-bracket expressions with the Hamiltonian function (11).
19. B. V. Chirikov, E. Keil, and A. M. Sessler, CERN Report ISR-TH/69-59 (CERN, Geneva, Switzerland; 15 October 1969).
20. For example, Ref. 12, pp. 42-43, where is also listed a quantity $\lambda_I = \lambda - 1$ for fixed point families that have rotation m/(4m+1).
21. John M. Greene, J. Math. Phys. 9, 760-768 (1968).
22. James H. Bartlett, "Stability of Area-Preserving Mappings", Paper III-3 of Ref. 3.
23. J. Ford and G. H. Lunsford, Phys. Rev. A1, 59-70 (1970).
24. C. Froeschlé, Astron. and Astrophys. 9, 15-23 (1970); C. Froeschlé and J. P. Scheidecker, Ibid. 22, 431-436 (1973); and other references cited therein.
25. This transformation, (14a,b), can be put into McMillan's form[11] by the change of variables $x = \sqrt{\sin\alpha}\, Y$, $y = (X - Y\cos\alpha)/\sqrt{\sin\alpha}$, with f(Y) then becoming $2Y\cos\alpha + Y^2 \sin^{3/2}\alpha$.

26. The curves of Fig. 16 are plots of

$$\nu_n \equiv \sum_{m=1}^{n} \frac{1}{m}\Psi_m \exp(-\frac{n-m}{\tau}) / \sum_{m=1}^{n} \exp(-\frac{n-m}{\tau}) \text{ with } 1/\tau = 0.015,$$

the sliding exponential factor being designed to provide some smoothing of the results (L. Jackson Laslett, unpublished LBL Report). Extended computations of this nature can present challenging problems with respect to computer accuracy.[27]

27. C. Froeschlé and J.-P. Scheidecker, J. Comp. Phys. 11, 423-439 (1973); Astrophys. & Space Sci. 25, 373-386 (1973).

28. I am deeply indebted to Paul J. Channell for many stimulating and helpful conversations concerning topics discussed here. Responsibility for the views expressed in this paper, however, remains exclusively my own.

AN ILLUSTRATIVE EXAMPLE OF THE CHIRIKOV CRITERION FOR STOCHASTIC INSTABILITY

Paul J. Channell
The Institute for Advanced Study, Princeton, New Jersey 08540

INTRODUCTION

Our conception of classical mechanics tends to be dominated by a few examples which are analytically soluble, for instance the harmonic oscillator and the two body problem with $1/r^2$ force law. However, according to a result of Siegel[1] integrability is very atypical.

It is now known that the phase space of a typical Hamiltonian system may contain trajectories with widely varying ergodic properties. If the system is very "near" an integrable system then the Kolmogorff, Arnold, Moser (KAM) Theorem[2,3] asserts that most (in the sense of measure theory) of the phase space is occupied by quasiperiodic trajectories. However, interspersed between the quasiperiodic trajectories is a set of very "random" trajectories[4]. The complete classification of trajectories according to ergodic type has still not been accomplished.

At the other extreme from integrable systems are systems which are known to possess very strong ergodic properties, i.e., to be very "random", such as two dimensional hard spheres[5] and geodesic flows on manifolds of negative curvature[2]. For a given dynamical system of interest one is thus led to ask whether it is integrable, "random" or somewhere in between. Unfortunately, no technique presently exists for deciding the answer to this question.

Numerical experiments[6] have indicated that as a system moves further[7] and further away from an integrable system, a threshold is reached at which the motion of the system suddenly becomes quite "random". For practical applications it can be very important to know when this threshold, or stochastic instability, will occur. An approximate criterion for calculating this threshold has been developed by Chirikov[8] and by Rosenbluth, et al.[9]. We will call this criterion the Chirikov Criterion.

The Chirikov Criterion has been applied to a wide variety of problems. It has been used to calculate the transition to statistical behavior of an anharmonic lattice[10]. It has also been used to compute the beam-beam limit for colliding beam accelerators[11,12]. Recently, Finn[13] and Stix[14] have used the Chirikov Criterion to explain certain features of the disruptive instability in Tokamaks. Rosenbluth[15] proposed the concept of superradiabaticity, which uses

ISSN: 0094-243X/78/248/$1.50

the Chirikov Criterion as limit, to explain the confinement of particles in magnetic mirrors in the presence of a spectrum of electrostatic fluctuations. Kaufman and Smith[16] and Timofeev[17] have used the Chirikov Criterion to predict heating of plasma particles in the presence of a wave. It is probable that a large number of other applications of the Chirikov Criterion will be found.

The purpose of this paper is pedagogical. I will present the Chirikov Criterion by way of an example of which it can be simply calculated. The particular system that I study can be written as a transformation and studied numerically with ease. I will present some results which show that in this case the Chirikov Criterion is reasonably successful. I will then discuss the limitations and shortcomings of the criterion.

THE MODEL AND THE CRITERION

In order to introduce the Chirikov Criterion, I will start with an explicitly integrable system and perturb away from it. In terms of action-angle variables the Hamiltonian for the model can be written as

$$H_0 = \alpha I + \beta I^m , \tag{1}$$

where I is the action variable and α, β, m are adjustable constants. The Hamiltonian equations of motion for this system are

$$\frac{dI}{dt} = - \frac{\partial H_0}{\partial \theta} = 0 \qquad\qquad \frac{d\theta}{dt} = \alpha + \beta m I^{m-1} \equiv \omega(I). \tag{2}$$

The solutions of these equations are

$$I = \text{constant} \qquad\qquad \theta = \omega(I)t + \theta_0. \tag{3}$$

For our perturbed Hamiltonian we choose

$$H = H_0 + \varepsilon H_1 = H_0 + \varepsilon \cos\theta \sum_{\ell=-\infty}^{\infty} \delta(t - \ell), \tag{4}$$

where ℓ assumes integer values and ε is an adjustable constant which measures the "size" of the perturbation. We thus have a periodic Hamiltonian with period one (in arbitrary units). Let us note that

$$\sum_{\ell=-\infty}^{\infty} \delta(t - \ell) = 1 + 2 \sum_{m=1}^{\infty} \cos(2\pi m t). \tag{5}$$

Let

$$\omega_0 = 2\pi \tag{6}$$

be the period of the perturbation. The Hamiltonian can now be written as

$$H = \alpha I + \beta I^m + \frac{\varepsilon\omega_0}{2\pi}\left(\cos\theta + \sum_{m=1}^{\infty}\left[\cos(\theta + m\omega_0 t) + \cos(\theta - m\omega_0 t)\right]\right). \tag{7}$$

For small ε we can use an iteration scheme in which we put unperturbed motion, Eq. (3), into the ε term of Eq. (7). Doing this, we find terms of the form

$$\cos\left[\omega(I)t \pm m\omega_0 t\right] . \tag{8}$$

Because $\omega(I)$ is a function of I [see Eq. (2)], for a given m there may be an I for which

$$\omega(I) = \pm m\omega_0 . \tag{9}$$

This is called a resonance. Near this value of I, one term in the sum (7) will remain constant and the other terms will oscillate in time so that their effects will tend to average away. Thus, near this value of I, for small enough ε, the system is dominated by a single resonance. Let us write an approximate Hamiltonian for the system near this resonance;

$$H_{app.} = \alpha I + \beta I^m + \frac{\varepsilon\omega_0}{2\pi}\cos(\theta - m\omega_0 t) . \tag{10}$$

The equations of motion which result from this Hamiltonian are

$$\frac{dI}{dt} = \frac{-\partial H_{app.}}{\partial\theta} = \frac{\varepsilon\omega_0}{2\pi}\sin(\theta - m\omega_0 t) \tag{11}$$

$$\frac{d\theta}{dt} = \frac{\partial H_{app}}{\partial I} = \alpha + \beta m I^{m-1} = \omega(I) . \tag{12}$$

Let us write

$$I = I_R + \Delta I, \tag{13}$$

where I_R is defined by

$$\omega(I_R) = m\omega_0 .$$

We then can write approximately,

$$\frac{d(\theta - m\omega_0 t)}{dt} = \omega(I) - m\omega_0 \simeq \Delta I \frac{d\omega(I_R)}{dI} \equiv \Delta I\ \omega'(I_R)\ . \tag{14}$$

If we define

$$\psi = \theta| - m\omega_0 t\ , \tag{15}$$

then the equations of motion can be written as

$$\frac{d\psi}{dt} = \omega'(I_R)\ \Delta I \tag{16}$$

$$\frac{d\Delta I}{dt} = \frac{\varepsilon\omega_0}{2\pi} \sin\psi\ . \tag{17}$$

We recognize these as the equations of motion of a pendulum which has a separatrix with width in ΔI that is given by

$$\delta I = 4\sqrt{\left|\frac{\varepsilon\omega_0}{2\pi\omega'(I_R)}\right|}\ . \tag{18}$$

This is called the width of the resonance.

The description of a system with only one resonance is thus fairly simple. Unfortunately, our system, Eq. (7), has an infinite set of resonances given by all values of m in Eq. (9). When all resonances are included the behavior of the system can be quite complicated. The Chirikov Criterion asserts that stochastic instability, i.e. "random" behavior, results when resonances overlap, i.e., when resonances take up all the phase space. More precisely, stochasticity occurs when the distance between resonances is less than the width of the resonances. Roughly speaking, the idea is that when resonances overlap trajectories move from one resonance region to another resonance region very rapidly.

In our model system, the distance between resonances in frequency is given by

$$\Delta\omega = \omega_0\ , \tag{19}$$

as can be easily seen from Eq. (9). The width in frequency of the resonances can be calculated from Eq. (18) as

$$\delta\omega \simeq \frac{d\omega}{dI}\ \delta I = 4\sqrt{\left|\frac{\varepsilon\omega_0\omega'(I_R)}{2\pi}\right|}\ . \tag{20}$$

The Chirikov Criterion is thus

$$s = \frac{\delta\omega}{\Delta\omega} = 4\sqrt{\left|\frac{\varepsilon\omega'(I_R)}{2\pi\omega_0}\right|} \geq 1, \qquad (21)$$

where s is called the stochasticity parameter.

Let us now mention two conditions which must be met in order for the Chirikov Criterion to be applicable. The first is that the unperturbed system must be truly non-linear, i.e. the frequency must vary [see Eqs. (2) and (9)]. This condition is necessary in order that the concept of a multiplicity of resonances make sense. Furthermore, the unperturbed system must be sufficiently non-linear that more than one resonance occurs in the region of interest; otherwise the system is dominated by linear resonance which can be of various types.

The second condition which must be applied is that the unperturbed system not be too non-linear. If the system is too non-linear then the approximation in Eq. (14) is no longer valid and the idea of resonance breaks down.

The two conditions mentioned above have been expressed by Chirikov[10] as

$$\varepsilon \ll \frac{I}{\omega}\frac{d\omega}{dI} \ll \frac{1}{\varepsilon}\,. \qquad (22)$$

Let us now write out the Chirikov Criterion and the validity criteria more explicitly for our model system. Using Eq. (2), the Chirikov Criterion, Eq. (21), can be written as

$$|\varepsilon| \geq \frac{2\pi\omega_0}{16\ \beta m(m-1)I^{m-2}}\ . \qquad (23)$$

Note that for m = 2 the criterion is independent of I, otherwise, Eq. (23) is satisfied only when I is sufficiently large. For given ε, the I beyond which stochasticity sets in is given by

$$I \geq \left|\frac{2\pi\omega_0}{16\ \beta m(m-1)\varepsilon}\right|^{1/m-2} \qquad (24)$$

assuming that m > 2. The validity criteria, Eq. (22), can be written as

$$\varepsilon \ll \frac{\beta m(m-1)I^{m-1}}{\alpha + \beta m I^{m-1}} \ll \frac{1}{\varepsilon}\,. \qquad (25)$$

Let us note that the system with the Hamiltonian of Eq. (4) can be written as a transformation as

$$\theta_1 = \theta + \alpha + \beta m I^{m-1} ,$$
$$I_1 = I + \varepsilon \sin \theta_1 . \tag{26}$$

It is easy to verify that the transformation, (26), is canonical.

NUMERICAL RESULTS

The transformation, (26), is very easy to iterate on a computer many times. For convenience in plotting we make the canonical transformation

$$q = \sqrt{2I} \sin \theta ,$$
$$p = \sqrt{2I} \cos \theta . \tag{27}$$

In the (q,p) plane, I is half of the square distance from the origin and θ is the angle from the p axis. The plots in Figs. 1-3 are in the (q,p) plane. Conditions (25) were satisfied in all runs.

From Eq. (24) we see that beyond a certain distance from the origin the trajectories should be random. As can be seen from Figs. 1-3, this is indeed the case. The random splatter of points outside the closed curves were all iterates of a single initial point. In each figure the value of I at which stochasticity set in is listed in Table I for a number of cases and is compared to theory, as given by Eq. (24). As can be seen, the agreement is fairly good.

If the regions with closed curves in Figs. 1-3 were enlarged sufficiently, then Zehnder's Theorem[4] assures us that small stochastic regions would appear. However, the measure of these regions is small and can be ignored (except in higher dimensions where Arnold Diffusion[18] may be important). In addition, these small stochastic regions are confined by the surrounding closed curves (but not in higher dimensions). Thus, for all practical purposes this example supports the Chirikov Criterion as a useful way of computing the stochasticity threshold.

Recently I have been informed[19] that Chirikov, after seeing the preliminary notes for this work, has derived a threshold which agrees better with the numerical results than Eq. (24).

DISCUSSION

Although fairly good agreement with the Chirikov Criterion was found, the model we chose lacks sufficient generality to exhibit all the difficulties which can occur. For example, suppose that the perturbing term in the Hamiltonian had the form

$$\varepsilon H_1 = \varepsilon \sum_{k=-\infty}^{\infty} A_k \cos k\theta \sum_{\ell=-\infty}^{\infty} \delta(t-\ell), \tag{28}$$

m	ε	Theoretical Threshold	Observed Threshold
10	.1	1.04	.81
5	.1	1.83	1.3
4	.1	3.21	2.19
4	.05	4.53	3.24
4	.02	7.17	4.84
4	.01	10.14	7.56

Table I - All runs were made with $\alpha = .5$, $\beta = .2$, where α, β, and m are defined in Eq. 1. ε is defined in Eq. 4.

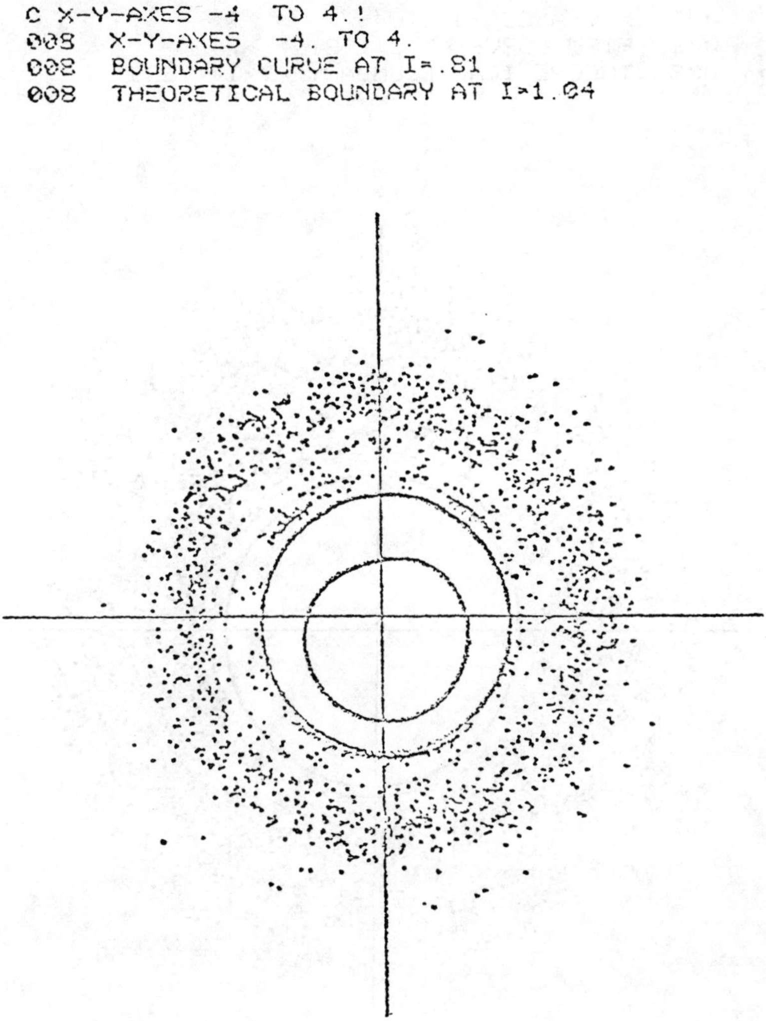

Fig. 1. Iteration of Eq. 26 with $m = 10$, $\varepsilon = .1$. All runs used $\alpha = .5$, $\beta = .2$.

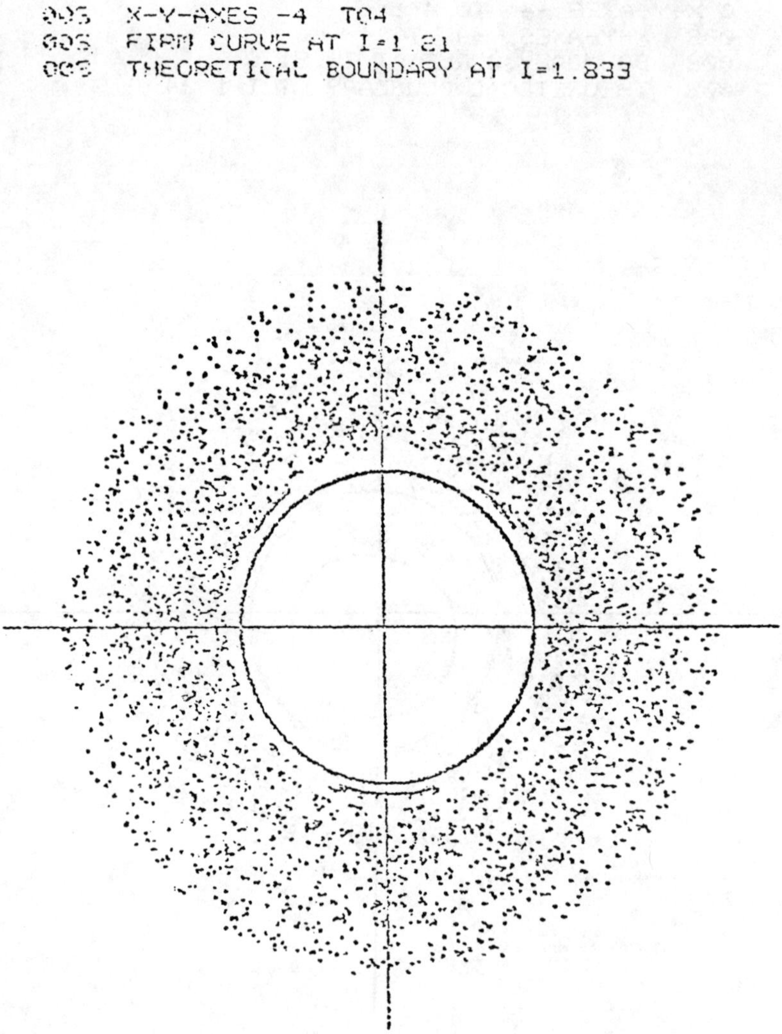

Fig. 2. Same as Fig. 1 with m = 5, ε = .1.

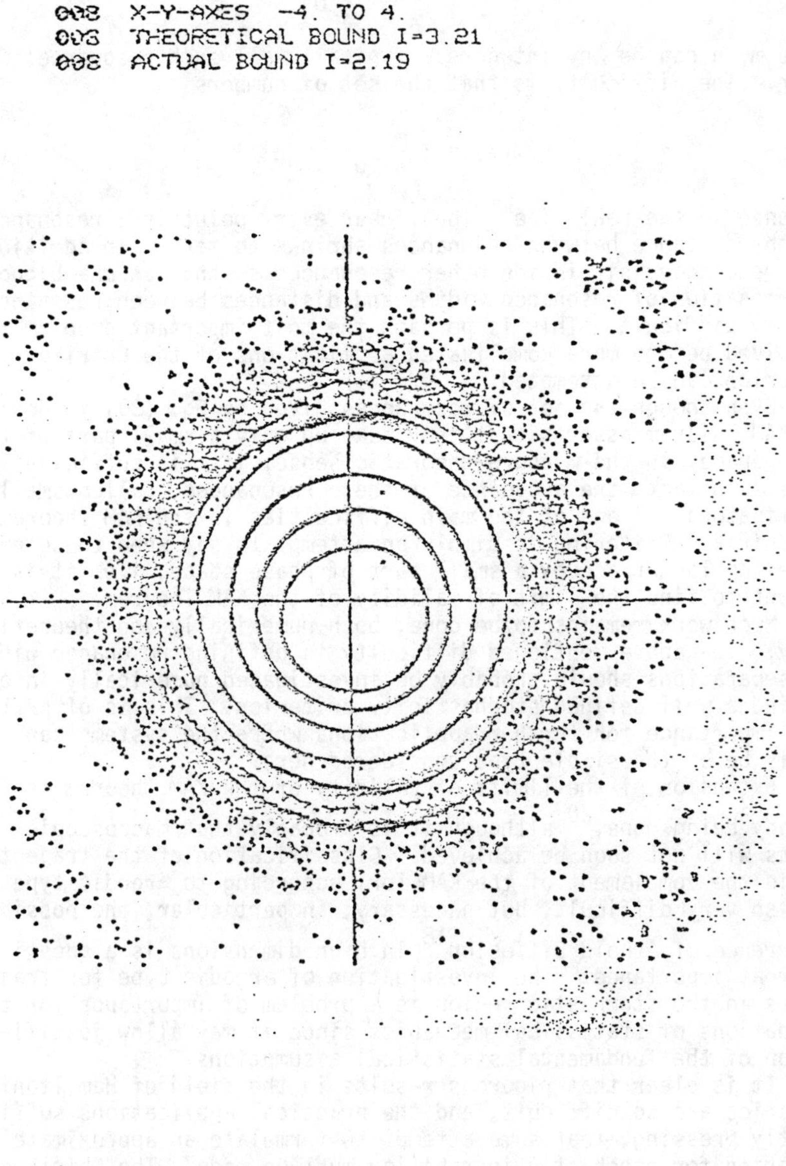

Fig. 3. Same as Fig. 1 with m = 4, ε = .1.

where A_k are constants. Then, instead of resonances given by Eq. (9), resonances would occur at all I such that

$$n\omega\ (I) = \pm m\omega_0, \tag{29}$$

where m, n can be any integers. Typically all such resonances will occur. The difficulty is that the set of numbers

$$\omega(I) = \pm \frac{m}{n}\,\omega_0, \tag{30}$$

is dense in the real line. Thus, near every point is a resonance, and the distance between resonances shrinks to zero. In addition, many resonances sit inside other resonances so that an unambiguous determination of resonance widths and distances between resonances is very difficult. This is perhaps the most important problem to be solved before more complicated applications of the Chirikov criterion can be attempted.

Even though the set of resonances given by Eq. (30) is dense, the KAM Theorem asserts that they take up only a small part of the phase space, in the measure-theoretic sense, if ε is sufficiently small. In fact, the avoidance of these resonances, called small denominators, is one of the main difficulties in the KAM Theorem. The Chirikov Criterion is simply an attempt to say when these resonances no longer occupy a small part of phase space, i.e. it is an attempt to find the limit of validity of the KAM Theorem.

Much work remains to be done, both numerically and theoretically. The above mentioned difficulty in defining resonance widths and separations should probably be investigated numerically in order to find a well defined stochasticity criterion. This is of particular importance for further applications where the systems can hardly be of the simple type considered here.

Extension of the limits of validity of the KAM Theorem is already being done,[20] although it is probably that macroscopic limits will not soon be achieved. Classification of the trajectories in the complement of the KAM Tori according to ergodic type is also very difficult, but necessary; in particular, the possible occurrence of Arnold Diffusion[18] in high dimensions is a question of great importance. The investigation of ergodic type for trajectories in the stochastic region is a problem of importance for the foundations of statistical mechanics since it may allow justification of the fundamental statistical assumptions.

It is clear that rigorous results in the field of Hamiltonian Mechanics are so difficult, and the practical applications sufficiently pressing, that some attempt to formulate an approximate criterion for stochastic instability must be made. The Chirikov Criterion, outlined above, is one such, reasonably successful, attempt.

ACKNOWLEDGEMENT

I want to thank Jackson Laslett for programming help, discussions, and especially encouragement of this work.

REFERENCES

1. C. L. Siegel, Math. Annalen. 128 (1954) p. 144.
2. Arnold, V. I. and Avez, A., Ergodic Problems of Classical Mechanics, Benjamin, New York, 1968.
3. J. Moser, Stable and Random Motions in Dynamical Systems, Princeton University Press, Princeton, 1973.
4. E. Zehnder, Comm. Pure Appl. Math. 26, 1973, p. 131.
5. Ya. Sinai, Dokl. Akad. Nauk. 153 No. 6 (1963). [Sov. Math. Dokl. 4 No. 6 (1963) p. 1818].
6. Hénon, M. and Heiles, C., Astron. J. 69, No. 1 (1964) p. 73.
7. Words such as "further" and "near" are, of course, very vague. In fact, they should be replaced by a specification of a topology on the space of Hamiltonian systems. Likewise, the word "random" is used very loosely; it should be replaced by the exact ergodic property intended, if possible.
8. B. V. Chirikov, Atomnaya Energiya 6, 630 (1959).
9. M. N. Rosenbluth, R. Z. Sagdeev, J. B. Taylor, and G. M. Zaslavskii, Nuclear Fusion 6, 297 (1966).
10. B. V. Chirikov, CERN Trans. 71-40 (1971).
11. B. V. Chirikov, E. Keil, and A. M. Sessler, CERN Report ISR-TH/69-59 (1969).
12. L. J. Laslett, Proc. 9th Intl. Conf. on High Energy Accel., SLAC p. 394 (1974).
13. J. Finn, Nuclear Fusion 15, 845 (1975).
14. T. H. Stix, Phys. Rev. Letters 36, 521 (1976).
15. M. N. Rosenbluth, Phys. Rev. Letters, 29, 408 (1972).
16. G. R. Smith and A. N. Kaufman, Phys. Rev. Letters, 34, 1613 (1975).
17. A. V. Timofeev, Fiz. Plazmy, 1, No. 1, p. 88 (1975). [Sov. J. of Plasma Phys. 1, 47 (1976).]
18. V. I. Arnold, Dokl. Akad. Nauk. USSR, 156, 9 (1964) [Sov. Math.-Doklady 5, 58 (1964).]
19. J. Ford, Private communication.
20. H. Rüssman, IN, Dynamical Systems, Theory and Applications, Ed. by J. Moser, Springer-Verlag, Berlin (1975).

SIMPLE PERIODIC ORBITS

Alan Weinstein
Department of Mathematics
University of California, Berkeley

We discuss existence theorems for periodic orbits of a Hamiltonian system near an equilibrium point. The averaging method is applied to an example of 1:2:4 resonance, and it is found that the three periodic orbits which appear are far from "orthogonal".

The purpose of this note is to illustrate with a simple example some general results about periodic orbits of a time-independent Hamiltonian system near a stable equilibrium point.

Choosing suitable canonical coordinates $(q_1,\ldots,q_n,p_1,\ldots,p_n)$ near the equilibrium point, we may assume that the Hamiltonian has the form

$$H = \frac{1}{2}\sum_{i=1}^{n} \omega_i(q_i^2 + p_i^2) + \text{(higher order terms)}, \qquad (1)$$

where $0 < \omega_1 \leqslant \omega_2 \leqslant \ldots \leqslant \omega_n$ are the fundamental frequencies of periodic solutions for the linearized system

$$\begin{aligned} \dot{q}_i &= \omega_i p_i \\ \dot{p}_i &= \omega_i p_i \end{aligned} \qquad (2)$$

whose Hamiltonian is $H_2 = 1/2 \sum_{i=1}^{n} \omega_i(q_i^2 + p_i^2)$. For any $E > 0$, the energy level $\{(q,p)|H(q,p) = E\}$ contains a small spherelike 2n-1 dimensional "surface" Σ_E surrounding the origin. By the conservation of energy, the surface Σ_E is invariant under the Hamiltonian system.

$$\begin{aligned} \dot{q}_i &= \frac{\partial H}{\partial p_i} \\ \dot{p}_i &= -\frac{\partial H}{\partial q_i} \end{aligned} \qquad (3)$$

The following theorem was proven by Liapunov[1] and Horn[2] for the case in which no ratio ω_i/ω_j is an integer, and extended by the author[3] to the general case.

ISSN: 0094-243X/78/260/$1.50

Theorem. There exists a positive number E_0 such that, for all E in the interval $(0,E_0)$, the equations (3) with Hamiltonian (1) have at least n periodic solutions on Σ_E. We may choose these solutions, say $(q_E^{(k)}(t), p_E^{(k)}(t))$, k = 1, 2, ... , n, so that for each k the limit

$$\lim_{E\to 0} (E^{-1/2}q_E^{(k)}(t), E^{-1/2}p_E^{(k)}(t)) \tag{4}$$

is a periodic solution of the linearized equations (2) with H_2 equal to 1.

We may call the periodic orbits given by this theorem simple. They are quite different from the very long period orbits of the type discussed by Helleman[4].

It would be interesting to know just how large E_0 can be. It has been proven that if Σ_E is convex[5], or even if each ray through the origin crosses Σ_E at exactly one point [6], there is at least one periodic orbit on Σ_E with a fairly short period. To show that, in fact, at least n orbits persist when E becomes large, it would be useful to know how far apart the orbits are when E is small.

Moser[7] has shown how to use the method of averaging to determine which periodic solutions of (2) occur as limits of the form (4) for periodic solutions of the full non-linear equations. These are called the bifurcation solutions. We shall apply his procedure to the case n = 3, $\omega_1 = 1$, $\omega_2 = 2$, $\omega_3 = 4$, i.e., three harmonic oscillators with frequencies in the ratio 1:2:4, to which a non-linear coupling has been applied.

In some joint work on the averaging method (unpublished) with Carl Simon, we found it useful to introduce complex coordinates $Z_i = q_i + \sqrt{-1}\, p_i$, as in reference 8. The solution of (2) may then be written

$$Z_i = e^{-\sqrt{-1}\,\omega_i t} Z_i(0) , \tag{5}$$

and the quadratic Hamiltonian is $H_2 = 1/2 \sum_{i=1}^{n} \omega_i Z_i \bar{Z}_i$.

To study periodic orbits of the non-linear system, we must average the cubic terms in H with respect to the solutions of (2). The general real cubic polynomial in $(q_1,q_2,q_3,p_1,p_2,p_3)$ can be written as a sum

$$H_3 = \sum_{|\underline{k}|+|\underline{\ell}|=3} a_{\underline{k\ell}} Z_1^{k_1} Z_2^{k_2} Z_3^{k_3} \bar{Z}_1^{\ell_1} \bar{Z}_2^{\ell_2} \bar{Z}_2^{\ell_3} \tag{6}$$

where $\underline{k} = (k_1,k_2,k_3)$, $\underline{\ell} = (\ell_1,\ell_2,\ell_3)$, $|\underline{k}| = \Sigma k_i$, $|\ell| = \Sigma \ell_i$, and the complex coefficients satisfy the "reality condition" $a_{\underline{k\ell}} = \bar{a}_{\underline{\ell k}}$. Substituting (5) into (6), integrating over t from 0 to 2π, and dividing by 2π, we obtain the averaged third-order terms

$$H_3^{av} = \sum_{\substack{|\underline{k}|+|\underline{\ell}|=3 \\ \underline{\omega}\cdot\underline{k} = \underline{\omega}\cdot\underline{\ell}}} a_{\underline{k\ell}} Z_1^{k_1} Z_2^{k_2} Z_3^{k_3} \bar{Z}_1^{\ell_1} \bar{Z}_2^{\ell_2} \bar{Z}_3^{\ell_3} ,$$

where $\underline{\omega}\cdot\underline{k} = \Sigma\omega_i k_i$ and $\underline{\omega}\cdot\underline{\ell} = \Sigma\omega_i \ell_i$. In fact, there are not many pairs $(\underline{k},\underline{\ell})$ which satisfy the necessary conditions, and we may write

$$H_3^{av} = \alpha Z_1^2\bar{Z}_2 + \bar{\alpha}\,\bar{Z}_1^2 Z_2 + \beta Z_2^2\bar{Z}_3 + \bar{\beta}\,\bar{Z}_2^2 Z_3 .$$

We will assume that α and β are both non-zero.

Now the bifurcation solutions of (4) may be determined by solving the equations:

$$\frac{\partial}{\partial \bar{Z}_i} (H_3(Z) + \lambda H_2(Z)) = 0,$$

and

$$H_2(Z) = 1 ,$$

where Z_i and $\bar{Z}_i$ are considered as independent variables, and the Lagrange multiplier λ is required to be <u>real</u>. (The value of λ tells how the periods of the solutions depend on E.) Our specific equations are

$$2\alpha\bar{Z}_1 Z_2 + \frac{\lambda}{2} Z_1 = 0$$

$$\alpha Z_1^2 + 2\beta\bar{Z}_2 Z_3 + \lambda Z_2 = 0$$

$$\beta Z_2^2 + 2\lambda Z_3 = 0 .$$

The simplest solution to these equations comes from setting $\lambda = 0$. Then $Z_1 = Z_2 = 0$, and we get the "fast" periodic solution of the linearized equations; the fact that periodic solutions of the non-linear equations bifurcate from this one follows from the work of Liapunov and Horn already mentioned.

Next, we can find two solutions with $Z_1 = 0$, $Z_3 = \pm 1/2(\beta/|\beta|)(Z_2/|Z_2|)Z_2$, and $\lambda = \pm |\beta|\ |Z_2|$. These are essentially the same as the solutions found by Duistermaat[9] in two degrees of freedom. This brings us to our quota of three periodic solutions, as guaranteed by the theorem. Whether there are any more solutions depends on the relative magnitudes of the coupling constants α and β.

If $|\alpha| > 1/4\ \beta$, then there are two more periodic solutions, with Z_1, Z_2, and Z_3 all non-zero, bringing the total number of simple periodic orbits up to five. If $|\alpha| < 1/4\ |\beta|$, there are no solutions other than the three already found, i.e., if the coupling between the two fast oscillators is sufficiently strong, then the slow oscillator is "shut out" of the simple periodic orbits. These orbits all lie near the space $Z_1 = 0$, in contrast to the orthogonal normal modes for a linear system.

Of course, we have only studied a few special orbits of our system. It would be interesting to study the general (non-periodic) orbits to see the extent to which their behavior is determined by the simple periodic orbits.

REFERENCES

1. M. A. Liapounov, Problème général de la Stabilité du Mouvement, Princeton Univ. Press, Princeton, 1949, pp. 375-392. (reprinted from Ann. Fac. Sci. Toulouse 9 (1907), and an earlier Russian version).
2. J. Horn, Beiträge Zur Thorie der Kleinen Schwingungen, Z. Math. Phys. 48 (1903), 400-434.
3. A. Weinstein, Normal Modes for Non-linear Hamiltonian Systems, Inventiones Math. 20 (1973), 47-57.
4. R. H. G. Helleman, these proceedings.
5. A. Weinstein, Periodic Orbits for Convex Hamiltonian Systems, (to appear).
6. P. Rabinowitz, Periodic Solutions of Hamiltonian Systems, U. of Wisconsin preprint.
7. J. Moser, Regularization of Kepler's Problem and the Averaging Method on a Manifold, Comm. Pure Appl. Math. 23 (1970), 609-636.
8. J. Moser, Lectures on Hamiltonian Systems, Memoirs of the Amer. Math. Soc. 81 (1968).
9. J. J. Duistermaat, On Periodic Solutions Near Equilibrium Points of Conservative Systems, Arch. Rat. Mech. Anal. 45 (1972) 143-160.

VARIATIONAL SOLUTIONS OF NON-INTEGRABLE SYSTEMS

Robert H. G. Helleman
School of Physics
Georgia Institute of Technology
Atlanta, Georgia 30332

ABSTRACT

There are few, if any, methods to construct the solutions of a "non-integrable"[26] (≈ non-separable) N-body Hamiltonian system [even a numerical integration cannot reliably follow certain orbits for very long].[8,5,26] Variational methods were equally unsuccessful in utilizing Hamilton's famous (variational) Principle. This is due to the fact that the stationary point of the action S is a *Saddlepoint* of an unknown type, under variations of the solution $\vec{x}(t)$, given the familiar initial conditions $\vec{x}(0)$, $\dot{\vec{x}}(0)$. Thus one cannot simply minimize S since for certain, a-priori unknown, variations it should have been maximized. Here we identify this class of variations for which the action S has a maximum [as well as the remaining one for which it has a minimum]. It will be shown that the solution can even be characterized by the type of its action saddlepoint [requiring 2N numbers for its definition also]. So, *we abandon the familiar initial value problem and, instead, specify the type of the action saddlepoint.* Hence it becomes feasible to devise a *convergent* variational procedure, à la Rayleigh-Ritz, since we now know where to maximize S and where to minimize it. We concentrate on periodic solutions of arbitrary period which in "most" Hamiltonian systems, are dense among all bounded solutions. The emphasis is on constructive methods which can be evaluated explicitly and rapidly [∿ "quadratic"-convergence].

CONTENTS

ISSN: 0094-243X/78/264/$1.50

1. INTRODUCTION

"... *it is easy to conjecture that the significance of the analogous variational principles of dynamics is largely formal.*"

G. D. BIRKHOFF[14]

Variational methods are employed successfuly throughout quantum mechanics, hydrodynamics, etc. to *construct* solutions of particularly difficult equations. Yet in classical mechanics variational constructions are seldom used to solve a *system* of nonlinearly interacting particles even though its dynamics is *derived* from a celebrated variational principle, Hamilton's Principle[13]. Moreover, in classical mechanics there is indeed a large supply of particularly difficult equations from the classic three-body problem of celestial mechanics[6,19,13] to the "non-integrable" N body problems of nonequilibrium statistical mechanics (where $N \simeq 10^{23}$).[26,8,7] There is no lack of "relevant" applications of classical mechanics either: to plasma physics, accelerators, satellites, chemistry, statistical mechanics, etc. So what is keeping us? Or is Hamilton's Principle a largely formal principle indeed! Below we describe the stumbling blocks and derive a method for overcoming them. This method converges rapidly (∿ "quadratically") even at extremely high nonlinearity in non-integrable systems and yields explicit local solutions "nearly everywhere". The essential ideas are discussed in sections 1.1 and 2.5.

Hamilton's Principle states that the action S is stationary, $\delta S = 0$, under first order variations away from the exact solution of the equations of motion. Thus "all" one has to do is find the "roots" of $\delta S = 0$ ... . The problem lies in the second order terms: *The stationary point of the action* S *is a saddlepoint, in general*[13,12,2,1], i.e. the second order change ΔS is *negative* for one class of variations but positive for all others. So one cannot simply minimize S, à la Rayleigh-Ritz[3,4], in order to obtain a solution. Moreover this class of variations (yielding $\Delta S < 0$) changes *drastically* going from one orbit to another, arbitrarily close to the first orbit. Hence the negative $-\Delta S$ class is not only unknown "a-priori", but changes *discontinuously* as one varies the trial solution!

Not even the "mini-max" - or more sophisticated variational constructions[12,4,3] can deal with such pathologies. Whence the dearth of variational (and other) solutions.

We find in sections 2.1-2.5 that this negative ΔS class consists of the variations of *all Fourier* - (or Stürm-Liouville) *coefficients "below" a certain frequency* ν_k, in the Fourier - (or St. L.) decomposition of the solution $x_k(t)$ [k = 1, ..., N]. The action saddlepoint and the solution $\vec{x}(t)$ are in fact *characterized* by these basis frequencies ν_1, ..., ν_N, cf. section 2.5 [plus N phase angles, δ, irrelevant at this stage]. On the other hand a solution is usually characterized by its initial conditions $\vec{x}(0)$, $\dot{\vec{x}}(0)$. However in this paper we specify the basis frequencies ν [and the δ's] instead of the "initial conditions". This has the important advantage of *fixing* the negative ΔS class of variations, i.e. we now know exactly for which Fourier (St. L.-) variables S has a maximum and for which it has a minimum. Moreover the neg. ΔS class can no longer change as we vary the trial solution and is known beforehand... . Fixing the neg. ΔS class enables us to construct a convergent variational method à la Rayleigh-Ritz, outlined below in section 1.1. The essential point of this paper therefore is that we abandon the familiar initial value problem and replace it with a specification of the "type and direction" of the action saddlepoint, cf. section 2.5. The existence of such saddlepoints, and their importance, was revealed by Jacobi:

> *"Diesen Satz, der übrigens für die Mechanik im engeren Sinne von gar keiner Wichtigkeit ist, habe ich im Crelleschen Journal bekannt gemacht."*
>
> C. G. J. JACOBI[32]

1.1 Construction, Sketch

To implement these ideas we decompose $\vec{x}(t)$ on a set of Fourier (or St. L.-) basis functions reducing the variational problem as usual to a *purely algebraic* one for the Fourier (or St. L.-) coefficients c_n and the action $\bar{S}(c)$, cf. section 2.6. Hamilton's Principle yields the set of coupled nonlinear (nonseparable[26]) algebraic equations,

$$0 = \partial\bar{S}/\partial c_n^*\Big|_{\hat{c}} \quad , \; n = 0, 1, 2, \ldots, \tag{1.1}$$

vanishing at the exact solution vector $\hat{c}$, but not at a trial vector c, in general. If this were a scalar root-finding problem Newton's method would be preferred[30,31]. In Newton's method one uses the derivative of (1.1) to calculate a correction Δc (bringing $c + \Delta c$ closer to the root $\hat{c}$) from:

$$\lambda_n(c)\Delta c_n = -\partial\bar{S}/\partial c_n^*\Big|_{c} \; , \qquad (n\neq m) \tag{1.2}$$

with

$$\lambda_n(c) \equiv \partial^2\bar{S}/\partial c_n \partial c_n^*\Big|_{c} \qquad (\lambda_0 \leqslant \lambda_1 \leqslant \lambda_2 \;\; \ldots) \tag{1.3}$$

The same holds here since the variables c (can be transformed to) diagonalize the matrix of second derivatives, at c [denumbered in order of increasing λ]. One solves for the correction Δc, obtains a new trial vector $c + \Delta c$, etc. In general any addition of the form $g(c)\Delta c$ can be allowed on the lhs of (1.1) as long as the iterations converge, since in that case $\Delta c \to 0$. The corresponding change in the action is therefore

$$\Delta\bar{S}_{(n)} \approx \partial\bar{S}/\partial c_n \, \Delta c_n = -|\partial\bar{S}/\partial c_n|^2/\lambda_n(c), \qquad (1.4)$$

per iteration. The above iterations would converge extremely rapidly ("quadratically") *if the starting vector* c *were "close enough" to the exact solution* $\hat{c}$ [30,31]. It is here that the classic problems of dynamics arise: Since $\bar{S}$ has a saddlepoint the λ's (1.3), being curvatures, change their sign, e.g. at n=m,

$$\lambda_n(\hat{c}) \begin{cases} < 0 \text{ for } n < m \\ = 0 \text{ for } n = m \text{ (section 2.5)} \\ > 0 \text{ for } n > m \end{cases} \qquad (1.5)$$

cf. sections 2.2-2.5. If these λ values, at the *exact* solution, were known beforehand and used in (1.2)-(1.4) one would be guaranteed:

$$\Delta\bar{S}_{(n)} \begin{cases} > 0 \text{ for } n < m \\ < 0 \text{ for } n > m \end{cases} \qquad (1.6)$$

[n = m is treated differently, cf. (1.8)]. In that imagined case the convergence of $\bar{S}$ to a maximum for n < m, and to a minimum for n > m, would be as simple as in Newton's scalar method. However the actual "levels" $\lambda_n(c)$ do change at each iteration and some may easily have the wrong sign compared to (1.5). Hence some of the Δc_n and $\Delta\bar{S}_{(n)}$ acquire the wrong sign, in (1.2), (1.4), and the new c diverges away from $\hat{c}$. We call this the "*Level Crossing Problem*" [after a similarly severe problem in the quantum mechanics of coupled oscillators]. In addition there are several levels with very small λ [for $\hat{c}$: at n = (m,) m ± 1, ..., cf. (1.5)] where a tiny error in c causes a large *relative* error in that λ and an enormous "correction" in its Δc and $\Delta\bar{S}$ according to (1.2)-(1.4). This is known as the "*Small Denominator Problem*"[6,5,10]. In the literature this name usually covers all divergence problems even though Moser points out that resolving the "smallness" problem itself does not guarantee convergence[5]. Both problems become more pathological the closer the levels λ are to each other [they come arbitrarily close if the solution is quasi-periodic]. As a result divergence is the rule rather than the exception[5,6,26,8] in non-integrable systems. Countless elegant but formal iterative-, perturbative-, series-, and variational schemes are invented each year. However the upshot of theorems due to Siegel and Poincaré is that, at best, they can only have a *zero* radius of convergence when applied to "most" Hamiltonian systems[26,5,6]. Our solutions, below, converge locally only, i.e. each new solution is good for one orbit only. But Poincaré and Siegel tell us that, in general, it is the best one can hope for.

Our remedy is to specify the index n = m at which the "levels" (1.5) change sign instead of specifying $\vec{x}(0)$, $\dot{\vec{x}}(0)$ as usual [the corresponding eigenvectors e_m, e_m^*, needed to diagonalize (1.3) require the N basis frequencies ν and the N δ's mentioned above for their specification, cf. section 2.5]. Hence we have a constraint $\lambda_m(c) = 0$ (1.5) on each trial vector instead of the familiar constraints imposed on c by the initial conditions $\vec{x}(0)$, $\dot{\vec{x}}(0)$, cf. section 2.6. At the solution $\hat{c}$ we have, for n = m,

$$\partial\overline{S}/\partial c_m^*\Big|_{\hat{c}} = 0 \qquad (1.7.a)$$

If we made an infinitesimal change to $\hat{c} + \delta c$ under our constraint (on δc, since $\hat{c}$ already satisfies it) we must again have

$$\partial\overline{S}/\partial c_m^*\Big|_{\hat{c}+\delta c} \approx 0 + \lambda_m(\hat{c})\ \delta c_m = 0 \qquad (1.7.b)$$

since $\partial^2\overline{S}$ is diagonal, cf. (1.3) and (1.5). We could ("analytically") continue forever adding more δc's since our new c's are all chosen to diagonalize $\partial^2\overline{S}$ and satisfy the constraint $\lambda_m(c) = 0$ [degeneracy is discussed in section 2.5]. Thus the "integrated" version is a *new constraint*

$$\partial\overline{S}/\partial c_m^*\Big|_{c} = 0, \qquad (1.8)$$

to be satisfied by each trial vector c, instead of (satisfying) the usual initial conditions [eq. (1.8) can be solved *linearly* for a scaling factor[10,16] or $|\Delta c_m|^2$, etc.]. If we then choose a starting vector c which satisfies (1.8) and proceeded with *infinitesimal* fractions of the Δc in (1.2) we would be guaranteed convergence! In practice we prefer to "initially" have larger corrections Δc than infinitesimal ones ..., and the constraint (1.8) may no longer yield $\lambda_m(c) = 0$ exactly (but small). Therefore we could "initially", i.e. during the first few iterations of (1.2), (1.8), still have some of the Problems mentioned before. They are avoided by modifying ["relocating"] Newton's Method (1.2) into:

$$[\lambda_n(c) - \lambda_m(c)]\, \Delta c_n = -\partial\overline{S}/\partial c_n^*\Big|_c , \qquad (1.9)$$

at $n \neq m$, plus (1.8) to obtain Δc_m. These deceptively simple remedies (1.8), (1.9) only make sense of course if one knows *beforehand* that the exact levels λ change sign at $n = m$ as we do by (our) definition of the orbit. Now the $\Delta\overline{S}_{(n)}$ produced by (1.9), (1.4) all have the correct sign (1.6) by definition, and the "Level Crossing Problem" is eliminated. As we iterate (1.9), (1.8) the $\overline{S}$ increases monotonically with Δc_n for $n < m$ and decreases monotonically for $n > m$, i.e. *the $\overline{S}$ approaches its saddlepoint monotonically* ...! In section 3.2 we show that the iterations (1.9), (1.8) converge. As the corrections Δc diminish we necessarily have: $\lambda_m(c) \to 0$, from (1.8), and our method (1.9) reduces to Newton's method (1.2) [the convergence is "quadratic" asymptotically].

There may still be an initial "Small Denominator Problem" in (1.9) at $n = m \pm 1$. It is not a serious problem and we merely avoid it here by considering a solution only over (arbitrarily) long but finite time intervals [one can show that this puts an "a-priori" lower bound[20-23] on the magnitude of the denominators arising at $n = m \pm 1$]. In particular it enables us to obtain periodic solutions of arbitrary (long) period. We confine ourselves, from section 2.5 on, to such (not necessarily simple[29]) periodic orbits, of which there are infinitely many: Already near every "stable" periodic orbit there are infinitely many other periodic orbits[8,5,14] with vastly different periods. In fact the periodic orbits are "*dense*" among the bounded solutions of "most" Hamiltonian systems [Pugh] and play a fundamental role in the study of such systems[13-29]. For the well behaved "integrable" systems the latter is not too surprising[26,8,5]. However for the more "stochastic" non-integrable systems[26,8] the role of the (dense) periodic solutions might not seem so obvious. Yet is was Poincaré, and others[19], who emphasized that "the periodic solutions might well be the only breach through which such (non-integrable) systems could be penetrated".

2. THE ACTION IS A SADDLEPOINT, NOT AN EXTREMUM

In this section the truth of the above title, under variations in $\vec{x}(t)$, is demonstrated for two elementary examples as well as for the Duffing equation and the Hénon-Heiles system[5,8,10]. The "Stürm-Liouville" arguments employed in sections 2.5 and 2.6 will be essential for section 3. No knowledge of this theory is required beyond what is commonly used in wave mechanics, i.e. the existence of eigenfunctions for the Schrödinger Equation.

The stationary point of the action $S(\tau)$ (2.1) is a minimum under variations in the functions $\vec{x}(t)$, if the time τ in $S(\tau)$ is short enough[1,2]. However for longer times τ, we shall see that, the stationary point becomes a saddlepoint, i.e. it is a maximum under certain variations and a minimum under other ones[13]. In the limit as

$\tau\to\infty$ [or τ = the period, if the solution is periodic] the action $\overline{S}$ still is a function of the Fourier-coefficients and-frequencies in the Fourier decomposition of the $x_k(t)$, for example. Equivalently, we find this $\overline{S}$ to have a minimum under variations in all coefficients whose frequency is larger than some particular ν_k and a maximum for those "below ν_k".

Knowing this and the values of the ν_k's it becomes feasible to devise a *convergent* variational procedure, à la Rayleigh-Ritz,[3,4] for obtaining (the Fourier decomposition of) the solution $\vec{x}(t)$ since we know then where to minimize $\overline{S}$ and where to maximize it, in section 3.

2.1 The Action is Not a Minimum After Some Time; Jacobi's Example

An elegant example, due to Jacobi[1,2] is that of a particle moving at *constant* velocity v along the surface of a sphere. *Its action*

$$S(\tau) \equiv \int_0^\tau \mathcal{L}(\vec{x}(t), \dot{\vec{x}}(t))\, dt = \tfrac{1}{2}v^2\tau = \tfrac{1}{2}vs \ , \qquad (2.1)$$

is proportional to the distance s travelled. The particle completes a great circle in some period T and returns to its original position. For $\tau < \frac{1}{2}T$ the actual orbit, along a great circle, is the shortest path between two points. So $S(\tau)$ has a local minimum under all variations of the orbit if $\tau < \frac{1}{2}T$. On the other hand when $\tau > \frac{1}{2}T$ there are many nearby paths on the sphere *shorter* than the actual orbit [e.g. the actual orbit goes from the north pole via the south pole to some further point Q; a shorter nearby path is obtained by taking another great circle nearby but making a short-cut to Q just before arriving at the south pole. Since the previous variations are still allowed there are *longer* paths also]. Hence $S(\tau)$ has a local maximum under some variations, and a minimum under other ones, if $\tau > \frac{1}{2}T$, i.e. *a saddle - "point" in general*.

Similar arguments could be made for an ellipsoid with main periods T_1 and T_2 in the y- and x-directions. The $S(\tau)$ then has a saddlepoint under variations in x if τ exceeds some fraction of T_1 and under variations in y if τ exceeds another fraction of T_2.

There exist a number of sufficient and/or necessary criteria for the $S(\tau)$ to remain a minimum over a short time τ. These are due to Jacobi, Legendre and Weierstrass and can be evaluated for some simple cases [12,3,4]. In such cases one can employ the more usual variational and optimization methods to minimize $S(\tau)$ everywhere and obtain the solution $\dot{x}(t)$ over these short (!) times. Since we want to obtain the solution for all times, e.g. over the full period $\tau = T$ and even for $T\to\infty$, we have developed other techniques.

2.2 The Action is Not a Minimum for the Lowest Fourier Coefficients

First we demonstrate this on a simple driven linear oscillator. The above properties of S as a function of the time τ are of course reflected in the properties of $\overline{S}$ as a function of the Fourier coefficients of x(t). Consider

$$\ddot{x} = -\omega_2^2 x \quad + b_1 \cos \omega_1 t + b_3 \cos \omega_3 t \,, \qquad (2.2)$$

with

$$0 < \omega_1 < \omega_2 < \omega_3 \,, \qquad (2.3)$$

a simple harmonic oscillator driven below and above its harmonic frequency ω_2. Its solution can always be expressed as

$$x(t) = \Sigma_{n=1}^{3} A_n \exp(i\omega_n t) + A_{-n} \exp(-i\omega_n t) \qquad (2.4)$$

The action $\overline{S}$ is easily obtained from the Lagrangian of (2.2), with (2.4),

$$\overline{S} \equiv \lim_{\tau\to\infty} \frac{S(\tau)}{\tau} = \sum_{n=1}^{3} [\omega_n^2-\omega_2^2]A_n A_{-n} + \tfrac{1}{2}b_1(A_1+A_{-1}) + \tfrac{1}{2}b_3(A_3+A_{-3}) \qquad (2.5)$$

This $\overline{S}$ is no longer a function of the time and the variational problem has become a *purely algebraic* one. At the solution of (2.2) the first partials vanish according to Hamilton's Principle

$$0 = \partial\overline{S}/\partial A_{-n} = [\omega_n^2-\omega_2^2]A_n + \tfrac{1}{2}(b_1\delta_{n,\pm 1} + b_3\delta_{n,\pm 3}) \,, \qquad (2.6)$$

for $n = \pm 1, \pm 2, \pm 3$, which is of course the same as taking the ω_n-Fourier component of the equation of motion (2.2). The second variation can now be calculated also:

$$\partial^2\overline{S}/\partial A_{-n}\partial A_p = \delta_{n,p}[\omega_n^2-\omega_2^2] \qquad (2.7)$$

Therefore the type of the stationary point is:

$$\overline{S}(\ldots,A_n,\ldots) \textit{ has a local} \begin{Bmatrix} \textit{minimum, for } n > 2 \\ \textit{maximum, for } n < 2 \end{Bmatrix} \,, \qquad (2.8)$$

under variations in A_n, with $n > 0$. *Hence* $\overline{S}$ *has a saddlepoint,* with relative maxima for all the Fourier terms "below" the harmonic one, and minima "above" the harmonic term.

2.3. The Action Saddlepoint for a Driven Nonlinear Oscillator

In this section and the next one the existence of an action saddlepoint is demonstrated for some *nonlinear* systems. The methods used are essential for later arguments.

If we add a cubic term to a driven linear oscillator we obtain a form of Duffing's Equation[10]

$$\ddot{x} = -\omega^2 x + Rx^3 + B \cos \nu_2 t \text{ , with} \tag{2.09}$$

$$\mathcal{L} = \tfrac{1}{2}\dot{x}^2 - \tfrac{1}{2}\omega^2 x^2 + Rx^4/4 + xB \cos \nu_2 t \text{ ,} \tag{2.10}$$

and $S(\tau)$ defined as usual (2.1). Consider a variation $\psi(t)$ about an exact solution $\hat{x}(t)$ of (2.09):

$$x(t) \equiv \hat{x}(t) + \psi(t) \text{ ,} \tag{2.11}$$

satisfying the same boundary conditions, at $t = 0$ and $t = \tau$, as $\hat{x}(t)$ itself[3,4]. Substitution of this new function in (2.09) yields, to first order in ψ,

$$L\psi \equiv - d^2\psi/dt^2 - (\omega^2 - 3R\hat{x}^2(t))\, \psi \text{ ,} \tag{2.12}$$

$\neq 0$, in general. As in the Schrödinger Equation we use the existence of a complete set of eigenfunctions ψ_n and eigenvalues λ_n, with

$$(L\psi_n =)\ -\ddot{\psi}_n - (\omega^2 - 3R\hat{x}^2(t))\psi_n = \lambda_n \psi_n \text{ , } n=0,1,2\ldots, \tag{2.13}$$

demonstrated in Stürm-Liouville theory[3,4], for $\hat{x}(t)$ bounded on $(0,\tau)$. The eigenvalues are ordered,

$$\lambda_0 \leqslant \lambda_1 \leqslant \ldots \leqslant \lambda_n \leqslant \ldots \leqslant \infty \tag{2.14}$$

[and have no clusterpoints if $\tau < \infty$]. Moreover:

$$\lim_{n\to\infty} \lambda_n \tau^2/n^2\pi^2 = 1 \text{ ,} \tag{2.15}$$

and $\psi_n(t)$ has precisely n zero's within $[0,\tau)$[3,4,17], whence $\lambda_n > 0$ for large enough n. The eigenfunctions can be characterized uniquely[17,3,4] by the number of zero's. They are orthonormal under the usual internal product

$$\int_0^\tau dt\ \psi_k \psi_n = \delta_{k,n} \text{ ,} \tag{2.16}$$

and the $\psi(t)$ of (2.11) can be expanded[3,4] as:

$$\psi(t) = \Sigma_{n=0}^{\infty} d_n \psi_n(t) \tag{2.17}$$

If we evaluate the action S using the function x(t), instead of the exact solution $\hat{x}(t)$, we can obtain the change in the action as

$$\Delta S(\tau) \equiv S\Big|_{\hat{x}+\psi} - S\Big|_{\hat{x}} \approx \int_0^\tau dt\, [\tfrac{1}{2}\dot{\psi}^2 - \tfrac{1}{2}\omega^2\psi^2 + R\binom{4}{2}\hat{x}^2\psi^2/4] \tag{2.18}$$

$$= \tfrac{1}{2}\int_0^\tau dt\, [-\psi\ddot{\psi} - \omega^2\psi^2 + 3R\hat{x}^2\psi^2] = \tfrac{1}{2}\int_0^\tau dt\, \psi L\psi = \tag{2.19}$$

$$= \tfrac{1}{2}\,\Sigma_{n=0}^{\infty} \lambda_n d_n^2\,, \tag{2.20}$$

to second order in ψ, where we used (2.13), (2.16) and (2.17) and the vanishing of the partial term since $\hat{x}$ and x(t) satisfy the same boundary conditions. The first order terms vanish of course [Hamilton's Principle]. In (2.20) *we have in fact "diagonalized"* S about its stationary point. Hence the type of the stationary point follows from the signs of the λ_n. From (2.18) - (2.20) we see that *the lowest eigenvalue* λ_0 *can be characterized variationally as*[3,4]

$$\lambda_0(\tau) = \mathrm{MIN}_\psi \left\{ [\int_0^\tau dt\, \psi L\psi]/\int_0^\tau dt\psi^2 \right\} =$$

$$= \mathrm{MIN}_\psi \left\{ [\int_0^\tau dt\, \dot{\psi}^2 - (\omega^2 - 3R\hat{x}^2(t))\psi^2]/\int_0^\tau dt\psi^2 \right\}, \tag{2.21}$$

where the minimum is over all functions $\psi(t)$ such that x(t) (2.11) has the same boundary conditions as $\hat{x}(t)$[3,4]. If there exist bounded solutions $\hat{x}(t)$ with $3R\hat{x}^2(t) > \omega^2$ for *all* t we clearly have $\lambda_0 > 0$ always. Otherwise we take a trial function which is nearly constant over most of the interval $(0,\tau)$ except for a matching to the boundary conditions. Then, as we let $\tau \to \infty$

$$\lambda_0(\infty) \leqslant -\omega^2 + 3R\overline{\hat{x}^2}^{\,\tau\to\infty} \leqslant 0 \tag{2.22}$$

[leaving out from (2.21) all positive terms we also see that $\lambda_0 \geqslant -\omega^2$, for $R > 0$]. *Hence* S *certainly has a saddlepoint* under variations in all solutions with $3R\overline{\hat{x}^2} < \omega^2$.

2.4. The Action Saddlepoint for the Hénon-Heiles System

Here we show that the action for a system of *two* coupled nonlinear oscillators has a distinct saddlepoint under variations in the solution, with the aid of standard Stürm-Liouville results, as in Sec. 2.3. Consider the well known Hénon-Heiles system

$$\ddot{x} = -x - 2xy \tag{2.23}$$

$$\ddot{y} = -y + y^2 - x^2 , \tag{2.24}$$

a non-integrable system, from most indications[5,27,8]. The Lagrangian is

$$\mathcal{L} = \tfrac{1}{2}(\dot{x}^2 + \dot{y}^2) - \tfrac{1}{2}(x^2+y^2) - x^2y + y^3/3 \tag{2.25}$$

Again we vary the exact solution, here of (2.23) - (2.24), to

$$x(t) \equiv \hat{x}(t) + \psi(t) \text{ and } y(t) \equiv \hat{y}(t) + \phi(t) \tag{2.26}$$

Substitution in (2.23) - (2.24) yields, to first order in ψ and ϕ, the eigenvalue problem

$$L\begin{pmatrix}\psi_n\\ \phi_n\end{pmatrix} \equiv \begin{pmatrix}-\ddot{\psi}_n - (1+2\hat{y})\psi_n - 2\hat{x}\phi_n\\ -\ddot{\phi}_n - (1-2\hat{y})\phi_n - 2\hat{x}\psi_n\end{pmatrix} = \lambda_n\begin{pmatrix}\psi_n\\ \phi_n\end{pmatrix}, \tag{2.27}$$

analogous to (2.12) - (2.15) [the inner product $\{\vec{\psi}\cdot L\vec{\psi}\}$ cf. (2.32), with $\vec{\psi}^T \equiv (\psi,\phi)$, has the required lower bound[3,4] if $\hat{x}(t)$ and $\hat{y}(t)$ are bounded]. Upper bounds for λ_0 can be calculated as in (2.21):

$$\lambda_0(\tau) = \mathrm{MIN}_{\psi,\phi}\left\{\frac{\int_0^\tau dt(\dot{\psi}^2+\dot{\phi}^2)-\psi^2(1+2\hat{y})-\phi^2(1-2\hat{y})-4\hat{x}\psi\phi}{\int_0^\tau dt\ (\psi^2+\phi^2)}\right\} \leqslant \tag{2.28}$$

$$\leqslant \mathrm{MIN}_{\pm}\,\mathrm{MIN}_z\left\{\left[\int_0^\tau dt\ \dot{z}^2 - z^2(1\pm2\hat{y})\right]\Big/\int_0^\tau dt\ z^2\right\}, \tag{2.29}$$

which may be checked taking $(\psi,0)$ and $(0,\phi)$ in (2.28). The latter can be estimated as in (2.21) - (2.22)

$$\lambda_0(\infty) \leqslant -(1 + 2|\hat{\bar{y}}|) < 0 \tag{2.30}$$

Expanding ψ,ϕ in eigenfunctions,

$$\vec{\psi} \equiv \begin{pmatrix} \psi \\ \\ \phi \end{pmatrix} = \Sigma_{n=0}^{\infty} d_n \begin{pmatrix} \psi_n \\ \\ \phi_n \end{pmatrix}, \qquad (2.31)$$

with (ψ_n, ϕ_n) normalized under the (scalar) inner product (2.16), the resulting change in the action S is found, as in (2.18) - (2.20) to second order in ψ, ϕ,

$$\Delta S(\tau) \equiv \tfrac{1}{2} \{\vec{\psi}\cdot L\vec{\psi}\} \equiv \tfrac{1}{2} \int_0^{\tau} dt\ [\vec{\psi}(t)\cdot L(t)\vec{\psi}(t)] = \tfrac{1}{2} \Sigma_n \lambda_n d_n^2 \qquad (2.32)$$

From (2.30) we see that the lowest λ's are negative while the higher ones are positive, cf. (2.15). Formula (2.32) therefore demonstrates that the action S of this system has a saddlepoint under variations.

The index n again denumbers the eigenvalues in increasing order. However the number of zero's of $\psi_n(t)$ or $\phi_n(t)$ need no longer be equal to n, as it was in the scalar case, [e.g. consider a system of two uncoupled equations whose two sets of λ's interleaf]. Thus the asymptotic estimate (2.15) is no longer valid in this vector case. The actual number of zero's in ψ_n and ϕ_n becomes important in the next section and in the process we obtain estimates (2.40) - (2.43) for the present λ's.

2.5 Characterizing Saddlepoints by their Basis Frequencies, ν_k

Of particular interest are the "<u>p</u>eriodic <u>b</u>oundary <u>c</u>onditions"[11,26]:

$$x(\tau) = x(-\tau) \quad , \quad \dot{x}(\tau) = \dot{x}(-\tau) \ , \qquad (2.33)$$

i.e. we only consider those functions x, $\hat{x}$, ψ, y, $\hat{y}$, ϕ which have some (minimal) period 2τ. For "most" Hamiltonian systems such closed orbits are *dense* among all bounded orbits [Pugh], whence their importance. In this p.b.c. case some of the variational eigenvalues, λ, do remain simple but many become degenerate[22,13,17-24]. Yet a *complete* set of independent eigenfunctions exists in general, cf. (2.50). In order to simplify the presentation we confine ourselves here to solutions $\hat{x}$, $\hat{y}$ which are even functions of t. This will be generalized in refs. 11, 15 and elsewhere.

In this p.b.c. case one knows at least one exact eigenvector $\vec{\psi}(t)$: for example in the Hénon-Heiles system it is constructed by differentiating a solution of (2.23), (2.24), with respect to t:

$$d^2\dot{\hat{x}}/dt^2 = -\dot{\hat{x}}(1+2\hat{y}) - 2\hat{x}\dot{\hat{y}}$$
$$d^2\dot{\hat{y}}/dt^2 = -\dot{\hat{y}}(1-2\hat{y}) - 2\hat{x}\dot{\hat{x}} \qquad (2.34)$$

Comparison with the variational equations (2.27) shows that

$$\psi_m = \dot{\hat{x}} \quad , \phi_m = \dot{\hat{y}} \quad , \text{ with } \lambda_m = 0 , \qquad (2.35)$$

are the components of an eigenvector $\vec{\psi}_m$ of (2.27), with zero eigenvalue[13,19]; apart from a normalization constant. Clearly the $\vec{\psi}_m$ satisfies periodic boundary conditions (2.33) if the exact solution does, cf. (2.23), (2.34). Since we consider an even $\hat{x}(t)$ this eigenvector is an odd function of t. We note that

> *An (even) periodic solution of a nonintegrable autonomous system is completely determined by the (odd) eigenvector* $\vec{\psi}_m(t)$, *with* $\lambda_m = 0$, *of the variational equations,* (2.36)

up to a possible degeneracy, due to some symmetry in the equations of motion: If this $\vec{\psi}_m$ were available one would recover the exact solution $\hat{x}(t)$, $\hat{y}(t)$ by integrating $\vec{\psi}_m(t)$ and solving for the integration constants $\hat{x}(0)$, $\hat{y}(0)$ from (2.34) or (2.23), (2.24). If these equations allow several $\hat{x}(0)$, $\hat{y}(0)$ values we find several exact solutions $\hat{x}(t)$, $\hat{y}(t)$ at once [e.g. eq. (2.12) allows $-x_0$ as well as x_0]. We characterize the eigenvector by the number of zero's in each component, i.e. by *the important integers:*

$$m_1 \equiv \text{"the number of zero's of } \dot{\hat{x}}(t) = \psi_m(t)\text{"}, \qquad (2.37)$$
$$m_2 \equiv \text{"the number of zero's of } \dot{\hat{y}}(t) = \phi_m(t)\text{"},$$

within $[0,\tau)$. The integers m_1, m_2 *are relative primes* since 2τ is the minimal period of the exact solution [the eigenvalue $\lambda_m = 0$ can be doubly degenerate itself but the two eigenfunctions will have the same numbers of zero's m_1, m_2, cf. (2.50) and refs. 20-23]. *In this paper the exact periodic solution of the equations of motion (2.23), (2.24), will itself be characterized by specifying the* m_1, m_2 *and* τ, rather than by the familiar initial-displacements and -velocities. Thus the (even) periodic solutions of (2.23), (2.24) are determined from their period 2τ and the number of zero's in their velocities [or, in an eigenvector with $\lambda = 0$]. Again, we may be specifying several ("symmetrical") solutions "at once" thanks to the several possible $\hat{x}(0)$, $\hat{y}(0)$ values, mentioned above .

While this is reminiscent of the "inverse (scattering) approach"[20-24] we shall call it a "backward approach"[10,9] [before someone else does]. Since we are dealing with periodic functions of period 2τ the two functions $\hat{x}(t)$, $\hat{y}(t)$ *may just as well be characterized by the two basis frequencies,*

$$\left.\begin{aligned} \nu_1 &\equiv m_1 \nu_r \qquad (2.38) \\ \nu_2 &\equiv m_2 \nu_r \qquad (2.39) \end{aligned}\right\}, \text{ with } \nu_r \equiv 2\pi/2\tau ,$$

[sin (t $n\pi/\tau$) and cos (t $n\pi/\tau$) have precisely n zero's within $[0,\tau)$]. The ν_r is the (Poincaré-) "Recurrence Frequency", the greatest common divisior of ν_1 and ν_2 since m_1 and m_2 are relative primes. From (2.37) we find that ν_1 is the frequency of the main Fourier component of $\dot{\hat{x}}(t)$ [hence a major component of $\hat{x}(t)$ itself] and ν_2 is the "main frequency" of $\dot{\hat{y}}(t)$.

Approximations for the λ's of the present vector-problem may be obtained by "uncoupling" the variational equations, e.g. by taking $(\psi,0)$ and $(0,\phi)$ in (2.27) as we did to obtain (2.30). The spectrum then separates into x- and y-parts, λ'_n and λ''_n. An estimate for the Hénon-Heiles system is

$$\lambda'_n \sim n^2\nu_r^2 + \lambda'_0\Big|_{\text{var. est.}} \approx n^2\nu_r^2 - (1+2\bar{y}) \tag{2.40}$$

$$\lambda''_n \sim n^2\nu_r^2 + \lambda''_0\Big|_{\text{var. est.}} \approx n^2\nu_r^2 - (1-2\bar{y}) \ , \tag{2.41}$$

using the + and - cases of (2.29). In general we do not know the $\bar{y}$ value but in our methods the m_1, m_2, ν_r are specified so we also have "a-priori" estimates

$$\lambda'_n \sim (n^2-m_1^2)\nu_r^2 \tag{2.42}$$

$$\lambda''_n \sim (n^2-m_2^2)\nu_r^2 \tag{2.43}$$

The eigenvalues of the vector problem are approximated by the union of (2.40) with (2.41) or (2.42) with (2.43)[20-23].

The eigenvalue $\lambda = 0$ can be more than doubly-degenerate when there exist additional (differentiable) integrals of the motion[19] [for example take a system consisting of N *uncoupled* Hamiltonian equations]. If there exists an additional (differentiable) integral one can reduce the number of degrees of freedom to again obtain a system with only one such integral[19,13]. However that procedure can be inconvenient and we shall, elsewhere, discuss a variant of our methods suitable for such "partially integrable" systems. If there are N such (independent) integrals the problem can be reduced to a scalar one. Hence this (completely) "integrable" problem can be solved by standard methods[26,5]. In this "integrable" case our "local" basis frequencies ν_k (2.39) equal the "global" ω_k in the "angle"-variables[26,7,8]. Here we take it that all reductions have been performed and that the remaining Hamiltonian system has no other (differentiable) integrals[26,8,5], as is the case with (2.23), (2.24), and is "completely non-integrable".

The Duffing equation (2.09) has an explicit time dependence and differentiation does *not* yield any eigenfunction, unlike (2.34). In general, a non-autonomous equation need not have one eigenvalue equal to zero, under the above variational approach [cf. ref. 13 for different variational principles]. However all we truly need in section 3 is the knowledge that two subsequent eigenvalues have different signs, established in section 2.3. A version of the method of section 3 is successfully applied to the Duffing equation in ref. 10, and is extended in ref. 11. One might embed the Duffing equation in an autonomous system, e.g.

$$\ddot{x} = -\omega^2 x + Rx^3 + y \qquad \ddot{y} = -\nu_2^2 y \, , \tag{2.44}$$

recovering (2.09) for $y(0) = B$, $\dot{y}(0) = 0$. However this also introduces an integral [the Hamiltonian of the linear equation] plus some degeneracies as mentioned before.

The *existence* of such stationary ("critical") points (1.1) in the action can be demonstrated for certain types of periodic solutions, also by less direct but more "global" methods[28,29]. This shows the *existence* of certain classes of "simple" periodic solutions "near" an equilibrium point of the potential, as Alan Weinstein has told us[29]. These (simple) periods are close to the ones of the harmonic approximation.

2.6 The Equivalent Algebraic Saddlepoint

Since we do not know the eigenfunctions beforehand, we choose some suitable orthogonal basis and diagonalize $\Delta\bar{S}$ to obtain the λ's, cf. (2.20), (2.32). In the periodic boundary case (2.33) we take the *whole* trial vector expanded

$$x(t) = \sum_{n=-\infty}^{\infty} A_n e^{in\nu_r t} \qquad y(t) = \sum B_n e^{in\nu_r t} \, , \tag{2.45}$$

cf. (2.38), (2.39), rather than just the deviation $\vec{\psi}$ from the exact solution $\hat{\vec{x}}$ as in (2.26). Often we prefer the more concise notation

$$\vec{x}(t) \equiv \begin{pmatrix} x(t) \\ y(t) \end{pmatrix} = \Sigma_p C_p \vec{f}_p, \text{ with } \vec{f}_{2n} \equiv \begin{pmatrix} e^{in\nu_r t} \\ 0 \end{pmatrix}, \vec{f}_{2n+1} \equiv \begin{pmatrix} 0 \\ e^{in\nu_r t} \end{pmatrix} \tag{2.46}$$

with

$$C_{-p} = C_p^*, \; C_{2n} = A_n \text{ and } C_{2n+1} = B_n \; (n \geqslant 0) \tag{2.47}$$

In order to obtain even x(t), y(t) we take A, B, C real in this paper. One could now evaluate the action, solely in terms of the Fourier coefficients (and τ),

$$\overline{S}(C) \equiv S(2\tau)/2\tau, \tag{2.48}$$

cf. (2.5), (2.25), where C is an infinite vector with components C_p, etc. and where we have integrated out the time dependence. Hamilton's Principle, or a variant of it[13], requires the first partials to vanish at the exact solution $\hat{C}$. Hence at another C we have

$$\partial\overline{S}/\partial C_p^*\Big|_C \neq 0, \qquad p=0,\pm1,\pm2,\ldots, \tag{2.49}$$

in general. The matrix of second derivatives can be evaluated again cf. (2.7), here for arbitrary C, and can be diagonalized

$$L_{k,p} \equiv \partial^2\overline{S}/\partial C_k^*\partial C_p = \Sigma_n\, U_{k,n}\, \lambda_n\, (U^\dagger)_{n,p}\ , \tag{2.50}$$

i.e. the second derivatives matrix L is diagonal when we transform to a new vector c with

$$c \equiv U\Big|_C\ C \tag{2.51}$$

The $\lambda_n(\hat{c})$ are the same as the Stürm-Liouville eigenvalues (2.27) [compare the $\Delta\overline{S}$, obtained from (2.50) with the $\Delta\overline{S}$ in (2.32)]. Of course eqs. (2.49) - (2.51) are more easily obtained by a direct Fourier decomposition of the equations of motion. In practice we truncate the Fourier vectors and "all" that is left to do, in section 3, is to find the exact solution $\hat{c}$ ... .

3. REACHING THE ACTION SADDLEPOINT WITHOUT SLIDING

In this section we locate those solutions of a non-integrable system[26], that satisfy periodic boundary conditions (2.33), through variational methods, i.e. we try to reach the saddlepoint of the action $\overline{S}$. The method used is a form of Newton's method (of root finding), modified to avoid the classic problems of dynamics discussed in section 1.1, i.e. we avoid "sliding off" the saddlepoint. Convergence is shown in section 3.2 .

Consider the trial solution $\vec{x}(t)$ expanded, with coefficients c_n over a (periodic) orthonormal basis on $(-\tau,\tau)$, cf. (2.45)-(2.51). The equations of motion for $\vec{x}(t)$ can then be decomposed into a set of algebraic nonlinear equations for c

$$\partial\overline{S}/\partial c_n^*\Big|_{\hat{c}} = 0 \quad , n=0,\pm1,\pm2,\ \ldots \tag{3.1}$$

cf. (2.6), (2.49), where $\bar{S}$ is the action (2.48) and $\hat{c}$ the coefficient vector of the exact solution of the equations of motion (2.23), (2.24). Equations (3.1) express Hamilton's Principle. We saw in section 2 that this stationary point of $\bar{S}(c)$ is a saddlepoint of unknown type and curvature, i.e. one usually does not know beforehand whether the curvature of $\bar{S}(\ldots, c_n, \ldots)$ is positive or negative (or zero). As discussed in section 1.1 we no longer specify the usual initial conditions. Instead we specify the index n=m at which the curvature of $\bar{S}(\ldots, c_n, \ldots)$ changes from negative to positive [zero curvature at n=m] and the period 2τ of the solution. This entails specifying *two* integers m_1, m_2 (in a *two* degree of freedom problem), cf. (2.37), since the $\bar{S}(\ldots, A_n, \ldots; \ldots)$ changes its curvature near $n=m_1$ while $\bar{S}(\ldots; \ldots, B_n, \ldots)$ does so near $n=m_2$, where the A, B are the (Fourier) coefficients of x(t), y(t), cf. (2.45). We replaced the usual constraints on the (Fourier)coefficients, $\Sigma A = \hat{x}(0)$ and $\Sigma B = \hat{y}(0)$, by the new constraint (1.8). According to (2.37) this constraint, (1.8), translates here into *the new constraints*:

$$\partial\bar{S}/\partial A^*_{m_1}\Big|_C = 0 \quad \text{and} \quad \partial\bar{S}/\partial B^*_{m_2}\Big|_C = 0\ , \tag{3.2}$$

with $C \equiv (A;B)$, cf. (2.47), *to be satisfied by each trial vector* C [these "resonance conditions"[10] (3.2) can usually be solved linearly and simultaneously, by introducing two scaling factors[16,10]].

Specifying m_1, m_2 and τ is equivalent with specifying *two* basis frequencies ν_1, ν_2 (2.38), (2.39). However two second order differential equations require *four* initial constants: We have in fact also specified two phase angles $\delta_1 = \delta_2 = 0$ in the exponentials (2.45) by taking real (Fourier-) coefficients C and thus confining ourselves to even solutions in this paper[11]. From the exact $\hat{C}$ we calculate $\hat{x}(0)$, $\hat{y}(0)$, a "backward approach" if there ever was one...!

3.1 Monotonic Methods of Newton Form

We arrived in section 1.1 at a method of Newton Form to approach the saddlepoint monotonically and eliminate the classic problems involved in solving (3.1). For the first few iterations we prefer a slightly modified version of (1.9)

$$[\lambda_n(c)-\bar{\lambda}(c)]\ \Delta c_n = -\ \partial\bar{S}/\partial c^*_n\Big|_c\ , \quad |n|\neq m, \tag{3.3}$$

with

$$\bar{\lambda}(c) \equiv \tfrac{1}{2}[\lambda_{m-1}(c) + \lambda_{m+1}(c)]\ , \tag{3.4}$$

close to $\lambda_m(c)$ as in (1.9). This extends the region of linear convergence [in practice we soon switch over to λ_m, instead of $\bar{\lambda}$, to take advantage of the resulting (asymptotic) quadratic convergence]. In addition there is the constraint (1.8) for the diagonal variables

$$\partial\bar{S}/\partial c_m^* = 0 \ , \tag{3.5}$$

cf. (3.2), to be satisfied before each iteration of (3.3) [e.g. by solving (linearly) for $|\Delta c_m|^2$, a scalar problem]. We usually start the iterations with the c_m (and one or two more coefficients) as the only nonvanishing ones, and solve their starting values from (3.5). As discussed below (2.39) that means we are by definition starting with the largest (Fourier) coefficients [a sensible idea in any variational approach] no matter how large the nonlinearity. Hence the simple (analytic) starting approximations are usually within 10% of the final results[10,25,9,15,16].

3.2 Convergence

All our methods (3.3), (1.9), (1.2) are of "Newton Form"[30,31],

$$g_n(c)\ \Delta c_n = -\ \partial\bar{S}/\partial c_n^*\Big|_c \ , \tag{3.6}$$

with different choices of g(c). Iterating (3.6) the error after j iterations is easily seen to decrease as[30,31]

$$\text{"error"} = o(\rho^j) \quad , \text{ where } \rho \equiv \text{MAX}_n\ |1 - \lambda_n(\hat{c})/g_n(\hat{c})| \ , \tag{3.7}$$

(with $|n|\neq m$) locally near $\hat{c}$. The spectral radius ρ for our method (3.3), (3.5) is

$$\rho = \text{MAX}_\pm\ |1 - \lambda_{m\pm1}/(\lambda_{m\pm1} - \bar{\lambda})| < 1 \ , \tag{3.8}$$

where $\lambda_{m\pm1}(\hat{c})$ are meant as the nearest nonzero λ's to λ_m [i.e. we ignore degeneracy in our *notation*]. The only thing needed to show (3.8) is $\lambda_{m-1} < 0$ and $\lambda_{m+1} > 0$. Thus the method converges locally already (near $\hat{c}$) in view of (1.5) and (3.7)[30,31]. It is clear that when we switch to λ_m and (1.9), instead of $\bar{\lambda}$ and (3.3), we have $\rho(\hat{c}) = 0$, i.e. quadratic convergence locally[30,31]. Actually in our use of the initial method (3.3) we *always ensure* $\rho(c) < 1$, for all c: We argued below (1.8) that $\lambda_m(c)$ can always be kept as close to zero as one desires by proceeding with small enough corrections Δc. Here we have the far less stringent requirement

$$\lambda_{m-1}(c) < 0 \quad \text{and} \quad \lambda_{m+1}(c) > 0 \ , \tag{3.9}$$

i.e. to "stay in the saddle" while ("analytically") continuing from c to $\hat{c}$ along the path we are constrained to by (3.5). The Δc needed to satisfy (3.9) need *not* be infinitesimal since $\lambda_{m+1} - \lambda_{m-1}$ has a finite lower bound in the case of periodic boundary conditions;[20-23] cf. (2.38) - (2.43). This is trivially accomplished by modifying (3.3) into

$$[\lambda_n(c) - \bar{\lambda}(c)]\,\Delta c_n = - w\partial\bar{S}/\partial c_n^*\Big|_c \ , \text{ with } 0 < w \leqslant 1 \ , \qquad (3.10)$$

plus (3.5) for the case $|n|=m$. In the theory of linear systems the w is known as a "relaxation parameter"[31,9,10]. In the case of periodic boundary condition $w \gtrsim |\bar{\lambda}/\lambda_0| \approx \nu_r^2/1$ suffices, cf. (2.38), (2.43), (2.22) and eq. (3.36) of ref. 9 [a factor w is to be inserted before the first $\lambda_{m\pm1}$ of (3.8]. Whence $\rho(c) < 1$ can always be ensured.

3.3 Practical Short-Cuts

In practice we need not continually transform to diagonal variables c but can stay in the original (Fourier) variables $C \equiv (A;B)$ (2.45). If in doubt how the subsequent short-cuts might alter the eventual convergence keep in mind that already the original Newton's Method (1.2) itself, with the constraint (1.8) will converge if we make the corrections small enough, i.e. the w small enough in

$$[\partial^2\bar{S}/\partial^2 C]\Big|_C \ \Delta C = - w\ \partial\bar{S}/\partial C^*\Big|_C \qquad (3.11)$$

$|n| \neq 2m_1,\ 2m_2+1$, cf. (2.47), in what I hope is an obvious notation, with constraints

$$\partial\bar{S}/\partial C_{2m_1}^*\Big|_C = 0 \quad \text{and} \quad \partial\bar{S}/\partial C_{2m_2+1}^*\Big|_C = 0 \qquad (3.12)$$

cf. (2.47). The corrections ΔC are solved from the linear system (3.11). Some of the subsequent short-cuts do require a smaller w, hence a larger number of iterations, and one has to weigh the labour of diagonalizing, to c, against the labor of iterating more often. Several short-cuts have analogues in the theory of linear systems[31]. A first simplification of our (1.9) is

$$\left\{\begin{bmatrix} \partial^2\bar{S}/\partial^2 A & \partial^2\bar{S}/\partial A\partial B \\ \partial^2\bar{S}/\partial B\partial A & \partial^2\bar{S}/\partial^2 B \end{bmatrix} - \begin{bmatrix} \lambda'_{m_1} I & \\ & \lambda''_{m_2} I \end{bmatrix}\right\}\Bigg|_C \begin{pmatrix} \Delta A \\ \Delta B \end{pmatrix} = - w \begin{pmatrix} \partial\bar{S}/\partial A^* \\ \partial\bar{S}/\partial B^* \end{pmatrix} \qquad (3.13)$$

where the λ' are the eigenvalues of the $\partial^2\overline{S}/\partial^2A$ matrix alone, and λ'' those of $\partial^2\overline{S}/\partial^2B$, with m_1 and m_2 defined in (2.37). A further reduction arises if we only consider the eigenvalues[31]:

$$(\lambda'_{n_1} - \lambda'_{m_1})\ \Delta A_{n_1} = -\ w\ \partial\overline{S}/\partial A^*_{n_1} \qquad (3.14)$$

$$(\lambda''_{n_2} - \lambda''_{m_2})\ \Delta B_{n_2} = -\ w\ \partial\overline{S}/\partial B^*_{n_2} \qquad (3.15)$$

In that case it would be helpful to at least diagonalize the linear parts of the equations of motion, as already done in (2.23), (2.24). Again this requires diagonalization. The ultimate reduction is to use the asymptotic estimates for λ' and λ'' which are known *beforehand* in our approach

$$(n_1^2 - m_1^2)\ \nu_r^2\ \Delta A_{n_1} = -\ w\ \partial\overline{S}/\partial A^*_{n_1} \qquad (3.16)$$

$$(n_2^2 - m_2^2)\ \nu_r^2\ \Delta B_{n_2} = -\ w\ \partial\overline{S}/\partial B^*_{n_2}\ , \qquad (3.17)$$

cf. (2.42), (2.43), (2.15). For all these "short-cut" methods the version corresponding to (3.10), (3.3) is obvious.

The simplest one to implement numerically is (3.16), (3.17), (3.2) requiring no diagonalization, but using smaller w's and thus more iterations. Practical results can be found, for the Hénon-Heiles system (2.23), (2.24) in ref 15 (and 11), for Hénon's mapping in ref. 9, for the Fermi-Pasta-Ulam Chain in ref. 16 and for the Duffing equation (2.09) in refs. 10, 11. Typically the iterations of (3.16), (3.17) converge rapidly at any nonlinearity up to infinity. Even in the so called "stochastic regions"[26] the rate of convergence is high[15]. The same holds for most *unbounded* orbits [in that case we take a real-exponential basis, i.e. ν_r imaginary in (2.45)] as well as the bounded orbits. However the rate of convergence of (3.16), (3.17) decreases very near to the points where the orbits change from bounded to unbounded, i.e. the "escape" points of (2.23), (2.24) or the "breaking" points of a chain. Yet even there good results are obtainable at the price of an exorbitant number of iterations.

Since we mostly employ the *linear* variational equations, cf. section 2, some new insights might be gained also from the linear Carleman - formalism discussed in ref. 25 by Elliott Montroll and the author. Finally we note that while it is quite common to characterize a scalar Stürm-Liouville eigenfunction by the number of its zero's the author believes it is less common to do so, in the vector case (2.37), and would appreciate further references.

4. ACKNOWLEDGEMENTS

It is a pleasure to thank Joe Ford, Alan Weinstein, Tassos Bountis and Elliott Montroll for their help and encouragement. I thank Dr. Danby for his comments in Como as well as sending me ref.

19, Alan Weinstein for pointing out ref. 28 and John Greene for reminding me of refs. 21, 22. I wish to compliment Adolf Hochstim for holding this interesting workshop.

5. REFERENCES

1. C. G. J. Jacobi, "Vorlesungen über Dynamik", ed. A. Clebsch, Reimer Verlag, Berlin (1866); espec. p. 48.
2. E. T. Whittaker, "Analytical Dynamics", Cambridge University Press, 4th ed. (1964); espec. chapt. 9.
3. R. Courant and D. Hilbert, "Methods of Mathematical Physics", Interscience Publ. Inc., N.Y. (1965); espec. chapters 5 and 6 of Vol. 1.
4. S. H. Gould, "Variational Methods for Eigenvalue Problems", Univ. of Toronto Press (1957); espec. chapters 2 and 3.
5. J. Moser, "Lectures on Hamiltonian Systems", Memoirs A.M.S., 81, 1-60 (1968);
6. J. Moser, "Stable and Random Motions in Dynamical Systems", Princeton Univ. Press (1973).
7. J. Moser, "Nearly Integrable and Integrable Systems", in this volume.
8. M. V. Berry, "Regular and Irregular Motion", in this volume.
9. R. H. G. Helleman, "On the Iterative Solution of a Stochastic Mapping", p. 343-370, in "Statistical Mechanics and Statistical Methods", ed. U. Landman, Plenum Publ. Co., N.Y. (1977); espec. section III.D. Misprints: (eq. (2.2)): $x_t \sin\alpha$; (eq. (2.24)) $\bar{\chi}_t =$; (captions of figures 3 and 5): $y_t \equiv [\chi_{t-1} - \cos\alpha\chi_t]/\sin^2\alpha$ and $x_t \equiv \chi_t/\sin\alpha$; (eq. (3.29)): $\chi_t\chi_{t-1}$ term left out, cf. (3.28).
10. C. R. Eminhizer, R. H. G. Helleman and E. W. Montroll, "On a Convergent Nonlinear Perturbation Theory without Small Denominators or Secular Terms", J. Math. Phys. 17, 121-140 (1976); espec. sections 4 and 5.
11. R. H. G. Helleman and T. Bountis, "Periodic Solutions of Arbitrary Period, Variational Methods", in Proc. of the "Volta-Memorial" Conference (Como, Italy, 1977) on "Stochastic Behavior in Hamiltonian Systems", eds. G. Casati and J. Ford, Springer, N.Y. (1978).
12. P. Funk, "Variationsrechnung und ihre Anwendung in Physik und Technik", 2nd Ed., Springer Verlag (1970); espec. chapters 2, 3 and 7.
13. L. A. Pars, "Analytical Dynamics", John Wiley, Inc. (1965); espec. chapt. 26 (26.3), chapt. 23 (23.5, 23.6).
14. G. D. Birkhoff, "Dynamical Systems", Am. Math. Soc., Providence R.I. (1927); revised edition (1966); espec. chapt. 2.
15. T. Bountis, Thesis, University of Rochester (1978).
16. C. R. Eminhizer, Thesis, University of Rochester (1975).
17. E. C. Titchmarsh, "Eigenfunction Expansions Associated with Second Order Differential Equations", Parts 1 and 2, Oxford University Press (1946), (1958).

18. A. H. Nayfeh and D. T. Mook, "Nonlinear Oscillations", J. Wiley (Interscience Series, Pure and Applied Math.) New York (1978).
19. J. M. A. Danby, "Qualitative Methods in Celestial Mechanics", Publications of the Institute of Mathematics of the University of São Paulo, Brasil (1971); espec. sections 4 and 5.
20. W. Magnus and S. Winkler, "Hill's Equation", Interscience Publ. New York (1966).
21. H. P. McKean and P. van Moerbeke, "The Spectrum of Hill's Equation", Inventiones Math. 30, 217-274 (1975).
22. H. P. McKean and E. Trubowitz, "Hill's Operator ...", Comm. Pure Appl. Math. 24, 143-226 (1976).
23. B. M. Levitan and I. S. Sargsjan, "Introduction to Spectral Theory", Translations of Math. Monographs, Vol. 39, Am. Math. Soc. Providence (1975).
24. H. Flaschka and D. W. McLaughlin, Prog. Theor. Phys. 55, 438-456 (1976).
25. E. W. Montroll and R. H. G. Helleman, American Institute of Physics Conf. Proc. 27, 75-110 (1976), A.I.P. New York; and the article by Montroll in this volume.
26. R. H. G. Helleman, "Dynamics Revisited, a Glossary", in this volume; espec. its references.
27. M. Hénon and C. Heiles, Astron. J. 69, 73-79 (1964).
28. P. H. Rabinowitz, "Periodic Solutions of Hamiltonian Systems", MRC preprint #1783, Math. Research. Center, Madison, Wisconsin (1977).
29. A. Weinstein, "Bifurcations and Hamilton's Principle", preprint, Math. Dept. Berkeley (1977) & refs.; "Normal Modes for Nonlinear Hamiltonian Systems", Inventiones Math. 20, 47-57 (1973); and his article in this volume; and J. Moser, "Periodic Orbits near an Equilibrium and a Theorem by Alan Weinstein", Comm. Pure Appl. Math. 29, 727-747 (1976).
30. W. C. Rheinboldt, "Methods for Solving Systems of Nonlinear Equations", S.I.A.M. Monographs (1974); espec. section 3.2.
31. J. M. Ortega and W. C. Rheinboldt, "Iterative Solution of Nonlinear Equations in Several Variables", Academic Press, N.Y. (1970).
32. "I have announced this theorem, which incidentally is of absolutely no importance to Mechanics-in the strict sense-, in the Crelle Journal", ref. 1, p. 48.

THE LIE TRANSFORM: A NEW APPROACH TO CLASSICAL PERTURBATION THEORY

Allan N. Kaufman
Physics Dept. and Lawrence Berkeley Laboratory
University of California - Berkeley, Calif. 94720

ABSTRACT

A survey is presented of the pre-history, discovery, development, and applications of the Lie transform in classical dynamics, with particular attention to its utility in plasma physics.

"Give the world the Lie [transform]."[1]

THE AWKWARDNESS OF CLASSICAL-CLASSICAL PERTURBATION THEORY

The classical (pre-Lie) approach[2] to perturbation theory in classical ($\hbar = 0$) Hamiltonian systems proceeds by the use of a generating function to perform a canonical transformation from old variables q, p to new variables Q, P. This function depends on both old and new variables, e.g., S(q, P, t); as a consequence, the transformation itself, e.g.

$$Q(q,P,t) = \partial S(q,P,t)/\partial P \ , \qquad (1)$$

appears in mixed form, whereas what we need is Q(q,p,t), i.e., the new variables in terms of the old, or vice versa. Again, the relation between the new Hamiltonian K(Q,P,t) and the old Hamiltonian h(q,p,t) appears in a mixed representation:

$$K(Q,P,t) = h(q,p,t) + \partial S(q,P,t)/\partial t \ ; \qquad (2)$$

it would be preferable to have a relation directly between the functions K and h, rather than between their values at corresponding points in phase space.

If the old Hamiltonian h is expressible as a power series in a small perturbation parameter ε, it is reasonably straightforward to obtain perturbation expansions for S and K. However, even in order ε^2, these expressions are lengthy, and no pattern is manifest. To order ε^3, not only is the amount of algebra disheartening, but any relationships which reside in the physics are hidden. Such relationships would provide new insights and results.

THE SUPERIORITY OF CLASSICAL-QUANTUM PERTURBATION THEORY

Even before the development of the powerful perturbation techniques of quantum electrodynamics in the fifties, quantum theory utilized a systematic perturbation theory[3] which led to a set of well-known "rules". The superiority of these rules was so evident, that it became quite popular in plasma physics to derive purely classical results by quantizing the system and letting $\hbar \to 0$ in the result. This approach was widely used both for Vlasov (collisionless) plasmas and for plasmas with discreteness (collisional) effects.[4-7]

Yet it was disconcerting to see that the most expeditious route to a classical result was via a detour through quantum theory. One had the feeling that a classical formalism equal in power to the quantum one was waiting to be discovered.

EVIDENCE FOR THE EXISTENCE OF A BETTER CLASSICAL FORMALISM

Several years ago, S. Johnston and I examined the problem of the interaction of three electromagnetic modes in an arbitrarily nonuniform Vlasov plasma. On the basis of experience with related problems, we were led to the oscillation-center theory, a classical (pre-Lie) canonical formalism developed earlier by R. Dewar[8] for the study of quasi-linear diffusion, and applied by Johnston[9] to the problem of induced scattering. After some fifty pages of algebra, utilizing the methods of conventional classical perturbation theory, we succeeded in obtaining a remarkably concise expression[10] for the three-wave coupling coefficient in terms of Poisson brackets (PB) of the single-particle perturbation Hamiltonian and its convective time-integral along unperturbed orbits. In the mixed representation (q,P), these PB did not appear naturally, for they are defined only in terms of conjugate variables. Rather, guided by esthetic principles, we had to extract them from the algebraic mess, and then show that all of the residue cancelled.

The beauty of the concise form thus obtained then led us to a new insight[11] as to the essence of the wave coupling and of the Manley-Rowe relations governing such interactions. Thus, it appeared that the interaction Hamiltonian of the three waves was simply the trilinear contribution to the single-particle (new) Hamiltonian K, summed over all the (non-resonant) particles. The Manley-Rowe conditions followed trivially from this expression.

It was then clear to us that a far simpler method must exist for obtaining such a simple result, and we struggled to find it.

THE DEVELOPMENT OF A BETTER CLASSICAL FORMALISM

This "struggle" was remarkably short, from our point of view, for the desired formalism had already been discovered by workers in celestial mechanics (Hori[12], Deprit[13], Kamel[14], and others), and had even been very clearly presented in the textbook by Nayfeh[15]. This formalism was based on the Lie transform, and its hallmark was the PB, as desired.

It had been used and generalized by Dragt and Finn[16] for the study of magnetic moment invariance in a dipole field. More recently, it has been applied by Howland[17] for Kolmogorov's superconvergent perturbation theory, and by McNamara[18] for the treatment of higher-order resonances.

FURTHER DEVELOPMENT OF THE LIE TRANSFORM

The power and beauty of the Lie formalism became even more evident in the non-perturbative Vlasov formulation developed by Dewar[19], who then applied it to renormalized plasma turbulence.

In essence, the idea is to transform the phase space $p,q \equiv \underline{z} \rightarrow \underline{Z}$ by means of a generating function $S(\underline{z},t)$, which depends only on the old variables; hence it can appear in PB. The function S is chosen so as to simplify the form of the new Hamiltonian K. The old and new Hamiltonians, h and K, each produce the time-evolution of the corresponding phase-space densities f and F, by the respective Liouville equations:

$$\frac{\partial f}{\partial t} + \{f,h\} = 0 \;, \; \frac{\partial F}{\partial t} + \{F,K\} = 0 \tag{3}$$

These densities are equal at corresponding points:

$$f(\underline{z};t) = F(\underline{Z};t) \; . \tag{4}$$

Now if K is a much simpler function than h, the evolution of F is correspondingly simpler than that of f. This simplification is accomplished by proper selection of S, which then has an evolution equation of its own to satisfy. Thus instead of one difficult equation for f, one now has two simpler equations for F and S. The physical content is of course the same, but it may be more apparent in the new variables. In the next sections, we demonstrate these ideas more explicitly.

Older averaging methods[20,21] also succeeded in simplifying the Hamiltonian. But it appeared that information was lost in the averaging process; and it was not clear (to us at least) how to proceed to higher order. In the Lie method, nothing is lost: what has disappeared from the Hamiltonian is stored in the generating function. Also, the procedure for successive orders appears to be well specified.

THE LIE FORMALISM

"After all, what is a Lie [transform]? 'Tis but the truth in masquerade".[22]

A full discussion of the mathematical apparatus has been prepared in a "Lie handbook"[23] by J. Cary. Here we quote the relevant formulae. We begin by associating, with any phase function $A(\underline{z})$, a Lie operator $\tilde{A}$, defined by

$$\tilde{A} \equiv \{A(\underline{z}),\ \} \equiv \sum_{\mu} \frac{\partial A}{\partial q_\mu}\frac{\partial}{\partial p_\mu} - \frac{\partial A}{\partial p_\mu}\frac{\partial}{\partial q_\mu} \ . \tag{5}$$

For example, the momentum $\underline{p}$ has the Lie operator $\tilde{\underline{p}} = -\,\partial/\partial\underline{q}$. This is strikingly similar to the association in quantum theory of a Hermitian operator with a classical phase function ($\underline{p} \to -\,i\hbar\partial/\partial\underline{q}$).

Consider now any phase function, possibly time-dependent and depending also on the perturbation parameter ε: $S(\underline{z},t,\varepsilon)$. The corresponding Lie operator $\tilde{S}$ then generates a transformation operator $T(\varepsilon)$ by

$$\partial T(\varepsilon)/\partial\varepsilon = \tilde{S}(\varepsilon)\, T(\varepsilon) \ . \tag{6}$$

The latter operator effects a canonical transformation:

$$\underline{Z}(\underline{z},t,\varepsilon) = T(\varepsilon)\underline{z} \ . \tag{7}$$

Corresponding phase functions, e.g., f and F in Eq. (4), are related by the inverse of T:

$$F = T^{-1}\, f \quad \text{or } f = T\, F \ . \tag{8}$$

The dynamical variables and densities evolve according to the respective Hamiltonians h and K which are related by

$$\frac{\partial K}{\partial \varepsilon} = \frac{\partial H}{\partial \varepsilon} - T^{-1}(\varepsilon)\,\frac{\partial S}{\partial t} \ , \tag{9}$$

with

$$h = TH \ . \tag{10}$$

PERTURBATION EXPANSION

We implement this bare outline with a perturbation expansion:

$$h(\varepsilon) = h_0 + \varepsilon h_1 + \varepsilon^2 h_2 + \ldots$$
$$\varepsilon S(\varepsilon) = 0 + \varepsilon S_1 + \varepsilon^2 S_2 + \ldots \tag{11}$$

Substituting into Eq. (6), we obtain

$$T(\varepsilon) = 1 + \varepsilon\tilde{S}_1 + \frac{1}{2}\,\varepsilon^2(\tilde{S}_1^2 + \tilde{S}_2) + \ldots \tag{12}$$

We then use Eq. (10) to find $H(\varepsilon)$, and Eq. (9) to find $K(\varepsilon)$:

$$K(\varepsilon) = h_0 + \varepsilon K_1 + \varepsilon^2 K_2 + \ldots \tag{13}$$

In particular, we have

$$K_1 = h_1 - \dot{S}_1 , \tag{14}$$

where the dot implies a convective derivative along unperturbed orbits:

$$\dot{A} \equiv \partial A/\partial t + \{A, h_0\} . \tag{15}$$

Now we choose S_1 so as to make K_1 vanish. By Eq. (14):

$$\dot{S}_1 = h_1 . \tag{16}$$

We then proceed to K_2, finding

$$K_2 = h_2 - \frac{1}{2}\dot{S}_2 - \frac{1}{2}\tilde{S}_1 h_1 . \tag{17}$$

Suppose h_2 vanishes in the problem studied; we then choose $S_2 \equiv 0$, and obtain

$$K_2 = -\frac{1}{2}\tilde{S}_1 h_1 , \tag{18}$$

with S_1 given by (16). One may easily proceed to third order, obtaining

$$K_3 = \frac{1}{3}\tilde{S}_1^2 h_1 ; \tag{19}$$

this neat result required lengthy algebra[10] by the old method.

OSCILLATION CENTER THEORY

The low frequency force produced non-linearly by a high frequency field has been known for some time[24]. Its current name is "ponderomotive force" (PMF), and the object on which it acts is called the "oscillation center" (OC). The particle oscillates at high frequency about the orbit of the OC. The concept is quite analogous to the gyration of a particle about its guiding center in a magnetic field.

The PMF concept has been very useful in the study of r.f. confinement[24], quasi-static density perturbations due to amplitude modulation of high-frequency waves[25], and parametric instabilities[26]. Its meaning, however, appeared nebulous until Dewar derived it from a canonical transformation[8]. His pre-Lie formalism led to lengthy algebraic calculations, and now has been superseded in turn by the Lie method, which is astonishingly simple.

Consider a monochromatic electric field

$$\underline{E}(\underline{x},t) = \tilde{\underline{E}}(\underline{x}) \exp(-i\omega t) + \text{c.c.} \tag{20}$$

acting on a particle in a static field represented by $h_0(\underline{z})$. The linear perturbation in the Hamiltonian is

$$h_1(\underline{z},t) = -\, c^{-1} \int d^3x\ \underline{j}(\underline{x}|\underline{z}) \cdot \underline{A}(\underline{x},t)\ , \tag{21}$$

where $\underline{j}(\underline{x}|\underline{z})$ is the current density at $\underline{x}$ due to a particle at $\underline{z}$, and $\underline{A}$ is the vector potential representation of $\underline{E}$. We insert (21) into (16) and (18), immediately obtaining a formula for the second-order OC Hamiltonian K_2, in terms of PB of the phase function $\underline{j}(\underline{x}|\underline{z})$ (at two positions $\underline{x},\underline{x}'$). We then recognize this PB as having appeared in the Kubo-type formula for the linear Vlasov susceptibility $\chi(\underline{x},\underline{x}';\omega)$ of an unperturbed distribution $f(\underline{z})$, representing its linear current response to the field (20). Comparing the expressions for K_2 and $\underset{\sim}{\chi}$, we read off the remarkable relation[28]

$$K_2(\underline{z}) = -\,(4\pi)^{-1} \int d^3x \int d^3x'\ \tilde{\underline{E}}^*(\underline{x})\ \tilde{\underline{E}}(\underline{x}') :\ \delta\underset{\sim}{\chi}'\ (\underline{x},\underline{x}';\omega)/\delta f(\underline{z}). \tag{22}$$

Hence the lengthy calculations of the past, to deduce second-order forces, are no longer necessary: all one needs is the (generally known) linear response kernel. (Note that $\underset{\sim}{\chi}$ is a linear functional of f, so that K_2 is of course independent of f. The prime on $\underset{\sim}{\chi}$ denotes Hermitian, or non-resonant,part. To include the resonant response is far more difficult[29].)

Relation (22) may be considered as a generalization of the well-known expression[30] for the second-order free energy of a dielectric in thermal equilibrium in a static electric field. However, it has a fascinating new feature, namely a momentum dependence, whereas older versions of K_2 ignored this feature.

Suppose that the unperturbed Hamiltonian is $\underline{q}$-independent: $h_0(\underline{p})$. Then the OC Hamiltonian reads

$$K(\underline{z}) = h_0(\underline{p}) + K_2\ (\underline{q},\underline{p})\ , \tag{23}$$

to second order. Note that $\underline{z}$ here represents the OC variables. Its dynamics are given by

$$\dot{\underline{p}} = -\,\partial K/\partial \underline{q} = -\,\partial K_2/\partial \underline{q} \ , \tag{24}$$

$$\dot{\underline{q}} = \partial K/\partial \underline{p} = \partial h_0/\partial \underline{p} + \partial K_2/\partial \underline{p} \ . \tag{25}$$

Eq. (24) states that the OC momentum is acted on by the PMF ($-\partial K_2/\partial \underline{q}$), which generalizes the standard formula for the latter. Eq. (25) distinguishes between the OC velocity $\dot{\underline{q}}$ and the OC momentum $\underline{p}$ (for $h_0 = p^2/2$, $m = 1$). The difference, $\dot{\underline{q}} - \underline{p} = \partial K_2/\partial \underline{p}$, represents the shift in the particle's average velocity, due to its being in the wave field; summed over particles, it represents the particle contribution to the wave momentum[31].

Another consequence of momentum-dependence in K_2 is that the PMF mimics a magnetic field[32]. Namely, if we expand K_2 to first order in $\underline{p}$:

$$K_2(\underline{q},\underline{p}) \equiv e\Phi(\underline{q}) - ec^{-1}\underline{p} \cdot \underline{A}(\underline{q}) + \mathcal{O}(p^2) \ , \tag{26}$$

the OC equation of motion reads ($m = 1$)

$$\ddot{\underline{q}} = e(\underline{\mathcal{E}} + c^{-1}\, \dot{\underline{q}} \times \underline{\mathcal{B}}) \ , \tag{27}$$

where the pseudo-potentials Φ, $\underline{A}$ are defined by Eq. (26), and

$$\begin{aligned} \underline{\mathcal{B}}(\underline{q}) &\equiv (\partial/\partial \underline{q}) \times \underline{A}(\underline{q}) \ , \\ \underline{\mathcal{E}}(\underline{q}) &\equiv -\,\partial\Phi/\partial \underline{q} - c^{-1}\, \partial \underline{A}/\partial t \ . \end{aligned} \tag{28}$$

The pseudofields $\underline{\mathcal{B}}$ and $\underline{\mathcal{E}}$ satisfy a Faraday law:

$$c\, \nabla \times \underline{\mathcal{E}} = -\,\partial \underline{\mathcal{B}}/\partial t \ . \tag{29}$$

All these quantities Φ, $\underline{A}$, $\underline{\mathcal{B}}$, $\underline{\mathcal{E}}$ are quadratic in the amplitude $\underline{\tilde{E}}$ of the true field.

If the unperturbed motion is in a true magneto-static field $\underline{B}_0$, the interpretation is different, but related. A dependence of K_2 on gyromomentum p_g implies a nonlinear shift in the OC's gyrofrequency $\Omega \equiv \dot{\phi}_g = \partial K/\partial p_g = \partial h_0/\partial p_g + \partial K_2/\partial p_g \equiv \Omega_0 + \delta\Omega$. Since $\Omega_0 = eB_0/mc$, the gyrofrequency shift $\delta\Omega$ may again be interpreted as being due to a magnetic pseudofield $\mathcal{B}$, quadratic in wave amplitude $\underline{\tilde{E}}$.

SELF-CONSISTENT FIELDS

To determine the Maxwell field $\underline{E}(\underline{x},t)$ self-consistently, we need the current density

$$\underline{j}(\underline{x},t) = \int d^6z \ \underline{j}(\underline{x}|\underline{z}) \ f(\underline{z};t) \tag{30}$$

as a source in the Maxwell equations (summation over species implied). It is essential to recognize that the OC current

$$\underline{j}^{OC}(\underline{x},t) \equiv \int d^6z \ \underline{j}(\underline{x}|\underline{z}) \ F(\underline{z};t) \tag{31}$$

constitutes only part of (30). To see their relation, we substitute (8) into (30):

$$\begin{aligned} \underline{j}(\underline{x},t) &= \int d^6z \ \underline{j}(\underline{x}|\underline{z}) \ TF(\underline{z},t) \\ &= \int d^6z \ F(\underline{z},t) \ T^{-1} \ \underline{j}(\underline{x}|\underline{z}) \\ &\equiv \int d^6z \ F(\underline{z},t) \ \underline{J}(\underline{x}|\underline{z}) \ . \end{aligned} \tag{32}$$

We interpret $\underline{J}(\underline{x}|\underline{z})$ as the current at $\underline{x}$ due to a particle whose OC is at $\underline{z}$. Thus $\underline{J}(\underline{x}|\underline{z})$ consists of the "bare" part $\underline{j}(\underline{x}|\underline{z})$, from the OC, and the "polarization" correction $(T^{-1} - 1)\underline{j}(\underline{x}|\underline{z})$, due to the particle's jitter motion. As we know from guiding center theory, these separate parts sometimes cancel!

Still guided by that analogy, we expect that the situation for charge density is simpler. Indeed,we find that the OC charge density $\rho(\underline{x}|\underline{z})$ and the slow part of the true charge density $T^{-1}\rho(\underline{x}|\underline{z})$ differ only by order $(\delta r/L)^2$, where $\delta r \sim eE/m\omega^2$ is the first-order displacement due to the wave field, and L is the scale length of the wave amplitude modulation. Hence,we may normally neglect the polarization charge density and use

$$\rho(\underline{x},t) = \int d^6z \ F(\underline{z},t) \ \rho(\underline{x}|\underline{z}) \tag{33}$$

for the slow charge density source in the slow Poisson equation.

On the basis of such considerations, we have recently derived a simple formula[33] for the quasi-static, quasi-neutral density perturbation caused by spatial modulation of a magnetoplasma wave of arbitrary polarization:

$$\delta n(\underline{x}) = - (4\pi)^{-1} \ (T_e + T_i)^{-1} \ [|\underline{\tilde{E}}(\underline{x})|^2 - |\underline{\tilde{B}}(\underline{x})|^2] \ . \tag{34}$$

This formula (within its range of validity) predicts that a "magnetic" wave ($\tilde{B} > \tilde{E}$) leads to density enhancement, in contrast to the well-known density depression produced by an "electric" wave ($\tilde{E} > \tilde{B}$). Thus, a magnetic wave may contribute to plasma confinement[34].

CONCLUSION

On the basis of the results obtained only in the last few years, the Lie approach to classical dynamics would appear to be most promising. It sheds new light on old results, reveals previously unobserved relationships, and leads to new results by extremely economical means. Because the Lie transform preserves Poisson brackets, it is particularly successful in dealing with Hamiltonian systems. Its relation to Lagrangian methods[21,35-37] has yet to be investigated. Finally, one may hope that dissipative systems[38] may be treated by a variant of the Lie method.

In summary, "a Lie [transform] is ... an ever-present help in time of need".[39]

ACKNOWLEDGEMENTS

My understanding of these matters has benefited tremendously from discussions with Robert Dewar, John Cary, and Robert Littlejohn, and from the always informed encouragement of Oscar Manley. The U. S. Department of Energy provided support.

REFERENCES

1. Sir Walter Raleigh, The Lie [Transform] (1595).
2. H. Goldstein, Classical Mechanics (Addison-Wesley, Reading, Mass. 1950); H. Corben & P. Stehle, Classical Mechanics (Wiley, New York, 1960).
3. L. I. Schiff, Quantum Mechanics (McGraw-Hill, New York, 1949).
4. E. Harris, in Advances in Plasma Physics, edited by A. Simon & W. B. Thompson (Interscience, New York, 1968), Vol. 3.
5. B. Coppi, M. Rosenbluth, & R. Sudan, Ann. Phys. 55, 207 (1969).
6. A. Kaufman & T. Nakayama, Phys. Fluids 13, 956 (1970).
7. V. N. Tsytovitch, Non-linear Effects in Plasma (Plenum, New York 1970); Theory of Turbulent Plasma (Consultants Bureau, New York, 1977).
8. R. Dewar, Phys. Fluids 16, 1102 (1973).
9. S. Johnston, Phys. Fluids 19, 93 (1976).
10. S. Johnston & A. Kaufman, in Plasma Physics, edited by H. Wilhelmsson (Plenum, New York, 1977), p. 159.
11. S. Johnston & A. Kaufman, Bull. Am. Phys. Soc. 21, 1094 (1976).
12. G. Hori, Publ. Astr. Soc. Japan 18, 287 (1966).
13. A. Deprit, Cel. Mech. 1, 12 (1969).
14. A. Kamel, Cel. Mech. 1, 190 (1969).
15. A. Nayfeh, Perturbation Methods (Wiley, New York, 1973), Sec.5.7.

16. A. Dragt & J. Finn, J. Math. Phys. 17, 2215 (1976).
17. R. Howland, Cel. Mech. 15, 327 (1977).
18. B. McNamara, UCRL-79843 (submitted to J. Math. Phys., 1978).
19. R. Dewar, J. Phys. A9, 2043 (1976).
20. E. Burshtein & L. Solov'ev, Sov. Phys. Doklady 6, 731 (1962).
21. G. Whitham, Linear and Non-linear Waves (Wiley, New York, 1974).
22. George Gordon, Lord Byron, Don Juan (1823).
23. J. Cary, LBL-6350 (1978); for an application to symmetry and invariants, see J. Cary, J. Math. Phys. 18, 2432 (1977).
24. H. Motz & C. Watson, in Adv. in Electronics and Electron Physics 23, 153 (1967).
25. P. Kaw, G. Schmidt, & T. Wilcox, Phys. Fluids 16, 1522 (1973).
26. J. Drake, P. Kaw, Y. Lee, G. Schmidt, C. Liu, & M. Rosenbluth, Phys. Fluids 17, 778 (1974).
27. L. Altshul & V. Karpman, Sov. Phys. JETP 20, 1043 (1965).
28. J. Cary & A. Kaufman, Phys. Rev. Lett. 39, 402 (1977).
29. R. Dewar, J. Phys. A 11,9 (1978).
30. L. Landau & E. Lifshitz, Electrodynamics of Continuous Media (Addison-Wesley, Reading, Mass. 1960), Sec. 11.
31. A. Kaufman, J. Plasma Phys. 8, 1 (1972).
32. J. Cary & A. Kaufman, LBL-6306 (1977).
33. A. Kaufman, J. Cary, & N. Pereira, LBL-7234, Proc. Third Topical Conference on RF Plasma Heating, Pasadena (Jan. 1978).
34. J. Cary and J. Hammer, LBL-7538 (1978).
35. J. Dougherty, J. Plasma Phys. 4, 761 (1970); 11, 331 (1974).
36. R. Dewar, Phys. Fluids 13, 2710 (1970).
37. J. Galloway & H. Kim, J. Plasma Phys. 6, 53 (1971).
38. B. Rosen, SIT-304 (1977).
39. Sen. John Tyler Morgan, quoted by D. McCullough in The Path Between the Seas (1977).

DISCUSSION OF SOME WEAKLY NONLINEAR SYSTEMS IN CONTINUUM MECHANICS

Jim Meiss and Kenneth Watson
Physics Department
and
Lawrence Berkeley Laboratory
University of California, Berkeley

1. INTRODUCTION

We are interested here in dynamical systems which are weakly nonlinear in the following sense. It is first supposed that such a system has a linear approximation by which it can be described as a superposition of sinusoidal waves, or, equivalently, as a set of uncoupled harmonic oscillators. In higher approximations, these oscillators are coupled by non-linear terms. We describe this non-linear coupling as "weak" if the perturbing effect on a given oscillator is very small during one period of oscillation.

There are many examples of physical systems which by this criterion may reasonably be described as weakly non-linear. This includes lattice vibrations, sound waves, laser beam interactions with matter, plasma waves, etc. Many important examples are also encountered in geophysics. These include plasma waves in the ionosphere, seismic waves, internal waves in the atmosphere, ocean surface waves, and internal waves in the ocean.

The discussion given in what follows of these phenomena will be rather general and somewhat formal. Our own current interest concerns the effect of non-linearities in the ocean internal wave field and reference will be made to this application. Two other examples, which are mathematically closely related, will be described. These are the interaction of a laser beam with a plasma and the interaction of a single ocean internal wave with a spectrum of surface waves.

In anticipation of our examples, we briefly review some properties of internal and surface waves. The spectrum[1,2] of vertical displacement of surface waves at horizontal wavenumber $\underset{\sim}{k}$ is written as $\psi(\underset{\sim}{k})$. This is so normalized that

$$< \zeta_S^2 > = \int \psi(\underset{\sim}{k})\, d^2k \, , \tag{1.1}$$

where $\zeta_S(\underset{\sim}{x},t)$ is the vertical displacement, due to surface waves, at time t and position $x = (x,y)$ on the (local) reference plane of the ocean surface ($\tilde{z} = 0$). In deep water the linear theory gives the angular frequency

$$\omega_k = \sqrt{gk} \, , \tag{1.2}$$

ISSN: 0094-243X/78/296/$1.50

where g is the acceleration of gravity. Hydrodynamic non-linearities lead to the transfer of excitation in wavenumber space and are thought to play a significant role in the evolution of the spectrum $\psi(\underset{\sim}{k})$.

A spectrum $\psi(\underset{\sim}{k},j)$ of vertical fluid displacement $\zeta(\underset{\sim}{x},z,t)$ (z = 0 corresponds to the ocean surface, z = -B to the bottom) was introduced by Garrett and Munk[3,4] to describe the internal wave field. Here $\underset{\sim}{k}$ represents the horizontal wavenumber vector and j (= 1,2,...) the vertical mode number for a linear internal wave. A simple WKB model used by Garrett and Munk has the angular frequency of the linear waves (if ω is large compared to the inertial frequency)

$$\omega_{kj} \cong \frac{kbN_0}{\left(j - \frac{1}{4}\right)\pi} . \qquad (1.3)$$

Here b $\sim$ 1200 m is a scale depth and $N_0 \sim 5 \times 10^{-3}\ \mathrm{sec}^{-1}$ is a Vaisala frequency parameter. [The expression (1.3) is intended to be applied only over restricted regions of the internal wave spectrum.] The normalization of ψ is chosen so that

$$< \zeta^2 > = \sum_{j=1}^{\infty} \int \psi(k,j) d^2k . \qquad (1.4)$$

2. MATHEMATICAL DESCRIPTION OF SYSTEM

The physical quantities of interest will be described by one or more real field quantities, $F(\underset{\sim}{x},t)$, representing velocity, stream function, potential function, displacement, etc. A finite volume, area, or length (depending on the dimension of the system) is chosen (often with periodic boundary conditions) in which to Fourier analyze F:

$$F(\underset{\sim}{x},t) = \sum_{\underset{\sim}{k}} U_{\underset{\sim}{k}} A_{\underset{\sim}{k}}(t)\, e^{i\underset{\sim}{k}\cdot\underset{\sim}{x}}, \quad A^*_{\underset{\sim}{k}} = A_{-\underset{\sim}{k}} . \qquad (2.1)$$

In general, F will represent a column matrix whose elements are velocity components, fluid pressure, etc., and the U_k represent the column matrix eigenvectors for the linearized system. These may be conveniently normalized, say, to unity. The A_k's are then time dependent coefficients specifying the amplitude of the $\underset{\sim}{k}$th mode in the wave field.

Equation (2.1) is typical, but may require modification for specific cases. For example, in the case of the internal wave field, we have

$$F(\underset{\sim}{x},t) = \sum_{\underset{\sim}{k}} \sum_{j=1}^{\infty} U_{\underset{\sim}{k}j} A_{\underset{\sim}{k}j}(t) W_{\underset{\sim}{k}j}(z) e^{i\underset{\sim}{k}\cdot\underset{\sim}{x}} . \qquad (2.2)$$

Here $\underset{\sim}{x} = (x,y)$ is a vector in the horizontal reference plane $z = 0$ and the $W_{\underset{\sim}{k}j}(z)$ represent the eigenfunctions describing the z-dependence of the linear modes. The matrices F and U are three dimensional, the elements corresponding (say) to pressure variation, vertical displacement, and potential function for horizontal displacement.

In writing (2.1) and (2.2) we have assumed that the modes for the linearized system provide a complete, orthogonal set of basis functions. Then (2.1) and (2.2) can be used to describe the non-linear system.

In this discussion we shall suppose that the dynamical system can be described by Hamilton's principle in terms of a Lagrangian. A Hamiltonian may then be obtained and the equations of motion expressed in Hamiltonian form. We thus exclude important physical systems from our analyses - for example, those with dissipation. We do not consider this a serious limitation, however, since the equations for non-Hamiltonian systems are anticipated to have a structure similar to that for Hamiltonian systems.

Specific Lagrangian-Hamiltonian formulations have been published for surface waves[5], internal waves[6], and surface wave-internal wave interactions.[7] The laser beam-plasma wave interaction was also formulated in Hamiltonian form.[8]

A. The Linearized System

We begin with a somewhat detailed description of the linearized system. The Lagrangian, corresponding to the system described by the field (2.1) will then have the form*

$$L = \frac{1}{2}\sum_{\underset{\sim}{k}} e_k \left[\dot{A}_{\underset{\sim}{k}} \dot{A}_{-\underset{\sim}{k}} - \omega_k^2 A_{\underset{\sim}{k}} A_{-\underset{\sim}{k}}\right] . \qquad (2.3)$$

Here the e_k are a set of real constant coefficients. Hamilton's principle provides the equations of motion for the A_k's by variation of

*For internal waves the mode label j and a sum over j are required in (2.3).

$$I = \int_{t_1}^{t_2} L \, dt \, . \tag{2.4}$$

These are

$$\ddot{A}_{\underset{\sim}{k}} = - \omega_k^2 A_{\underset{\sim}{k}} \, , \tag{2.5}$$

where $\dot{A} \equiv dA/dt$, etc.

Real variables may be introduced with the transformation

$$A_{\underset{\sim}{k}} = (2\, e_{\underset{\sim}{k}})^{-1/2} \left[e^{i\frac{\pi}{4}} q_{\underset{\sim}{k}} + e^{-i\frac{\pi}{4}} q_{-\underset{\sim}{k}} \right] \, . \tag{2.6}$$

In terms of the $q_{\underset{\sim}{k}}$'s the Lagrangian (2.3) becomes

$$L = \sum_{\underset{\sim}{k}} \frac{1}{2} \left[\dot{q}_{\underset{\sim}{k}}^2 - \omega_{\underset{\sim}{k}}^2 q_{\underset{\sim}{k}}^2 \right] . \tag{2.7}$$

Canonical momenta may be introduced with the relation

$$p_{\underset{\sim}{k}} = \frac{\partial L}{\partial \dot{q}_{\underset{\sim}{k}}} = \dot{q}_{\underset{\sim}{k}} \, . \tag{2.8}$$

The Hamiltonian for the linear system is then

$$\begin{aligned} H_0 &= \sum p_{\underset{\sim}{k}} \dot{q}_{\underset{\sim}{k}} - L \\ &= \sum \frac{1}{2} \left[p_{\underset{\sim}{k}}^2 + \omega_{\underset{\sim}{k}}^2 q_{\underset{\sim}{k}}^2 \right] . \end{aligned} \tag{2.9}$$

In practice it is convenient to transform to variables which relate the Fourier label $\underset{\sim}{k}$ to the propagation wave number vector. This is accomplished with the introduction of the <u>action amplitude</u> variables

$$\begin{aligned} a_{\underset{\sim}{k}} = \frac{1}{2\sqrt{f_k}} \Big[& e^{i\frac{\Pi}{4}} (p_{\underset{\sim}{k}} - i\omega_{\underset{\sim}{k}} q_{\underset{\sim}{k}}) + \\ & e^{-i\frac{\Pi}{4}} (p_{-\underset{\sim}{k}} - i\omega_{\underset{\sim}{k}} q_{-\underset{\sim}{k}}) \Big] \, . \end{aligned} \tag{2.10}$$

Here the $f_{\underset{\sim}{k}}$ are real coefficients chosen to appropriately scale the $a_{\underset{\sim}{k}}$'s. The Poisson bracket relations for the $a_{\underset{\sim}{k}}$'s are *

$$(a_{\underset{\sim}{k}}, a_{\underset{\sim}{k}'}) = (a^*_{\underset{\sim}{k}}, a^*_{\underset{\sim}{k}'}) = 0, \qquad (2.11)$$

$$(a_{\underset{\sim}{k}}, a^*_{\underset{\sim}{k}'}) = -i\,(\omega_k/f_k)\,\delta_{\underset{\sim}{k}-\underset{\sim}{k}'}.$$

The Hamiltonian (2.9) when expressed in terms of the $a_{\underset{\sim}{k}}$'s has the form

$$H_0 = \sum_{\underset{\sim}{k}} f_k\, a^*_{\underset{\sim}{k}}\, a_{\underset{\sim}{k}}, \qquad (2.12a)$$

and the equations of motion are

$$\dot{a}_{\underset{\sim}{k}} = (a_{\underset{\sim}{k}}, H_0) = -i\omega_k a_{\underset{\sim}{k}}. \qquad (2.12b)$$

Integration gives

$$a_{\underset{\sim}{k}}(t) = a_{\underset{\sim}{k}}(0)e^{-i\omega_k t}. \qquad (2.13)$$

The Fourier amplitudes $A_{\underset{\sim}{k}}$ are expressed in terms of the $a_{\underset{\sim}{k}}$ as

$$A_k = \frac{i}{\omega_k}\sqrt{\frac{f_k}{2e_k}}\,[a_{\underset{\sim}{k}} - a^*_{-\underset{\sim}{k}}] \;. \qquad (2.14)$$

The field quantities (2.1) have the corresponding form

$$F(x,t) = \sum_{\underset{\sim}{k}} \frac{i}{\omega_k}\sqrt{\frac{f_k}{2e_k}}\, U_{\underset{\sim}{k}} a_{\underset{\sim}{k}}(0)\exp[i(\underset{\sim}{k}\cdot\underset{\sim}{x}-\omega_k t)] + \text{c.c.} \qquad (2.15)$$

We see from (2.14) that $\underset{\sim}{k}$ now appears as a wavenumber vector for an advancing wave.

Action-angle variables $J_{\underset{\sim}{k}}$, $\theta_{\underset{\sim}{k}}$ are also very convenient for studying the nonlinear problem. These are expressed in terms of the $a_{\underset{\sim}{k}}$'s as

*The reader will recognize that the $a_{\underset{\sim}{k}}$'s and $a^*_{\underset{\sim}{k}}$'s correspond to the annihilation and creation operators of quantum field theory.

$$a_{\underset{\sim}{k}} = [\omega_k J_{\underset{\sim}{k}}/f_k]^{\frac{1}{2}} \exp(-i\theta_{\underset{\sim}{k}}). \tag{2.16}$$

Substitution into Eq. (2.12) gives

$$H_0 = \sum_{\underset{\sim}{k}} \omega_k J_{\underset{\sim}{k}}. \tag{2.17}$$

The $J_{\underset{\sim}{k}}, \theta_{\underset{\sim}{k}}$ are canonical variables. The equations of motion for the linear theory are

$$\dot{J}_{\underset{\sim}{k}} = -\frac{\partial H_0}{\partial \theta_{\underset{\sim}{k}}} = 0, \quad \dot{\theta}_{\underset{\sim}{k}} = \frac{\partial H_0}{\partial J_{\underset{\sim}{k}}} = \omega_k. \tag{2.18}$$

B. Nonlinear Theory

We assume that for the weakly nonlinear system the Hamiltonian may be expressed in terms of the variables $a_{\underset{\sim}{k}}$ and in the form*

$$H = H_0 + V ,$$

$$V = \sum_{\underset{\sim}{k},\underset{\sim}{\ell},\underset{\sim}{n}} \Big\{ \delta_{\underset{\sim}{k}-\underset{\sim}{\ell}-\underset{\sim}{n}} [a^*_{\underset{\sim}{k}} a_{\underset{\sim}{\ell}} a_{\underset{\sim}{n}} + a_{\underset{\sim}{k}} a^*_{\underset{\sim}{\ell}} a^*_{\underset{\sim}{n}}] C^{\underset{\sim}{k}}_{\underset{\sim}{\ell}\underset{\sim}{n}} + \delta_{\underset{\sim}{k}+\underset{\sim}{\ell}+\underset{\sim}{n}} [a_{\underset{\sim}{k}} a_{\underset{\sim}{\ell}} a_{\underset{\sim}{n}} + a^*_{\underset{\sim}{k}} a^*_{\underset{\sim}{\ell}} a^*_{\underset{\sim}{n}}] \Big\} C_{\underset{\sim}{k}\underset{\sim}{\ell}\underset{\sim}{n}} + 4\underline{th} \text{ order} + \ldots \tag{2.19}$$

If V contains terms above third order in the a's, we interpret the condition of <u>weak nonlinearity</u> to imply that successively higher orders decrease rapidly in numerical magnitude (for example, in the rms sense). In practice, only terms of the lowest, or the two lowest orders, are usually kept in V. The equations of motion are readily obtained using the Poisson bracket relations (2.11) and the equations of motion

$$\dot{a}_{\underset{\sim}{k}} = \left(a_{\underset{\sim}{k}}, H\right) . \tag{2.20}$$

We may also express H in terms of the action-angle variables of the linear theory, but no longer "action-angle" variables. (To avoid a cumbersome label, we shall continue to refer to these as "action-angle" variables.) This has the form [derived from (2.19)],

---

* Somewhat more generally, we might encounter complex coefficients C. This would lead to constant phase shifts in the cosine terms in (2.21), but would not significantly modify our discussion.

$$H = \sum_{\underset{\sim}{k}} J_{\underset{\sim}{k}} \omega_{\underset{\sim}{k}} + \sum_{\underset{\sim}{k}_1,\underset{\sim}{k}_2,\underset{\sim}{k}_3} \Big[V_3\left(\underset{\sim}{k}_1,\underset{\sim}{k}_2,\underset{\sim}{k}_3\right) \times \cos\left(\theta_{\underset{\sim}{k}_1} - \theta_{\underset{\sim}{k}_2} - \theta_{\underset{\sim}{k}_3}\right) + V_3'\left(\underset{\sim}{k}_1,\underset{\sim}{k}_2,\underset{\sim}{k}_3\right) \times \cos\left(\theta_{\underset{\sim}{k}_1} + \theta_{\underset{\sim}{k}_2} + \theta_{\underset{\sim}{k}_3}\right)\Big] \left[J_{\underset{\sim}{k}_1} J_{\underset{\sim}{k}_2} J_{\underset{\sim}{k}_3}\right]^{1/2}$$

$$+ \sum_{\underset{\sim}{k}_1,\underset{\sim}{k}_2,\underset{\sim}{k}_3,\underset{\sim}{k}_4} \left[J_{\underset{\sim}{k}_1} J_{\underset{\sim}{k}_2} J_{\underset{\sim}{k}_3} J_{\underset{\sim}{k}_4}\right]^{1/2} \Big[V_4\left(\underset{\sim}{k}_1,\underset{\sim}{k}_2,\underset{\sim}{k}_3,\underset{\sim}{k}_4\right) \times \cos\left(\theta_{\underset{\sim}{k}_1} + \theta_{\underset{\sim}{k}_2} - \theta_{\underset{\sim}{k}_3} - \theta_{\underset{\sim}{k}_4}\right)$$

$$+ V_4'\left(\underset{\sim}{k}_1,\underset{\sim}{k}_2,\underset{\sim}{k}_3,\underset{\sim}{k}_4\right) \cos\left(\theta_{\underset{\sim}{k}_1} + \theta_{\underset{\sim}{k}_2} + \theta_{\underset{\sim}{k}_3} + \theta_{\underset{\sim}{k}_4}\right)$$

$$+ V_4''\left(\underset{\sim}{k}_1,\underset{\sim}{k}_2,\underset{\sim}{k}_3,\underset{\sim}{k}_4\right) \cos\left(\theta_{\underset{\sim}{k}_1} - \theta_{\underset{\sim}{k}_2} - \theta_{\underset{\sim}{k}_3} - \theta_{\underset{\sim}{k}_4}\right)\Big]$$

$$+ \text{5th order} + \ldots . \tag{2.21}$$

Referring to (2.19) we see that V_3 is expressed in terms of $C^{\underset{\sim}{k}_1}_{\underset{\sim}{k}_2,\underset{\sim}{k}_3}$ and V_3' in terms of $C_{\underset{\sim}{k}_1,\underset{\sim}{k}_2,\underset{\sim}{k}_3}$:

$$V_3\left(\underset{\sim}{k}_1,\underset{\sim}{k}_2,\underset{\sim}{k}_3\right) = 2\left[\omega_{\underset{\sim}{k}_1} \omega_{\underset{\sim}{k}_2} \omega_{\underset{\sim}{k}_3} \Big/ \left(f_{\underset{\sim}{k}_1} f_{\underset{\sim}{k}_2} f_{\underset{\sim}{k}_3}\right)\right]^{1/2} \times C^{\underset{\sim}{k}_1}_{\underset{\sim}{k}_2,\underset{\sim}{k}_3} \delta_{\underset{\sim}{k}_1-\underset{\sim}{k}_2-\underset{\sim}{k}_3}$$

$$V_3'\left(\underset{\sim}{k}_1,\underset{\sim}{k}_2,\underset{\sim}{k}_3\right) = 2\left[\omega_{\underset{\sim}{k}_1} \omega_{\underset{\sim}{k}_2} \omega_{\underset{\sim}{k}_3} \Big/ \left(f_{\underset{\sim}{k}_1} f_{\underset{\sim}{k}_2} f_{\underset{\sim}{k}_3}\right)\right]^{1/2} \times C_{\underset{\sim}{k}_1 \underset{\sim}{k}_2 \underset{\sim}{k}_3} \delta_{\underset{\sim}{k}_1+\underset{\sim}{k}_2+\underset{\sim}{k}_3},$$

$$\ldots . \tag{2.22}$$

The coefficient $V_4\left(\underset{\sim}{k}_1,\underset{\sim}{k}_2,\underset{\sim}{k}_3,\underset{\sim}{k}_4\right)$ vanishes unless $\underset{\sim}{k}_1+\underset{\sim}{k}_2 = \underset{\sim}{k}_3+\underset{\sim}{k}_4$, etc. The equations of motion for the $J_{\underset{\sim}{k}}$ and $\theta_{\underset{\sim}{k}}$ and

$$\dot{J}_{\underset{\sim}{k}} = - \frac{\partial H}{\partial \theta_{\underset{\sim}{k}}} \;, \; \dot{\theta}_{\underset{\sim}{k}} = \frac{\partial H}{\partial \theta_{\underset{\sim}{k}}} \;. \qquad (2.23)$$

It is evidently convenient for our discussion, and essential for numerical computation, to truncate the Fourier series (2.1) at a finite number, say N, of modes. For the remainder of this paper we shall always suppose that this has been done and that our phase space is of 2N dimensions.

3. PRELIMINARY DISCUSSION

The equations of motion (2.20) or (2.23) are of first order. Numerical integration appears (from past experience) to be relatively straight forward. The sums over wavenumbers, especially for fourth and higher order terms, can be costly in computer time, however. This provides a strong incentive to use the available "lore" relating to these equations to simplify the calculation*.

The essential idea in simplifying the equations of motion is that energy transfer among modes occurs only where there are "resonances". Terms not associated with resonances may be transformed away by a sequence of canonical transformations. These transformations lead to a slight (and presumably unimportant) re-definition of the normal modes for the linear system described by H_0. Expressed somewhat more formally, a given orbit of the linearized system is confined to an N-dimensional torus embedded in the 2 N-dimensional phase space. The KAM theorem leads us to anticipate that the effect of non-resonant terms will be to slightly distort this torus. The effect of the resonant terms, on the other hand, is expected to destroy the torus, leading to transfer of excitation among the modes.

The nice and hoped for implication of this pretty picture is that one can simply omit from the $\underset{\sim}{k}$-sums in (2.22) all terms not close to a resonance. Since this includes the vast majority of terms, a great economy is achieved in doing calculations. In this context we note that the principal technique currently available for studying these phenomena uses a Boltzmann-type transport equation derived from the random phase approximation[11]. The only contribution to this equation comes from terms precisely at resonance. [In the quantum mechanical version for a gas this corresponds to energy conservation at each binary collision.]

To transform away the non-resonant terms, we first average over all phases $\theta_{\underset{\sim}{k}}$ (so that secular terms are not introduced) obtaining

---

*An excellent reference for techniques described here is the review article by Chirikov [Ref. (10)].

$$K(\underset{\sim}{J}) \equiv \langle H \rangle_\theta$$

$$= \sum_{\underset{\sim}{k}} J_{\underset{\sim}{k}} \omega_k + \sum_{\underset{\sim}{k}_1, \underset{\sim}{k}_2} J_{\underset{\sim}{k}_1} J_{\underset{\sim}{k}_2} \left[V_4\left(\underset{\sim}{k}_1, \underset{\sim}{k}_2, \underset{\sim}{k}_1, \underset{\sim}{k}_2\right) \right.$$

$$\left. + V_4\left(\underset{\sim}{k}_1, \underset{\sim}{k}_2, \underset{\sim}{k}_2, \underset{\sim}{k}_1\right)\right]$$

$$+ \sum_{\underset{\sim}{k}_1, \underset{\sim}{k}_2, \underset{\sim}{k}_3} J_{\underset{\sim}{k}_1} J_{\underset{\sim}{k}_2} J_{\underset{\sim}{k}_3} [\ldots] + \ldots \tag{3.1}$$

The "perturbation" is taken to be

$$V \equiv H - K = V_3 + V_4 + V_5 + \ldots . \tag{3.2}$$

with $V_i \equiv V_i - \langle V_i \rangle_\theta$ so that $\langle V \rangle_\theta = 0$. The Hamiltonian is now written as

$$H = K(\underset{\sim}{J}) + \sum_{\underset{\sim}{k}_1, \underset{\sim}{k}_2, \underset{\sim}{k}_3} \left[J_{\underset{\sim}{k}_1} J_{\underset{\sim}{k}_2} J_{\underset{\sim}{k}_3} \right]^{1/2}$$

$$\times \left[V_3\left(\underset{\sim}{k}_1, \underset{\sim}{k}_2, \underset{\sim}{k}_3\right) \cos\left(\theta_{\underset{\sim}{k}_1} - \theta_{\underset{\sim}{k}_2} - \theta_{\underset{\sim}{k}_3}\right) \right.$$

$$\left. \times V_3'\left(\underset{\sim}{k}_1, \underset{\sim}{k}_2, \underset{\sim}{k}_3\right) \cos\left(\theta_{\underset{\sim}{k}_1} + \theta_{\underset{\sim}{k}_2} + \theta_{\underset{\sim}{k}_3}\right)\right]$$

$$+ V_4 + V_5 + \ldots \tag{3.3}$$

If the perturbation V were neglected, the equations of motion would be

$$\dot{J}_{\underset{\sim}{k}} = - \frac{\partial K}{\partial \theta_{\underset{\sim}{k}}} = 0$$

$$\dot{\theta}_k = \frac{\partial K}{\partial J_{\underset{\sim}{k}}} \equiv \Omega_{\underset{\sim}{k}} (\underset{\sim}{J}) \tag{3.4}$$

To continue, we perform a canonical transformation from variables $J_{\underset{\sim}{k}}$, $\theta_{\underset{\sim}{k}}$ to $I_{\underset{\sim}{k}}$, $\phi_{\underset{\sim}{k}}$ in order to remove V_3 in (3.3). The generating function for this transformation is written in the notation of Goldstein[12] as

$$F_2(\underset{\sim}{I}, \underset{\sim}{\theta}) = \sum_{\underset{\sim}{k}} I_{\underset{\sim}{k}} \theta_{\underset{\sim}{k}} + R(\underset{\sim}{I}, \underset{\sim}{\theta}) , \tag{3.5}$$

with

$$J_{\underset{\sim}{k}} = \frac{\partial F_2}{\partial \theta_{\underset{\sim}{k}}} = I_{\underset{\sim}{k}} + \frac{\partial R}{\partial \theta_{\underset{\sim}{k}}} ,$$

$$\phi_{\underset{\sim}{k}} = \frac{\partial F_2}{\partial I_{\underset{\sim}{k}}} = \theta_{\underset{\sim}{k}} + \frac{\partial R}{\partial I_{\underset{\sim}{k}}} . \tag{3.6}$$

To eliminate third order terms from H we must define

$$\begin{aligned} R = - \sum_{\underset{\sim}{k}_1, \underset{\sim}{k}_2, \underset{\sim}{k}_3} \Bigg[& V_3\left(\underset{\sim}{k}_1, \underset{\sim}{k}_2, \underset{\sim}{k}_3\right) \sin\left(\theta_{\underset{\sim}{k}_1} - \theta_{\underset{\sim}{k}_2} - \theta_{\underset{\sim}{k}_3}\right) \Big/ \left(\Omega_{\underset{\sim}{k}_1} - \Omega_{\underset{\sim}{k}_2} - \Omega_{\underset{\sim}{k}_3}\right) \\ & + V_3'\left(\underset{\sim}{k}_1, \underset{\sim}{k}_2, \underset{\sim}{k}_3\right) \sin\left(\theta_{\underset{\sim}{k}_1} + \theta_{\underset{\sim}{k}_2} + \theta_{\underset{\sim}{k}_3}\right) \Big/ \left(\Omega_{\underset{\sim}{k}_1} + \Omega_{\underset{\sim}{k}_2} + \Omega_{\underset{\sim}{k}_3}\right) \Bigg] \\ & \times \left[I_{\underset{\sim}{k}_1} I_{\underset{\sim}{k}_2} I_{\underset{\sim}{k}_3} \right]^{1/2} \end{aligned} \tag{3.7}$$

The above process can be repeated. Averaging the new Hamiltonian over phases leads to a new K(J) and a second canonical transformation can be made to eliminate terms through the fifth order. A third canonical transformation will eliminate terms through the seventh order, etc. This process is called "super convergent" by Chirikov.[10]

The difficulty with the above is that for continuum mechanics resonances occur and the perturbation theory fails. If, for example, we can find a set of wave numbers $\underset{\sim}{k}_1$, $\underset{\sim}{k}_2$, $\underset{\sim}{k}_3$ such that

$$\Omega_{\underset{\sim}{k}_1} - \Omega_{\underset{\sim}{k}_2} - \Omega_{\underset{\sim}{k}_3} = 0 , \tag{3.8}$$

corresponding to a "resonant triad", the expression (3.7) is meaningless. Even when triad resonances are not found, higher order resonances can be expected. For surface gravity waves, for example, the first resonance encountered is at fourth order ("resonant quartet").

This problem can be avoided (but at a cost!) as follows. We simply exclude from the $\underset{\sim}{k}$-sums in (3.7), and in subsequent transformations, terms "close to resonance". How "close to resonance" may be determined in a way described below. We now have remaining in the Hamiltonian, of course, those terms in V_3, etc., which were not included in R.

What have we gained? If we are going to do numerical computation we have gained a great deal. Most of the terms in our equations of motion (those not close resonance) have been eliminated. Our Hamiltonian is now of the form

$$H = K(\underset{\sim}{I}) + \sum_{\underset{\sim}{k}_1,\underset{\sim}{k}_2,\underset{\sim}{k}_3}^{R} \left[I_{\underset{\sim}{k}_1} I_{\underset{\sim}{k}_2} I_{\underset{\sim}{k}_3}\right]^{1/2}$$

$$\times \left[V_3\left(\underset{\sim}{k}_1,\underset{\sim}{k}_2,\underset{\sim}{k}_3\right) \cos\left(\theta_{\underset{\sim}{k}_1} - \theta_{\underset{\sim}{k}_2} - \theta_{\underset{\sim}{k}_3}\right)\right]$$

$$+ \dots , \qquad (3.9)$$

Where $\sum^{R}$ implies a sum over only those terms "near resonance".

To determine how "close to resonance" is required, we can add terms further and further from resonance, until the added terms have a negligible effect on our calculations.

4. DISCUSSION OF SIMPLE EXAMPLES

We begin our discussion of the Hamiltonian (3.9) for the simple case of three modes only. This represents an integrable system.

The Hamiltonian is of the form

$$H = K(I_1,I_2,I_3) + G\sqrt{I_1 I_2 I_3} \cos(\phi_1 - \phi_2 - \phi_3) . \qquad (4.1)$$

We first transform to new variables P,ψ with the generating function

$$F_2(P_2,\underset{\sim}{\phi}) = P_1(\phi_1 - \phi_2 - \phi_3) + P_2\phi_2 + P_3\phi_3 . \qquad (4.2)$$

Thus,

$$\psi_1 = \frac{\partial F_2}{\partial P_1} = \phi_1 - \phi_2 - \phi_3 ,$$

$$\psi_2 = \phi_2 ,$$

$$\psi_3 = \phi_3 ,$$

$$I_1 = \frac{\partial F_2}{\partial \phi_1} = P_1 ,$$

$$I_2 = P_2 - P_1 ,$$

$$I_3 = P_3 - P_1 . \tag{4.3}$$

The new Hamiltonian is

$$H = K(P_1, P_2 - P_1, P_3 - P_1) + G[P_1(P_2 - P_1)(P_3 - P_1)]^{1/2} \cos \psi_1 . \tag{4.4}$$

We note first that

$$\dot{P}_2 = - \frac{\partial H}{\partial \psi_2} = 0 , \quad \dot{P}_3 = - \frac{\partial H}{\partial \psi_3} = 0 ,$$

so

$$P_2 = \text{const}, \quad P_3 = \text{const} . \tag{4.5}$$

Also,

$$\dot{\psi}_1 = \frac{\partial H}{\partial P_1} = (\Omega_1 - \Omega_2 - \Omega_3) + \frac{G}{2} \left\{ \left[\frac{(P_2 - P_1)(P_3 - P_1)}{P_1} \right]^{1/2} - \left[\frac{P_1(P_3 - P_1)}{P_2 - P_1} \right]^{1/2} - \left[\frac{P_1(P_2 - P_1)}{P_3 - P_1} \right]^{1/2} \right\} \cos \psi_1 ,$$

$$\dot{P}_1 = - \frac{\partial H}{\partial \psi_1} = G[P_1(P_2 - P_1)(P_3 - P_1)]^{1/2} \sin \psi_1 . \tag{4.6}$$

We can assume without loss of generality that $P_3 > P_2$. Since the I's must be positive we have

$$0 < P_1 < P_2 . \tag{4.7}$$

It is evident from (4.6) that P_1 will oscillate between limits P_{min} and P_{max}.

Let us first suppose that

$$\Omega_1 - \Omega_2 - \Omega_3 \equiv \Delta G = \text{const} . \tag{4.8}$$

The Hamiltonian is then

$$\frac{H}{G} = \Delta P_1 + [P_1(P_2 - P_1)(P_3 - P_1)]^{1/2} \cos \psi_1$$

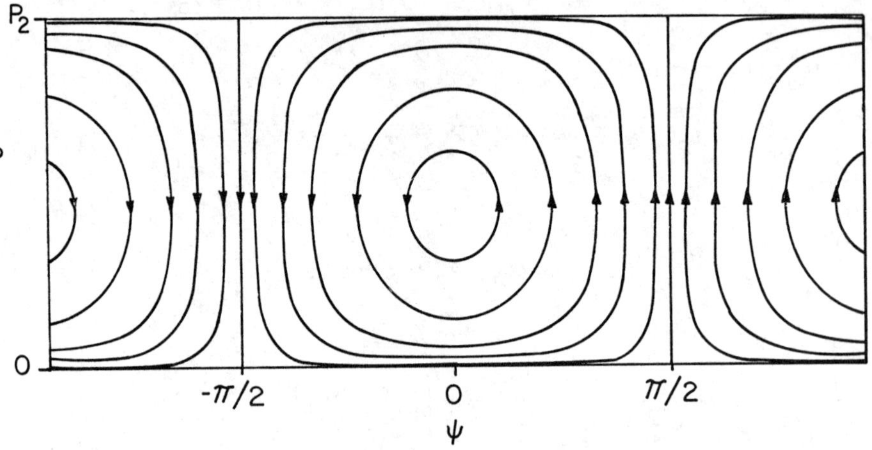

Figure 1. Illustration of Orbits for Eq. (4.9) when $\Delta = 0$, $P_2 = 1$, $P_3 = 3$. Fixed points are at $P_1 \cong P_2/2$. The central orbit is for $E = .75$, while the rectangular orbit is for $E = 0$.

$$= E = \text{const} . \qquad (4.9)$$

We consider first the case of exact resonance, corresponding to $\Delta = 0$. Typical phase space orbits in the $P_1 - \psi_1$ plane are shown in Fig. 1 . It is seen that all orbits correspond to a trapped phase ψ_1. For $-\pi/2 < \psi_1 < \pi/2$, $E > 0$; for $\pi/2 < \psi_1 < 3\pi/2$, $E < 0$. For those orbits corresponding to near maximum excursions for P_1 the phase ψ_1 tends to "switch" discontinuously from $-\pi/2$ to $\pi/2$. This is illustrated in Fig. 2 where P_1 and ψ_1 are shown as functions of time. This "switching" gives the maximum coupling among the oscillators, as is seen from (4.6), since then $\sin \psi_1 \cong \pm 1$.

Let us next consider the case that $\Delta > 0$. Orbits in the $P_1 - \psi_1$ plane are illustrated in Fig. 3 . A separatrix is observed at $E = \Delta P_2$ and at $E = 0$. For $0 < E < \Delta P_2$ the phase ψ_1 is not "trapped" but increases indefinitely. For $E > \Delta P_2$ we again have a "trapped" ψ_1. As Δ becomes very large the "trapped" domain becomes very small. Examples of the motion are shown in Fig. 4 .

Another case of interest is that for which

$$\Omega_1 - \Omega_2 - \Omega_3 = \alpha G P_1 \qquad \alpha = \text{const.}$$

This corresponds to exact resonance when P_1 is small. The system "detunes" as P_1 becomes larger. The orbits for this case are somewhat similar to those in Fig. 3 , the central fixed point being displaced toward the P_2-line. This effect is termed "non-linear stabilization" by Chirikov.[10]

Hasselmann[13] has given a frequently quoted "stability condition" for a single resonant triad. For the case that

$$\Omega_1 = \Omega_2 + \Omega_3,$$

he shows that for*

$$I_1 \gg I_2, I_3, \text{ system unstable}$$

$$I_2 \gg I_1, I_3 \text{ or } I_3 \gg I_1, I_2, \text{ system stable} \qquad (4.10)$$

(amplitudes purely oscillatory)

*Hasselmann expressed this in terms of the action amplitudes a_1, a_2, a_3.

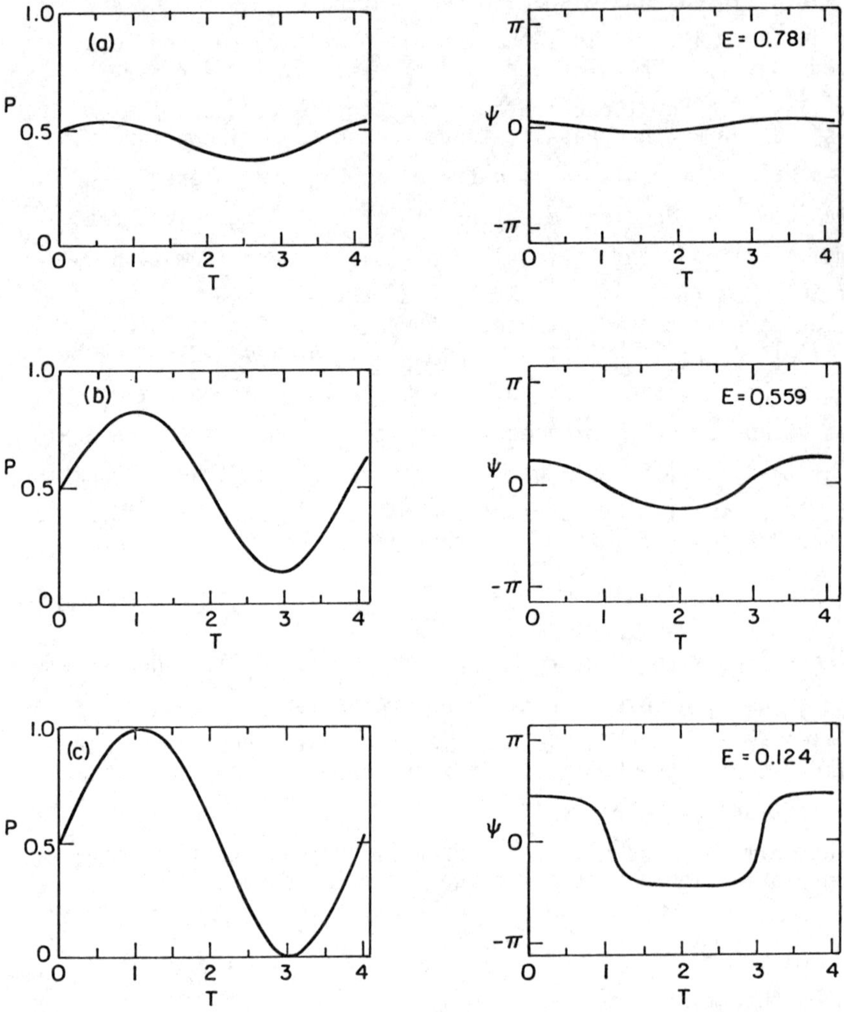

Figure 2. Time dependence of P_1 and ψ_1 for Hamiltonian (4.9) when $\Delta = 0$, $P_2 = 1$, $P_3 = 3$. (a) E = .781, (b) E = .559, (c) E = .124.

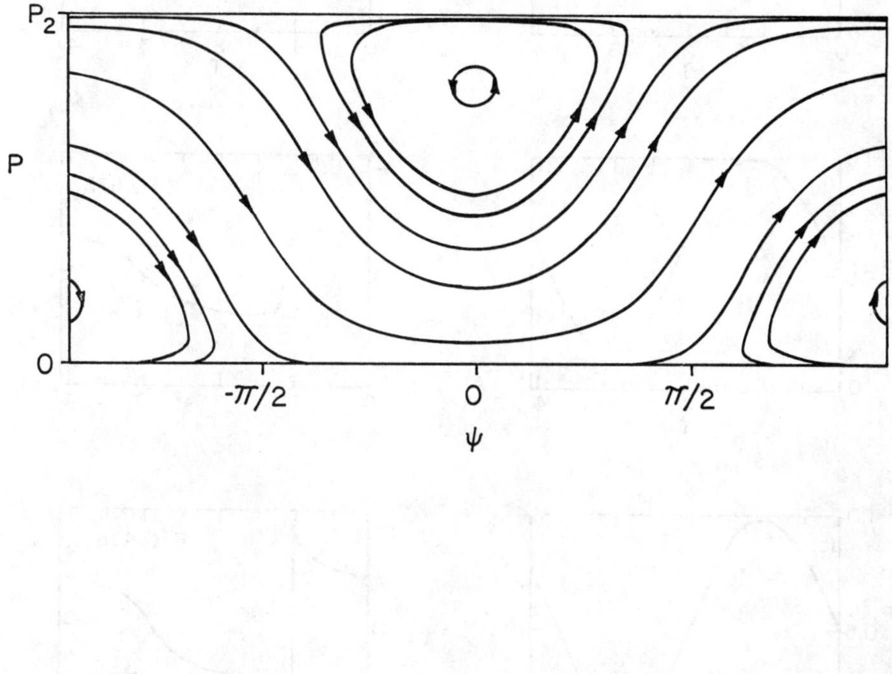

Figure 3. Illustration of orbits for Eq. (4.9) when $\Delta > 0$. Separatrices are curves for $E = \Delta P_2$ and $E = 0$. As Δ increases central fixed point moves up from approximately $P_2/2$ to P_2.

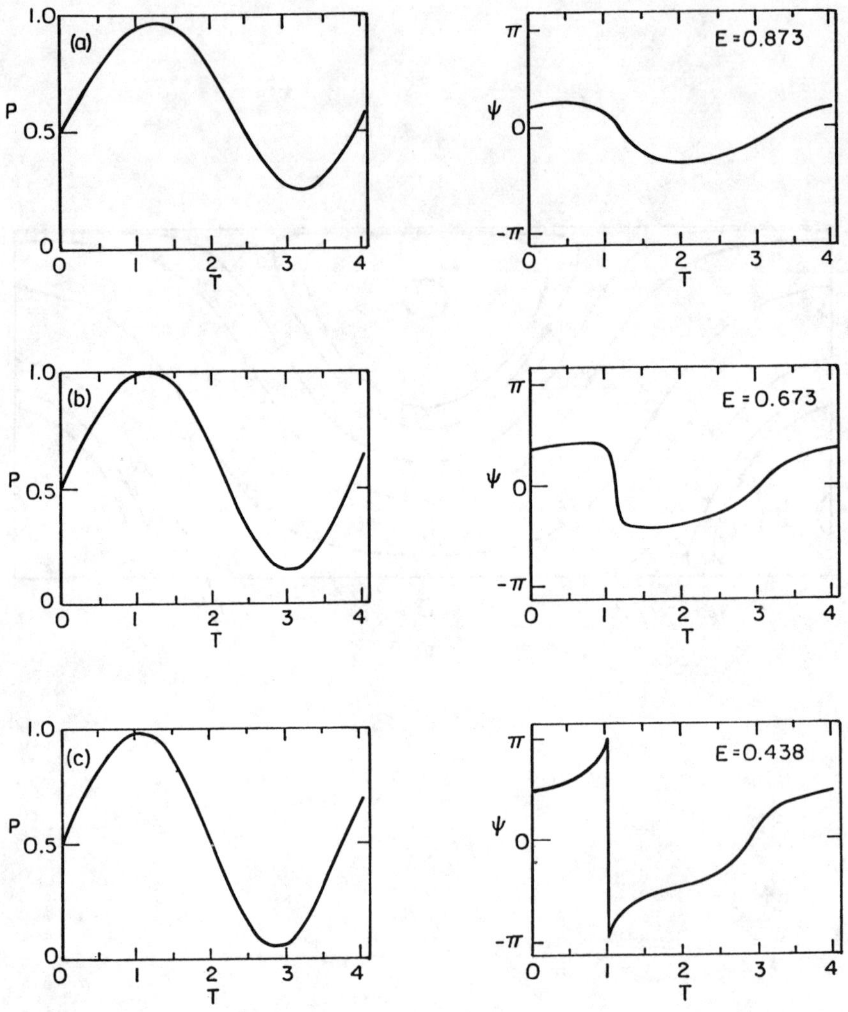

Figure 4. Time dependence of P_1 and ψ_1 for Hamiltonian (4.9) when $\Delta = 0.628$, $P_2 = 1$, $P_3 = 3$. (a) E = 1.095; (b) E = 0.673; (c) E = 0.438. In (c) the phase was reduced by 2π on the plot when it became greater than π.

Hasselmann's conditions can be easily understood from the phase diagram of Fig. 1 . The unstable case corresponds to an initial position in the upper left corner of the central rectangle in Fig. 1 . The exponential growth of J_2 and J_3 corresponds to moving down along the line $\psi_1 = \pi/2$. The stable case corresponds to an initial position near the elliptic fixed point. We shall later describe a generalization of the Hasselmann stability condition.

As a slightly more complex example, we consider three "high frequency" oscillators coupled to one of lower frequency for which

$$H = J\omega_0 + \sum_{n=1}^{3} J_n\omega_n + G[\sqrt{JJ_1J_2}\cos(\theta_2 - \theta_1 - \theta_0) + \sqrt{JJ_2J_3}\cos(\theta_3 - \theta_2 - \theta_0)] . \tag{4.11}$$

We suppose that

$$\omega_2 - \omega_1 - \omega_0 = 0$$

$$\omega_3 - \omega_2 - \omega_0 = \Delta \text{ is small.} \tag{4.12}$$

Thus "0", "1", and "2" are in exact resonance and "0", "2", and "3" are near resonance.

The equations of motion obtained from (4.11) were integrated for a model of a single internal wave mode interacting with three surface wave modes[7] ("0" corresponding to the internal wave mode"). In Fig. 5 we see the time dependence of the wave amplitudes (proportional to $\sqrt{J}$) in arbitrary units. The quantity Δ was chosen as

$$\Delta^{-1} = 1 ,$$

in the plotted time units. Mode "3" is effectively decoupled because Δ is too large. In Fig. 6 , on the other hand, Δ was chosen as

$$\Delta^{-1} = 5$$

in the same units. When the amplitude J becomes large enough to give effective coupling, mode "3" now participates in the exchange of energy among the oscillators.

The above example illustrates that one can remove from consideration, as was done in (3.9), terms far enough from resonance. Some numerical experimentation can be done to determine how "far" this should be.

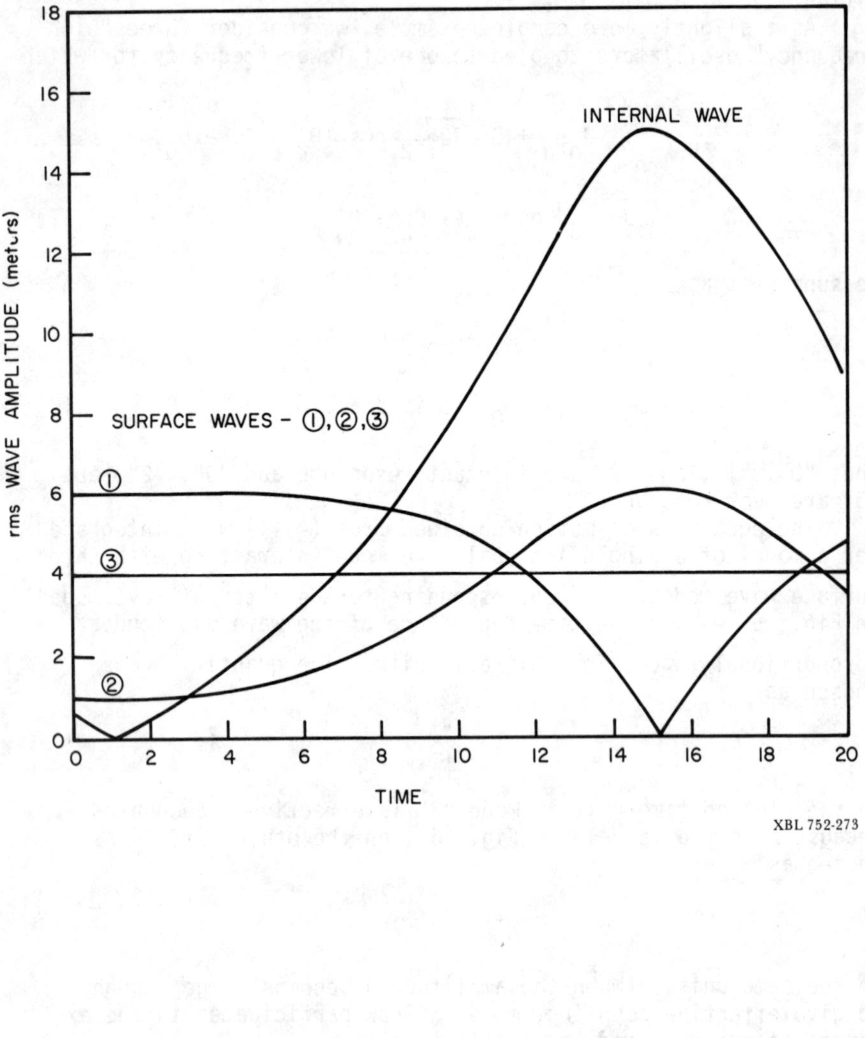

Figure 5. Illustration of motion for the Hamiltonian (4.11) for the slightly off resonance case for mode 3.

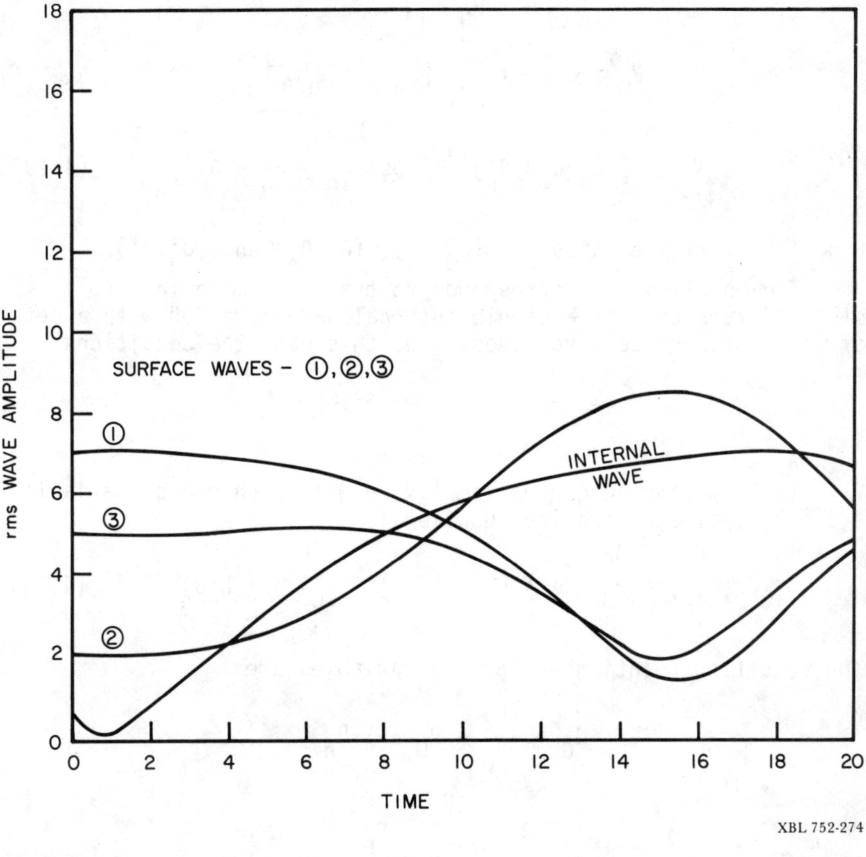

Figure 6. Illustration of motion for the Hamiltonian (4.11) for the case of resonance for mode 3.

As a somewhat more complex example, we consider a single "low frequency" mode "0" coupled to M pairs of "high frequency" oscillators. The Hamiltonian is

$$H = H_0 + V ,$$

$$H_0 = J_0\omega_0 + \sum_{n=1}^{M} [J_n\omega_n + J_n'\omega_n']$$

$$V = - \sum_n G_n[J_0 J_n J_n']^{1/2} \cos (\theta_n - \theta_n' - \theta_0) . \qquad (4.13)$$

Canonical variable pairs are (J_0,θ_0), (J_n,θ_n) and (J_n',θ_n').

The model (4.13) corresponds to the study made in Ref. 7 of the interaction of a single internal wave mode "0" with a set of pairs of surface wave modes. For this case the conditions

$$\omega_0 << \omega_n , \qquad \omega_0 << \omega_n'$$

obtain.

The Hamiltonian can be simplified with a change of variables $\underset{\sim}{J},\underset{\sim}{\theta} \rightarrow \underset{\sim}{P},\underset{\sim}{\psi}$. The generating function is

$$F_2(\underset{\sim}{P},\underset{\sim}{\theta}) = P_0\theta_0 + \sum_n P_n(\theta_n - \theta_n' - \theta_0) + \sum_n P_n'\theta_n' . \qquad (4.14)$$

The resulting relations among the variables are

$$J_0 = \frac{\partial F_2}{\partial \theta_0} = P_0 - \sum_n P_n ,$$

$$J_n = \frac{\partial F_2}{\partial \theta_n} = P_n$$

$$J_n' = P_n' - P_n$$

$$\psi_0 = \frac{\partial F_2}{\partial P_0} = \theta_0$$

$$\psi_n = \frac{\partial F_2}{\partial P_n} = \theta_n - \theta_n' - \theta_0 ,$$

$$\psi_n' = \theta_n' . \qquad (4.15)$$

The new Hamiltonian is

$$H = P_0\omega_0 + \sum_n P'_n\omega'_n + \sum_n P_n(\omega_n - \omega'_n - \omega_0)$$

$$- \sum_n^M G_n\left[\left(P_0 - \sum_\ell^M P_\ell\right) P_n(P'_n - P_n)\right]^{1/2} \cos \psi_n . \tag{4.16}$$

We note that

$$P_0 = \text{constant}$$

$$P'_n = \text{constant} . \tag{4.17}$$

Also,

$$\dot{P}_n = - G_n\left[\left(P_0 - \sum_\ell^M P_\ell\right) P_n(P'_n - P_n)\right]^{1/2} \sin \psi_n$$

$$\dot{\psi}_n = (\omega_n - \omega'_n - \omega_0) - \frac{G_n}{2}\left\{\left[\frac{\left(P_0 - \sum_\ell^M P_\ell\right)(P'_n - P_n)}{P_n}\right]^{1/2} - \left[\frac{\left(P_0 - \sum_\ell^M P_\ell\right) P_n}{(P'_n - P_n)}\right]^{1/2}\right\} \cos \psi_n$$

$$+ \sum_m^M \frac{G_m}{2}\left[\frac{P_m(P'_m - P_m)}{P_0 - \sum_\ell^M P_\ell}\right]^{1/2} \cos \psi_m . \tag{4.18}$$

From (4.15) and (4.17) it follows that

$$\dot{J}'_n = -\dot{J}_n = -\dot{P}_n$$

$$\dot{J}_0 = \sum_n^M \dot{J}'_n \tag{4.19}$$

We shall assume here that the coefficients G_n are all positive.

Hasselmann's stability condition (4.10) leads us to anticipate that the excitation transfer rate will be greatest when the high frequency mode amplitudes are largest. To study this we first consider the limit that

$$J_0 \ll J'_n \ll J_n , \tag{4.20}$$

and exact resonance obtains; that is, $\omega_n - \omega'_n - \omega_0 = 0$. We might guess that this case is similar to that of Eqs. (4.6) for those orbits in Fig. 1 close to the boundaries of the rectangle. In that case the sin ψ_1 in (4.6) tended to switch between the values of ± 1. This leads us to look for a solution with

$$\psi_n = \frac{\pi}{2} + \eta_n, \qquad |\eta_n| \ll 1, \qquad \sin \psi_n \cong 1. \tag{4.21}$$

Using the expressions (4.20) and (4.21), we rewrite (4.18) in the appropriate form

$$\dot{J}'_n \cong G_n[J_0 J_n J'_n]^{1/2} , \qquad \dot{J}_0 = \sum_n^M \dot{J}'_n ,$$

$$\dot{\eta}_n \cong - \sum_m^M \frac{G_m}{2} \left[\frac{J_m J'_m}{J_0}\right]^{1/2} \eta_m . \tag{4.22}$$

The second of these equations has the solution

$$\eta_n(t) = \eta_n^0 e^{-\Gamma t} + c_n, \; \Gamma \equiv \sum_{m=1}^M \frac{G_m}{2} \left[\frac{J_m J'_m}{J_0}\right]^{1/2} . \tag{4.23}$$

The c_n here are integration constants subject to the self consistency condition

$$\sum_n \frac{G_n}{2} \left[\frac{J_n J'_n}{J_0}\right]^{1/2} c_n = 0. \tag{4.24}$$

We shall see in a moment that Γ is a constant, and we observe that

$$\Gamma > 0 .$$

Equations (4.23) indicate that in this first approximation the phases η_n are not strongly driven by the interaction. They decay to the weighted mean of zero indicated by (4.24). To further study their variation we must look at the next order term in the η_n equations [the middle term on the right in (4.18)]. This has the proper sign to lead to decay of the η_n's toward zero, a conclusion which depends on the condition (4.20) that $J'_n \ll J_n$.

The first two of equations (4.22) can be combined in the form

$$J_0 = \sum_m G_m [J_0 J_m J'_m]^{1/2} .$$

A solution valid near t = 0 may be obtained by setting

$$J_m \cong B_m, \; J'_m \cong B'_m, \text{ all constants,}$$

and

$$J_0 = \gamma(t + t_0)^2 ,$$

$$\gamma = [1/2 \sum_m G_m (B_m B'_m)^{1/2}]^2 . \qquad (4.25)$$

The parameter t_0 is chosen to fix the initial value of J_0.

The second of equations (4.22) implies that J_0 grows more rapidly than do the J'_m (unless M = 1!). When J_0 has reached a value such that $J_0 \simeq \sum_m J'_m$, the exponential growth model described in ref. (7) may be expected to be valid if the relation $J_n \gg J_0$ is still valid. In this case,

$$J_0 \sim e^{\alpha t} , \; \alpha = \left[\sum_m G_m^2 B_m \right]^{1/2} . \qquad (4.26)$$

Numerical calculations (to be reported elsewhere) show consistency with the above conclusions. The relation (4.21) that $\sin \psi_m \approx 1$ appears to valid for a much wider range of initial conditions than our arguments would suggest. The growth of J_0 implied by (4.25) and (4.26) is also observed in numerical calculations.

The interaction of a single small amplitude internal wave mode with a surface wave field was studied in Ref. (7). The purpose was to investigate the rate at which the surface wave field could regenerate a depleted internal wave field. The wave number vector for the surface waves of each triad are related by

$$\underset{\sim}{k}'_n = \underset{\sim}{k}_n - \underset{\sim}{k}_0 . \qquad (4.27)$$

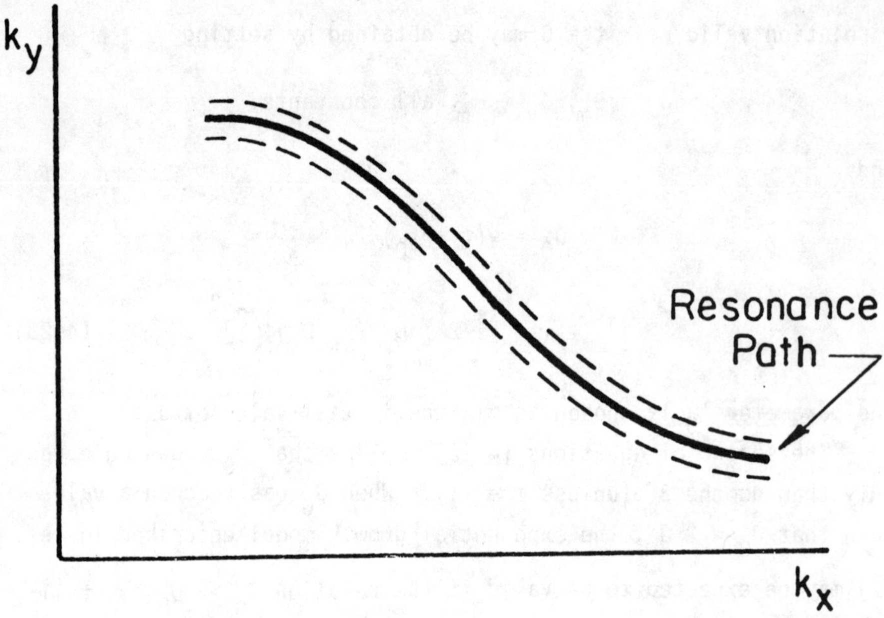

Figure 7. The resonance curve in the $\tilde{k}$-plane, as determined from Eqs. (4.27) and (4.28), is illustrated.

The resonance condition is

$$\omega_k = \omega_{k'} + \omega_0 \; . \tag{4.28}$$

If the $\underset{\sim}{k}'_n$ are determined by (4.27), we see that resonance occurs along a curve in the k-plane, as is illustrated in Fig. 7. This curve is spread into a path by the finite resonance width. Numerical calculations with as many as M = 72 surface wave pairs were reported in Ref. (7). Although the J'_n were comparable in magnitudes to the J_n (continuity in the mean of the surface wave spectrum), the results agreed surprisingly well, when J_0 was small, with the solution (4.25). The initial growth rate of J_0 tended to increase until the phases "locked" on the values implied by (4.21). A "plateau" was then reached in the growth rate, consistent with the value implied by (4.26). Finally, the growth rate of J_0 began to decrease and J_0 itself would presumably have begun to decrease had the calculation been continued.

In the work reported in Ref. (7) the effect of detuning from resonance was studied. This was done by sequentially adding more modes further from the resonance line until the effect of the added modes had no noticeable effect on the growth rate. Not surprisingly, this occured when

$$|\omega_n - \omega'_n - \omega_0|$$

was several times α.

We describe one final model taken from Ref. (8). The process considered is the radiation of a plasma with an intense laser beam of frequency $\omega_1 \gg \omega_0$, the plasma frequency. Quanta of energy $\hbar\omega_0$ are exchanged from beam to plasma leading to a second beam of frequency $\omega_2 = \omega_1 - \omega_0$. This continues leading to more beams of frequencies $\omega_3 = \omega_2 - \omega_0$, etc. The Hamiltonian to describe this, using classical theory, is of the form

$$H = J_0\omega_0 + \sum_{n=1}^{M} J_n\omega_n - \sum_{n=1}^{M-1} G_n[J_0J_nJ_{n+1}]^{1/2} \cos(\theta_n - \theta_{n+1} - \theta_0) \; . \tag{4.29}$$

A constant of the motion of this case is

$$J_{tot} = \sum_{n=1}^{M} J_n \; . \tag{4.30}$$

Numerical integration, beginning with the energy concentrated in mode "1", showed that the excitation was passed from mode-to-mode, reminiscent of a row of falling dominoes. The "locked phase" approximation appeared to agree reasonably well with the numerical calculations for this case also.

In this case, also, a generalized form of Hasselmann's stability condition is encountered. When the initial energy was concentrated in Mode M, there was no comparable flow of excitation up the chain--only a weak excitation of the two or three neighboring modes.

When a slight frequency mismatch is introduced at an intermediate point in the chain a build-up of excitation at the mismatch was found. A weak "ripple" of excitation was reflected back toward higher frequencies, but propagated only a few modes.

5. CONJECTURES AND QUESTIONS

We have discussed a number of simple examples and should properly return to the general case described by (2.21) with a definitive discussion. Our present understanding of these phenomenon is unfortunately so limited that we can only describe some crude calculational models, raise questions, and make conjectures.

Available numerical studies and theoretical analyses, such as the KAM theorem, suggest that the truncated Hamiltonian (3.9) is probably quite adequate. The "wealth of low order resonances" encountered for the continuum systems we have in mind suggests that (3.9) can be further truncated at a low order. Thus, we can anticipate that numerical integration for some hundreds (or more) of coupled modes is practical. For quasi-two-dimensional systems, such as ocean surface and internal waves, this may be quite realistic.

The "wealth of resonances" mentioned above, forming a network connecting all regions of wave-number space, invites one to conjecture "ergodicity". We note, however, the recent report[14,15] of something like a Fermi-Pasta-Ulam recurrence cycle following the Benjamin-Feir[16] instability of surface waves.

Following the work of Hasselmann (see Ref. (11) and earlier papers) a Boltzmann-type, or radiative transfer, equation has been used to discuss the mean properties of weakly non-linear continuum systems. The mean intensity over an ensemble is introduced as, for example,

$$F(\underset{\sim}{k}) = \text{const} \times \langle |a_k|^2 \rangle \tag{5.1}$$

[see (2.13)]. The random phase approximation is invoked to deduce a transport equation for the evolution of $F(\underset{\sim}{k})$ in time. The only exchange of excitation occurs at resonance.

The derivation of such transport equations depends in an essential manner on the random phase (or "quasi-Gaussian") assumption that the phases $\theta_{\underset{\sim}{k}}$ in (2.13) behave in a stochastic manner. For the simple examples which we have discussed here this has not been the case. The "trapping", or locking of phases, has been important in determining the rate at which excitation is transferred. For more complex systems, it is not obvious what will happen. Different resonances may "compete" for a given mode phase, causing it to "jump" from one domain to another. If this happens on a time scale short compared to that for significant transfer of excitation between modes, one might expect the random phase approximation to be valid.

The reader interested in an application of the transport equation might refer to Ref. (6), where energy transfer in an internal wave field is studied.

REFERENCES

1. See, for example, O. M. Phillips, The Dynamics of the Upper Ocean (Cambridge Univ. Press, London, 1966).
2. A recent review is given by T. P. Barnett and K. E. Kenyon Rep. Prog. Phys. 38, 667 (1975).
3. C. J. R. Garrett and W. H. Munk J. Geophys. Res. 80 (3), 291 (1975), and preceding papers.
4. Stanley Flatté, Editor, Sound Transmission Through a Fluctuating Ocean (Cambridge Univ. Press, London, 1978).
5. J. W. Miles, J. Fluid Mech. 83, 153 (1977); see also D. M. Milder, J. Fluid Mech. 83, 159 (1977).
6. C. H. McComas and F. P. Bretherton, J. Geophys. Res. 82(3), 1397 (1977).
7. K. Watson, B. West and B. Cohen, J. Fluid Mech. 77, 185 (1976).
8. B. Cohen, A. Kaufman and K. Watson, Phys. Rev. Lett. 29, 581 (1972).
9. The a_k variables were used, for example, by B. West, K. Watson and A. Thomson, Phys. Fluids 17, 1059 (1974), to describe surface waves.
10. B. V. Chirikov (1977), "A Universal Instability of Many-Dimensional Oscillator Systems".
11. K. Hasselman, Rev. Geophys. 4, 1 (1966).
12. H. Goldstein Classical Mechanics (Addison-Wesley, Inc., Reading Mass., 1959).
13. K. Hasselmann, J. Fluid Mech. 30, 737 (1967).
14. B. M. Lake, H. C. Yuen, H. Rungaldier, and W. E. Gerguson, J. Fluid Mech. 83, 49 (1977).
15. B. M. Lake and H. C. Yuen, J. Fluid Mech. 83, 75 (1977).
16. T. B. Benjamin and J. E. Feir, J. Fluid Mech. 27, 214 (1967).

ORDER AND DISORDER IN NONLINEAR FLUID MOTIONS

Greg Holloway
Physical Dynamics, Inc.
La Jolla, California

ABSTRACT

Most commonly encountered flows and most pressing fluid engineering and environmental questions concern motions which are neither weakly nonlinear nor strongly nonlinear, neither simply ordered nor chaotically disordered. In this note, I very briefly sketch some recent efforts to provide a unifying theoretical description which ranges continuously from small amplitude perturbation theory to statistical turbulence theory. Results to date are in agreement with numerical-hydrodynamical simulations. Extension of these ideas may describe a flow which, for short time, is solved deterministically while, for times longer than an inherent predictability time, is given only statistically.

INTRODUCTION

Everyday experience conditions us to the complex role played by nonlinear terms in the equations of fluid flow, e.g. by the advection of momentum and of buoyancy in the air around us. Life as we know it would not exist but for this complicated flow.

The methodology of linear partial differential equations proves largely ineffective for such real flows. Perturbation methods may describe the weakly nonlinear flow but usually only over a span of time which scales as some inverse power of the relative strength of nonlinearity. In some cases finitely nonlinear flows can be solved, e.g. the distortion of a steady, laminar flow. A good deal of interest currently fastens on systems which support soliton solutions essentially laminar, finite amplitude solutions with the ability to propagate and to survive collisions with other solitons. The latter remarkable property suggests that complicated flows might be described as a collection of solitons (a "soliton gas") plus an incoherent background ("radiation"). This picture is purely speculative. For strongly nonlinear flows, we expect the appearance of chaotic motion. Idealized as homogeneous, isotropic turbulence, such flows have been described approximately by dimensional scaling rules and by various statistical "closure" hypotheses. Analytical results are almost unavailable, mainly because the understood perturbative methods for weak nonlinearity here produce strongly divergent series whose divergences cannot yet be "removed" in any systematic way.

Characteristically, reality provides flows that are not so weak as to be close to a solvable linear flow, that do not seem to support or to exhibit solitons, and that are not so chaotic as to suggest any simple statistical approximation. Advances in the size and speed of modern computers have made numerical modelling a popu-

ISSN: 0094-243X/78/324/$1.50

lar means for investigating these flows of intermediate complexity. However, direct simulation, for example of the three-dimensional oceanic internal wave field from length scales of kilometers to dissipation scales of centimeters, is far beyond any foreseeable computation. The model must consider interactions among a smaller number, perhaps a few tens or tens of thousands, of modes. This leaves undiminished the theoretical question which here concerns the relation between the interactions, however complicated, among a "sparse" set of modes and the interaction among the infinitely many modes of the continuum.

I will describe an approach which may begin to bridge the separation between perturbation theories of weakly nonlinear motions and statistical approximations for strong, chaotic motions. The discussion will be rather general but one may think, for example, of the oceanic internal wave field. In this example, large scales propagate substantially as linear waves interacting only over many wave periods while smaller scales interact more strongly in times which may be comparable to wave periods. For a sufficiently energetic field, still smaller scales of motion may become disorganized until finally suppressed by viscosity. While the large scales may fairly be described as weakly nonlinear, their motion is not, and cannot be, isolated from motions on intermediate scales and so on to dissipation scales. I will sketch a means by which some selected motion could be treated explicitly, perhaps by numerical modelling, while the interaction with other scales is given statistically.

EQUATIONS OF MOTION

Suppose that the equations of a fluid motion can be expressed by an infinite set of coupled, first order, ordinary differential equations for mode amplitudes $a_\ell(t)$ where a vector field $\underline{y}(\underline{x},t)$ is represented on an orthogonal set of basis functions

$$\underline{y}(\underline{x},t) = \sum_{\ell \in S} a_\ell(t)\, \underline{F}_\ell(\underline{x}) .$$

Components of $\underline{y}$ may be the fluid velocity, density, temperature or whatever. $\underline{y}$ and $\underline{F}$ are defined on some finite spatial domain and satisfy given boundary conditions. Then S is a denumerably infinite index set. Assume the equations of motion are

$$\{\dot{a}_\ell + \sum_m b_{\ell m} a_m + \sum_{m,n} c_{\ell mn} a_m a_n + f_\ell = 0;\ (\ell,m,n) \in S \times S \times S\} \quad (1)$$

where $\dot{a}_\ell \equiv d/dt\, a_\ell$, $b_{\ell m}$ and $c_{\ell mn}$ are constant matrices and $f_\ell(t)$ is a mode amplitude of any external force, body heating, etc.

Equation (1) is taken to be first order in time and quadratically nonlinear after continuity equations for mass and momentum. Although we will begin with the assumed form (1), it should be recognized that the ability to write (1) explicitly usually depends upon an exceedingly nice specification, particularly of boundary conditions. In the spatial domain, the equation of motion involves linear and nonlinear integro-differential operators. Integral operators occur because it is often convenient to remove acoustic effects by approximating the flow as incompressible with the consequence that information propagates to any $\underline{x}$ from distant $\underline{x}'$ instantly. Nonlinear or inhomogenous boundary conditions are usually intractable unless considered perturbatively in small amplitude. The condition that the free surface of an ocean be a boundary at constant pressure generates nonlinearity of all degrees.

With such caveats in mind, we return to (1). Linear effects arise from molecular viscosity or conduction and from the cross coupling of components of $\underline{y}$ by imposed mean shearing or by buoyancy effects, resulting in wave propagation. Nonlinear effects are due to advection. We do not consider, e.g., free surface effects. Complex $\underline{F}_\ell$ may be chosen so that $b_{\ell m}$ is diagonal. Reality of $\underline{y}$ then imposes certain conjugate symmetries on the complex set $\{a_\ell\}$. Still the infinitely dimensioned system (1) is intractable. However all but a finite number of modes correspond to motion on scales so short as to be wholly dissipated by viscosity or conduction. Thus we may retain only a finite set of modes indexed by $S_1 \subset S$ provided S_1 includes modes which are sufficiently strongly dissipative. Then

$$\{\dot{a}_\ell + (i\omega_\ell + \nu_\ell)a_\ell + \sum_{m,n} c_{\ell mn} a_m a_n + f_\ell = 0;\ (\ell,m,n) \in S_1 x S_1 x S_1\} \tag{2}$$

where ω_ℓ and ν_ℓ are the real frequency and dissipation rates in mode ℓ.

Although finite, S_1 is still quite large with typically 10^6 to 10^{12} or more members so that (2) remains practically incalculable. Elsewhere in these proceedings, Professor Watson has suggested that it may be interesting to consider a model interaction among the members of a very much smaller set S_2. For members of S_2, molecular dissipation may be entirely negligible. Then, without forcing, the motion on S_2 may be described by an autonomous Hamiltonian so that methods of classical mechanics, e.g. contact transformations, and recent advances in mechanics, e.g. the Kolmogorov-Arnold-Moser Theorem, could lead to some understanding. Actually there is no basis for isolating motions on S_2 from the motions on $S_1 - S_2$. Indeed, for an autonomous Hamiltonian system we should expect to see, over time average, evolution towards an equipartition state which is very unlike fluid motion. Failure of a Hamiltonian system to evolve toward equipartition would be more the occasion for remark, e.g. as the Fermi-Pasta-Ulam problem. Later I will indicate a way by which motion on S_1 may be projected, in the sense of typical realizations of the motion, onto S_2 but only by introducing non-Hamiltonian effects on S_2. Such effects would appear even if S_1 were governed by a Hamiltonian.

The problem of solving system (2) is not just a problem of practical calculability. Although unique solutions certainly exist for all $t>0$, given any initial set $\{a_\ell(0)\}$, it is usually the case that solutions are not of uniformly bounded sensitivity in the initial conditions. Thus it is inappropriate to solve (2) beyond some time $t=t_p$, the "predictability time". For $t<t_p$, one could apply, e.g., perturbative methods or assume, e.g., Hamiltonian evolution on a very small set S_2. Often our interest concerns flows which have been evolving for a long time in a more or less statistically stationary way. Then it is appropriate to consider an ensemble of realizations of the motion and to predict only the evolution of the ensemble averaged statistics of the motion. Finally we could aspire to a calculation which, for $t<<t_p$, solves a particular realization of the motion but which gradually loses definition until, for $t>>t_p$, only statistical information is available.

STATISTICAL FORMULATION

At first I will treat the entire flow field statistically. In many applications this is just the result we seek, e.g. predictions of average spectral evolution of energy, of stress tensor, of heat or buoyancy flux, etc. Denoting ensemble average by angle braces, we have from (2)

$$\frac{d}{dt}\langle a_\ell\rangle + (i\omega_\ell+\nu_\ell)\langle a_\ell\rangle + \sum_{m,n} c_{\ell mn}\langle a_m a_n\rangle = 0 \qquad (3a)$$

and we omit external forcing for the present. The motion of $\langle a_\ell\rangle$ cannot be calculated since we do not know the motion of $\langle a_m a_n\rangle$. Continuing, we may write

$$\frac{d}{dt}\langle a_\ell a_j\rangle + (i(\omega_\ell+\omega_j) + \nu_\ell+\nu_j)\langle a_\ell a_j\rangle + \sum_{m,n}(c_{\ell mn}\langle a_m a_n a_j\rangle + c_{jmn}\langle a_m a_n a_\ell\rangle) = 0 \qquad (3b)$$

$$\frac{d}{dt}\langle a_k a_\ell a_j\rangle + (i(\omega_k+\omega_\ell+\omega_j) + \nu_k+\nu_\ell+\nu_j)\langle a_k a_\ell a_j\rangle + \sum_{m,n}(c_{\ell mn}\langle a_m a_n a_j a_k\rangle \qquad (3c)$$

$$+ c_{jmn} \langle a_m a_n a_\ell a_k\rangle \tag{3c cont.}$$

$$+ c_{kmn} \langle a_m a_n a_\ell a_j\rangle) = 0 \quad ,$$

and so on in an unclosed, hence unsolvable, hierarchy of moment equations. It is this difficulty which has engendered a variety of "closure" hypotheses.

Firstly we observe that an nth order moment may be reduced to a sum of products of lower order moments plus an irreducible term or nth "cumulant", i.e.

$$\langle a_{i_1} \ldots a_{i_n}\rangle = \sum_{j=1}^{n-1} \langle a_{i_1} \ldots a_{i_j}\rangle\langle a_{i_{j+1}} \ldots a_{i_n}\rangle + \langle a_{i_1} \ldots a_{i_n}\rangle^c \quad . \tag{4}$$

Most closure schemes correspond to setting some nth cumulant and all higher cumulants identically to zero. The choice $\langle aaa\rangle^c = 0$ (i.e. $\langle aaa\rangle = \Sigma \langle a\rangle\langle aa\rangle$) is sometimes called the quasilinear hypothesis. For flow without mean motion, i.e. $\langle a\rangle = 0$, a choice $\langle aaaa\rangle^c = 0$ (i.e. $\langle aaaa\rangle = \Sigma \langle aa\rangle\langle aa\rangle$) is called the quasinormal hypothesis. This is also called the random phase hypothesis since, if the phase of each mode a_ℓ is statistically independent of all other modes, then $\langle aaaa\rangle^c = 0$ and, over spatial average, statistics of $\underline{y}$ approach a normal distribution. The hypothesis is termed quasinormal because, strictly, it implies $\langle aaa\rangle^c = 0$ while, in fact, the object usually is just to calculate the value of $\langle aaa\rangle^c \neq 0$.

Such cumulant discard hypotheses often don't work very well. They may predict such unrealizable statistics as $\langle |a|^2\rangle < 0$. For the case of infinitesimally weak wave interactions, some success has been claimed for the quasinormal closure taken together with an updating scheme (Hasselman, 1962) or, what is really the same thing, a multiple time scaling (Benney and Saffman, 1966). For finite amplitude waves, such calculations are expected to fail in ways that cannot be estimated quantitatively. Indeed, for the case of planetary wave interactions, these calculations already fail at infinitesimal amplitude, c.f. discussion around eq. (8) below.

ALTERNATIVE TO CUMULANT DISCARD CLOSURE

The approach I wish to discuss is an alternative to cumulant discard. Suppose we write a cumulant hierarchy up to the equation of motion of the (n)th cumulant, which contains a term in the (n+1)st cumulant. We will substitute for the (n+1)st cumulant a term linear in the (n)th cumulant with a forefactor which is an unknown function of lower cumulants. This function or factor of proportionality is then the main theoretical object. Rather than discarding the (n+1)st cumulant we discard something else: the difference between the (n+1)st cumulant and some linear function of the (n)th cumulant.

Why? We could suppose that for a system close to random phase, i.e. cumulants close to zero, that we are taking the power series expansion of the (n+1)st cumulant in the small departure of the (n)th cumulant from zero. Signs of terms will be such as to tend to restore equilibrium, i.e. random phase. In fact, we don't have a direct prescription for computing coefficients of the series. In <u>principle</u> one can interpret such evolution statistically since departure of an (n)th cumulant from zero reduces overall entropy ($E = -\int P(\underline{a}) \ln P(\underline{a}) d\underline{a}$ where $P(\underline{a})$ is the joint probability density on state vectors $\underline{a} = \{a_\ell, \ell \varepsilon S_1\}$) until checked by evolution of the (n+1)st cumulant as well as by effects of viscous or wave dispersive relaxation. Departure of the (n+1)st cumulant from zero likewise entails a reduction in E, and so on. While one could imagine maximizing E over the distribution of cumulants, a computable method does not seem to follow. Rather, the point in this is to introduce the model of a higher cumulant as linear relaxation of a lower cumulant. The actual forefactor in this term is evaluated by a consistency argument given below.

Suppose we wish to describe second moments (energy, heat flux) in a nonlinear wave field according to (3b). Consider the case of a spatially statistically homogeneous wave field, whence $\langle a \rangle = 0$ and $\langle a_k a_j^* \rangle = \langle a_k a_k^* \rangle \, \delta_{kj}$. To evaluate (3b) we need $\langle aaa \rangle$ from (3c). Factoring fourth moments in (3c) according to (4), and replacing the fourth cumulant with a linear function $\eta_{k\ell j} \langle a_k a_\ell a_j \rangle$ we have

$$
\begin{aligned}
\frac{d}{dt} \langle a_k a_\ell a_j \rangle + (i(\omega_k + \omega_\ell + \omega_j) + \nu_k + \nu_\ell + \nu_j) \langle a_k a_\ell a_j \rangle \\
+ c_{\ell jk} \langle a_j a_j^* \rangle \langle a_k a_k^* \rangle \\
+ c_{jk\ell} \langle a_k a_k^* \rangle \langle a_\ell a_\ell^* \rangle \\
+ c_{k\ell j} \langle a_\ell a_\ell^* \rangle \langle a_j a_j^* \rangle \\
+ \eta_{k\ell j} \langle a_k a_\ell a_j \rangle = 0
\end{aligned}
\tag{5}
$$

where $\eta_{k\ell j}$ is a real, non-negative but otherwise unspecified function of second moments. If second moments are evolving slowly relative to the response time of higher cumulants, then substitution of the stationary, real solution of (5) into (3b) yields

$$\frac{d}{dt} A_k + 2\nu_k A_k = 2 \sum_{\ell,j} \Theta_{k\ell j} C_{k\ell j} A_\ell (A_j - A_k) \quad , \tag{6}$$

where

$$A_k \equiv \langle a_k a_k^* \rangle \quad ,$$

and

$$\Theta_{k\ell j} = \text{Real } \{(i(\omega_k + \omega_\ell + \omega_j) + \nu_k + \nu_\ell + \nu_j + \eta_{k\ell j})^{-1}\} \quad ,$$

and $C_{k\ell j}$ is obtained algebraically from the $c_{k\ell j}$ coefficients. Since $\eta_{k\ell j}$ is undetermined, so is $\Theta_{k\ell j}$ in (6). The simple form of (6) results from having considered an especially simple case, namely the homogeneous wave field for which basis functions $F(\underline{x})$ may be chosen as Fourier bases exp i $\underline{k}\cdot\underline{x}$.

It remains to specify $\eta_{k\ell j}$. From (5) we observe that $\eta_{k\ell j}$ is a rate at which triple moments relax. This rate is distinct from either phase shifting in the off-resonant ($\omega_k+\omega_\ell+\omega_j \neq 0$) wave triad or dissipative decay due to $\nu_k+\nu_\ell+\nu_j \neq 0$. Thus $\eta_{k\ell j}$ is a triple moment relaxation or decorrelation rate which specifically accounts for effects of nonlinear interaction.

In (6) we are attempting to approximate the statistical evolution of (2). But it may also be seen that (6) yields exactly the statistical evolution of a system of model variables $\{\tilde{a}_k, k\varepsilon S,\}$ which satisfy a linear stochastic differential equation of Langevin type:

$$\frac{d}{dt} \tilde{a}_k + i\omega_k \tilde{a}_k + (\nu_k + \eta_k) \tilde{a}_k = \tilde{f}_k \quad , \tag{7a}$$

where

$$\eta_k = \sum_{\ell,j} \Theta_{k\ell j} C_{k\ell j} A_\ell \quad , \tag{7b}$$

and

$$\langle \tilde{f}_k(t) \tilde{f}_k^*(t') \rangle = F_k \, \delta(t-t') \quad ,$$

with

$$F_k = \sum_{\ell,j} \Theta_{k\ell j} C_{k\ell j} A_\ell A_j \quad . \tag{7c}$$

(7) implies (6), but not conversely. Like (6), (7) is not fully

specified since $\eta_{k\ell j}$, hence $\Theta_{k\ell j}$, is not specified. However, if we form the coherence time or integral time scale for a triad of model variables $\tilde{a}_k\tilde{a}_\ell\tilde{a}_j$, this is

$$\tilde{\Theta}_{k\ell j} = \text{Real } \{(i(\omega_k+\omega_\ell+\omega_j) + \nu_k+\nu_\ell+\nu_j+\eta_k+\eta_\ell+\nu_j)^{-1}\} \ .$$

Then the <u>conjecture</u> is that we identify $\eta_{k\ell j} = \eta_k+\eta_\ell+\eta_j$, i.e. $\Theta_{k\ell j} = \tilde{\Theta}_{k\ell j}$.

In a sense this is the simplest <u>consistent</u> prescription insofar as the assumption of an unknown relaxation rate $\eta_{k\ell j}$ in (5) leads to an equation (6) which contains a relaxation rate (7b) which is then assumed for the rate in (5).

Clearly this is science by conjecture and could be phrased in other ways. Let me just mention that these manipulations with cumulants correspond to "renormalizations" (after methods from quantum field theory) of a series formed by averaging a formal iterative solution of (2) in a form $a_k = {}_0a_k + {}_1a_k + {}_2a_k + \cdots$. Here ${}_0a_k$ solves the linearized equation of motion and successive ${}_na_k$ are obtained by substitution of lower order iterates in (2). Multiplying together series for a_k and averaging yields a "bare" series expression for A_k which diverges after a time which varies inversely with the strength of nonlinearity. Renormalization attempts to remove the divergence by mixing orders in the iteration series. It turns out that (6) is close to a simplest "line" renormalization (Holloway, 1978b). However, unlike the case of quantum electrodynamics, for fluid interactions analytical properties of renormalizations have not been obtained so the choice of one scheme over another is conjectural. My preference, simply, is to conjecture about cumulants.

WAVY TURBULENCE OR TURBULENT WAVES, A UNIFYING TREATMENT

Some immediate consequences of $\Theta = \tilde{\Theta}$ may be remarked. Consider scales of motion for which molecular dissipation is negligible. Then as the amplitude of the motion becomes small, so that $\eta_k/\omega_k \to 0$,

$$\Theta_{k\ell j} \to \pi\delta(\omega_k+\omega_\ell+\omega_j).$$

We recover the resonant interaction approximation after weak wave theories, e.g. Hasselman. On the other hand, for motion of large amplitude such that wave propagation is negligible, i.e. $\omega_k/\eta_k \to 0$,

$$\Theta_{k\ell j} \to (\eta_k+\eta_\ell+\eta_j)^{-1}$$

which corresponds to a model (Edwards, 1964) of homogeneous turbulence. The theory of Edwards is close to turbulence theories of Kraichnan (1959) and Herring (1964), c.f. a review by Leslie (1973).

The weak wave limit is unambiguous and, in many cases, asymptotically exact. Dependence in (6) on our conjectures concerning $\eta_{k\ell j}$ vanishes more rapidly than η_k/ω_k. Recently McComas and Bretherton (1977) have computed resonant interaction rates for the oceanic internal wave field. However, these calculations already indicate relaxation rates which are not small compared to wave frequencies, hence the calculation is not consistent with the resonant interaction hypothesis for internal waves of typically observed amplitudes.

Much more ambiguity attends turbulence theory. Earlier theories of Kraichnan, of Edwards and of Herring are known systematically to underestimate energy transfer processes among small scales of motion. The problem has been discussed by Kraichnan (1965) and, in context of plasma turbulence, by Kadomtsev (1965). In the format of the present paper the difficulty comes from associating $\Theta_{k\ell j}$ with the phase correlation time $\tilde{\Theta}_{k\ell j}$ of random variables $\tilde{a}_k\tilde{a}_\ell\tilde{a}_j$. In short length scales, actual variables $a_k a_\ell a_j$ will be phase-advected rather more coherently by motion on long scales so that we should expect $\Theta_{k\ell j} > \tilde{\Theta}_{k\ell j}$. A number of methods have been devised which remove this spurious advective effect though all somewhat arbitrarily so.

While we may not be entirely satisfied with weak wave theories nor, certainly, with turbulence theories, it is clear that we can begin to calculate the statistical evolution of arbitrary amplitude waves. Sometimes we must. Planetary waves include the zonal flow as a zero frequency wave. The arbitrarily small amplitude, say, of zonal shear is already infinitely strong compared to wave frequency. Employ a specification

$$
\begin{aligned}
\Theta_{k\ell j} &= \text{Real } \{(i(\omega_k+\omega_\ell+\omega_j) + (\mu_k+\mu_j+\mu_\ell) \\
\mu_k &= \nu_k + \sum \Theta_{k\ell j}\, \hat{C}_{k\ell j}\, A_\ell \,.
\end{aligned}
\tag{8}
$$

I have made some calculations on the evolution of a planetary wave field which agree with results of a numerical-hydrodynamical simulation (Holloway and Hendershott, 1977). Here $\hat{C}_{k\ell j}$ is a coefficient like $C_{k\ell j}$ but modified after Kraichnan (1971) so that μ_k is independent of advection due to A_ℓ as the length scale in mode ℓ becomes infinite.

A similar program might be carried out for internal waves. On large length scales, a weak wave theory would be recovered. On shorter scales, nonlinearity would be treated more faithfully. Indeed, for a sufficiently energetic wave field, the spectrum of smaller scale motion would tend to Kolmogorov inertial scaling: $E(k) \propto k^{-5/3}$. Oceanic internal waves usually are not so energetic.

My point is not so much to suggest any particular calculation but to propose beginnings of bridges between theories of weak, nearly linear motions and theories of strong, often chaotic motions. A model like (8) is an example which both provides a basis for new calculations and also suggests limits of validity of previous, nearly linear calculations. Thus contours of $\Theta_{k\ell j}$ projected onto a wavevector plane will resemble the "resonance path" sketched in Meiss and Watson's Figure 7 (these proceedings). The width of the path roughly is $\mu_k|\partial\omega/\partial k|^{-1}$. This finite width will give errors in a resonant interaction prediction when the energy spectrum E(k) varies on a scale such that

$$\left|\frac{\partial\ \ell n\ E(k)}{\partial k}\right| > \frac{1}{\mu_k}\left|\frac{\partial\omega}{\partial k}\right| \quad ;$$

$|\partial\omega/\partial k|$ is the wave group speed c_g. Roughly, $|\partial\ \ell n\ E/\partial k|$ is a group length L_g. Then the resonant interaction calculation errs when

$$L_g > \frac{1}{\mu} C_g \quad , \qquad (9)$$

i.e. when wave groups interact in a time μ^{-1} short compared to the time L_g/C_g in which a group propagates over a few group lengths. Usually wave groups or packets and group speeds are thought to be more "real" than phases and phase speeds, yet the group speed does not appear naturally in theories of resonant wave interaction. Another interpretation of (9) is that information contained in initial conditions or boundary conditions is carried by waves along wave characteristics or rays at the group speed C_g but is degraded by nonlinear interaction at a rate μ so that C_g/μ is an information propagation length in the flow field. Indeed, if L_B is an external length scale imposed by boundary conditions, a stochasticity criterion $L_B > C_g/\mu$ is just a fluid mechanical version of Chirikov's overlapping resonances criterion.

A MIXED DETERMINISTIC/STOCHASTIC FLOW MODEL

Treated statistically, flow fields have been described including nonlinear effects that may range from arbitrarily weak to arbitrarily strong. Although limited and tentative certainly, some "success" may be claimed in describing the disordered state of fluid flow. However, the reader is reminded that we are here ignoring such possibilities as solitons. Moreover, numerical calculations have tested predictions against statistically homogeneous flows, real examples of which are rare even should one set out to make them. Despite these limitations, we may hope to explore connections between statistical predictions and more conventional efforts to solve or numerically to simulate particular fluid motions.

The possibility of collecting orderly and disorderly as well as linear and nonlinear motion in one scheme is apparent in comparing the deterministic equation (2) with the stochastic equation (7). Previously we have used (7) only to predict the statistical evolution of (2), with these predictions verified against direct numerical integrations of (2) on S_1 of about 10^4 members. Now the suggestion is just to introduce directly Langevin terms η and $\tilde{f}$ from (7) into calculations of a specific realization of the motion. In part this is already familiar to numerical hydrodynamicists who employ "eddy viscosities" quite unrelated to kinematic viscosity. I think we will obtain a more systematic expression of such "viscosities".

The more novel suggestion is the inclusion of a field of random agitation, perhaps along the lines of an "eddy noise" suggested by Rose (1977). To nonlinear deterministic equations of motion, we add stochastic driving terms so that no particular solution is reproducible. We compute only "typical" realizations. This is a natural result if we consider, as before, a motion on a very large basis set S_1 which we want to model on a much smaller set S_2. A great deal of information must be lost in this projection and the appearance of "eddy noise" is just our expression of this uncertainty. What I believe is that the physics of most "typical" realizations of a stochastic motion on S_2 can be made more faithful to the motion on S_1 than could any deterministic model motion on S_2. The hard theoretical questions deal with the formulation of Langevin terms. I'll just outline some difficulties.

Statistical (spectral) calculations have proven practicable for statistically homogeneous motion. Any specific realization of a motion is necessarily everywhere inhomogeneous. Recently however I've had some encouraging success with a statistical treatment of such an everywhere-inhomogeneous problem (Holloway, 1978a). The problem was to compute the effect of irregular bottom topography on ocean currents. Given a particular realization of the topography, numerical simulations of currents were in very good agreement with predictions of a statistical theory which, in principle, averages over realizations of the topography. Surely this agreement depended on choosing a complicated, highly irregular (but realistically so!) topography. What this suggests is that we take S_2 large enough that the motion on S_2 will be sufficiently complicated so that interactions with S_1 can be treated statistically. I suppose an algorithm that goes like this:

Define an initial set of amplitudes on S_2. Fit a smoothed variance spectrum E_2 to the amplitudes on S_2, e.g. by a spline fit with far fewer modes than the length of S_2. Assume a variance spectrum E_{12} for the members of S_1-S_2 so that the overall variance spectrum is $E_1 = E_{12} + E_2$. Compute Langevin terms based on E_{12} then include a realization of these terms in an explicit calculation of the evolution of amplitudes on S_2. Compute the statistical evolution of E_1 incorporating the explicit evolution of E_2. Except for direct roles of forcing or dissipation, the calculation should conserve energy, momentum, helicity or other invariants of the deterministic motion but here only on ensemble average.

If the initial excitation is entirely in S_2, then the early evolution is deterministic. Weakly nonlinear motion will remain predictable for long time. Stronger motion will be realized only stochastically after a predictability time t_p. Yet even after t_p, S_2 may exhibit "typical" flow structures, e.g. particular shear configurations or intermittancy, which will be more realistic since, approximately, realistic spectral transfer is taking place.

REFERENCES

Benney, D. J. and P. G. Saffman (1966) Nonlinear interaction of random waves in a dispersive medium. Proc. Roy. Soc. A289, 301.

Edwards, S. F. (1964) The statistical dynamics of homogeneous turbulence. J. Fluid Mech. 18, 239.

Hasselmann, K. (1962) On the nonlinear energy transfer in a gravity wave spectrum. Part 1. General theory. J. Fluid Mech. 12, 481.

Herring, J. R. (1964) Self-consistent-field approach to turbulence theory. Phys. Fluids 8, 2219.

Holloway, G. (1978a) A spectral theory of nonlinear barotropic motion above irregular topography. J. Phys. Ocean 8 (4).

Holloway, G. (1978b) On spectral transfer among strongly interacting waves. Geophys. and Astrophys. Fluid Dyn. (to appear).

Holloway, G. and M. C. Hendershott (1977) Stochastic closure for nonlinear Rossby waves. J. Fluid Mech. 82, 747.

Kadomtsev, B. B. (1965) Plasma Turbulence, London, Academic Press. (c.f. Chapter 3).

Kraichnan, R. H. (1959) The structure of isotropic turbulence at very high Reynolds number. J. Fluid Mech. 5, 497.

Kraichnan, R. H. (1965) Lagrangian-history closure approximation for turbulence. Phys. Fluids 8, 575.

Kraichnan, R. H. (1971) An almost-Markovian Galilean-invariant turbulence model. J. Fluid Mech. 47, 513.

Leslie, D. C. (1973) Developments in the Theory of Turbulence, Oxford Clarendon Press.

McComas, C. H. and F. P. Bretherton (1977) Resonant interaction of oceanic internal waves. J. Geophys. Res. 82, 1397.
Rose, H. A. (1977) Eddy diffusivity, eddy noise and subgrid scale modelling. J. Fluid Mech. 81, 719.

ON THE SOLUTION OF NONLINEAR RATE EQUATIONS BY MATRIX INVERSION†

by

Elliott W. Montroll*
Physical Dynamics, Inc.
La Jolla, California 92038

ABSTRACT

In this paper an idea of T. Carleman is used to imbed a finite set of nonlinear rate equations into an infinite set of linear equations. The matrix operator associated with the linear equations is generally a triangular matrix whose matrix elements are matrices. Since an explicit expression is available for the inverse of such a matrix, an algorithm can be found to yield the solution of the original set of nonlinear equations.

I. INTRODUCTION

In 1931 Torsten Carleman[1,2] showed that the nonlinear differential equations of the classical dynamics of a system of n interacting particles can formally be used to derive an infinite set of linear differential equations from which the solutions of the nonlinear set can be extracted. Since certain existence proofs are easily derived for sets of linear equations, he was able to immediately apply his idea to the derivation of existence proofs in nonlinear dynamics[2]. He does not seem to have exploited his ideas to give explicit solutions to special equations. The aim of this paper is to present several algorithms for the generation of formal solutions of several broad classes of nonlinear rate equations and broad classes of nonlinear models in classical dynamics.

In a previous paper Robert Helleman and the author[3] used Carleman's ideas to imbed the nonlinear dynamical equations of an anharmonic oscillator into an infinite set of linear equations. These equations were solved by employing a certain set of linear recurrance formulae rather than by direct matrix inversion. In this paper the direct matrix inversion method will be introduced.

Carleman's general ideas are introduced by a consideration of the nonlinear rate equation

$$dx/dt = -x+x^2 \qquad (1)$$

---

*Permanent Address, Institute for Fundamental Studies, Physics Department, University of Rochester, Rochester, New York 14627.

†This research was partially supported by the Fluid Dynamics Branch of the Office of Naval Research.

ISSN: 0094-243X/78/337/$1.50

It is easy to show by separation of variables and direct integration that

$$x(t) = \frac{x(0)}{x(0) + [1-x(0)]\exp(t)} . \qquad (2)$$

Equation (1) can also be solved (by a much longer calculation) in terms of an infinite set of linear equations as follows. Following the scheme proposed by Carleman

$$y_1 \equiv x,\ y_2 \equiv x^2, \ldots\ldots,\ y_n \equiv x^n . \qquad (3)$$

Then

$$dy_n/dt = dx^n/dt = n\, x^{n-1} dx/dt , \qquad (4)$$

so that from Eq. (2) $y_1(0) \equiv x(0) \equiv C, y_n(0) \equiv C^n$

$$dy_n/dt = - ny_n + ny_{n+1} . \qquad (5)$$

One can then use his favorite method for solving this set of equations. The method of Laplace transforms will be emphasized here because it can be generalized immediately for the discussion of coupled nonlinear equations in several variables. Let

$$Y_n(\lambda) = \int_0^{\infty} e^{-\lambda t}\, y_n(t)dt . \qquad (6)$$

Then, since

$$\int_0^{\infty} e^{-\lambda t}\, (dy_n/dt)dt = -C^n + \lambda Y_n(\lambda) , \qquad (7)$$

eq. (5) becomes

$$(\lambda+n)Y_n - nY_{n+1} = C^n , \qquad (8)$$

or in matrix form

$$AY = C \qquad (9)$$

$$\begin{pmatrix} (\lambda+1) & -1 & 0 & 0 & 0 & . \\ 0 & (\lambda+2) & -2 & 0 & 0 & . \\ 0 & 0 & (\lambda+3) & -3 & 0 & . \\ 0 & 0 & 0 & (\lambda+4) & -4 & . \\ . & . & . & . & . & . \end{pmatrix} \begin{pmatrix} Y_1 \\ Y_2 \\ Y_3 \\ . \\ . \end{pmatrix} = \begin{pmatrix} C \\ C^2 \\ C^3 \\ . \\ . \end{pmatrix} \quad (10)$$

It is easy to construct the inverse

$$A^{-1} = \begin{pmatrix} \frac{1}{\lambda+1} & \frac{1}{(\lambda+1)(\lambda+2)} & \frac{2.1}{(\lambda+1)(\lambda+2)(\lambda+3)} & \frac{3.2.1}{(\lambda+1)(\lambda+2)(\lambda+3)(\lambda+4)} & . \\ 0 & \frac{1}{\lambda+2} & \frac{2}{(\lambda+2)(\lambda+3)} & \frac{3.2}{(\lambda+2)(\lambda+3)(\lambda+4)} & . \\ 0 & 0 & \frac{1}{\lambda+3} & \frac{3}{(\lambda+3)(\lambda+4)} & . \\ 0 & 0 & 0 & \frac{1}{\lambda+4} & . \\ . & . & . & . & \end{pmatrix} \quad (11)$$

so that

$$Y_1 = \frac{C}{(1+\lambda)} + \frac{C^2}{(2+\lambda)(1+\lambda)} + \frac{2!C^3}{(3+\lambda)(2+\lambda)(1+\lambda)} + \frac{3!C^4}{(4+\lambda)(3+\lambda)(2+\lambda)(1+\lambda)} + \ldots \quad (12)$$

$$= \sum_{n=1}^{\infty} \Gamma(n)\ \Gamma(\lambda+1)C^n/\Gamma(\lambda+2) \quad (13)$$

$$= \sum_{n=1}^{\infty} C^n\ B(n,\lambda+1)\ ,$$

$B(n,\lambda+1)$ being the beta function defined by

$$B(n,m) = \int_0^1 x^{n-1}(1-x)^{m-1}dx = \Gamma(n)\Gamma(m)/\Gamma(n+m).$$

Hence

$$Y_1 = \int_0^\infty (1-cx)^{-1}(1-x)^\lambda dx \ . \tag{14}$$

Now let $(1-x) = \exp(-t)$.Then

$$\begin{aligned} Y_1(\lambda) &\equiv \int_0^\infty y_1(t)e^{-\lambda t} dt \\ &= \int_0^\infty \frac{C e^{-\lambda t} dt}{C+(1-C)\exp t} \ , \end{aligned} \tag{15}$$

so that

$$x(t) \equiv y_1(t) = C/[C+(1-C)e^t] \ , \tag{16}$$

which is equivalent to (2).

Notice that the matrix A of equations (9) and (10) is a triangular matrix with no elements below the diagonal. If the right hand side of (1) were the polynomial

$$f(x) = x+a_2x^2 + \dots +a_nx^n, \tag{17a}$$

the matrix A would still be triangular but with n-1 filled diagonal lines above the central diagonal. Since the expression for the elements of the inverse of a triangular matrix has a simple determinental form an explicit series expansion can be constructed for the solution of the equation

$$dx/dt = f(x) \ . \tag{17b}$$

We show in the next sections that formal solutions of the infinite set of linear equations resulting from sets of coupled nonlinear equations can be constructed from the inverse of triangular matrices.

Equation (5) was derived from eq. (1) by Bellman and Richardson[4] as were analogous equations for two coupled variables (similar to those considered in the next section), however they did not discuss the construction of the solutions of the equations. They used the equations to investigate the errors resulting from truncating the infinite set by making it finite. The truncation technique has of course been used in discussing many body problems as well as hydrodynamic ones.

If we write A as a matrix with elements a_{ij}, the i and j running through indices 1,2,3,... with 1 corresponding to 10, 2 to 01, 3 to 20, 4 to 11, 5 to 02 etc., and if we use the notation

$$a_{ij} \equiv (i,j) \equiv ij \text{ and } ij{:}k\ell \equiv (ij)(k\ell) \equiv a_{ij}\, a_{k\ell}, \tag{18}$$

the inverse of our triangular matrix

$$A = \begin{pmatrix} (11) & (12) & (13) & (14) & \cdots \\ 0 & (22) & (23) & (24) & \cdots \\ 0 & 0 & (33) & (34) & \cdots \\ 0 & 0 & 0 & (44) & \cdots \\ . & . & . & . & \cdots \end{pmatrix} \tag{19}$$

is

$$A^{-1} = \begin{pmatrix} \frac{1}{11} & -\frac{12}{11:22} & \frac{\det 123}{11:22:33} & -\frac{\det 1234}{11:22:33:44} & \cdots \\ 0 & \frac{1}{22} & -\frac{23}{22:33} & \frac{\det 234}{22:33:44} & \cdots \\ 0 & 0 & \frac{1}{33} & -\frac{34}{33:44} & \cdots \\ 0 & 0 & 0 & 0 & \\ . & . & . & . & \cdots \end{pmatrix} \tag{20}$$

where we define det 1 ≡ 1, det 12 ≡ 12

$$\det 123 \equiv \begin{vmatrix} 12 & 13 \\ 22 & 23 \end{vmatrix} \qquad \det 1234 \equiv \begin{vmatrix} 12 & 13 & 14 \\ 22 & 23 & 24 \\ 0 & 33 & 34 \end{vmatrix}, \tag{21a}$$

$$\det 12345 \equiv \begin{vmatrix} 12 & 13 & 14 & 15 \\ 22 & 23 & 24 & 25 \\ 0 & 33 & 34 & 35 \\ 0 & 0 & 44 & 45 \end{vmatrix}. \tag{21b}$$

Then, if C of eq.(9) has components

$$(C_1,C_2,C_3,C_4,\ldots) \equiv (c_1,c_2,c_1^2,c_1c_2^2,c_1^3,c_1^2c_2,c_1c_2^2,c_2^3,\ldots) \quad (22)$$

we find

$$Y_1 \equiv Z_{10} = \sum_1^{\infty} \frac{(-1)^{j+1}C_j \det(123\ldots j)}{11{:}22{:}33{:}\ldots{:}jj} \quad , \quad (23a)$$

$$Y_2 \equiv Z_{01} = \sum_2^{\infty} \frac{(-1)^{j+1}C_j \det(2,3,\ldots,j)}{22{:}33{:}\ldots{:}jj} \quad , \quad (23b)$$

$$Y_3 \equiv Z_{20} = \sum_3^{\infty} \frac{(-1)^{j+1}C_j \det(3,4,\ldots,j)}{33{:}44{:}\ldots{:}jj}, \text{ etc.} \quad (23c)$$

The Laplace inverse of these expressions can then be found yielding the required solution of our differential equations (2).

These ideas can be generalized for application to the solution of coupled rate equations. While the direct Laplace transform approach presented in Section II for coupled nonlinear rate equations may lead to secular terms and may not be appropriate for some problems, it does indicate the type of matrix inversion required for the solution of coupled equations generally.

II. TWO VARIABLE COUPLED RATE EQUATION WITH QUADRATIC NONLINEARITIES

Let us consider first two coupled nonlinear rate equations with quadratic nonlinearities,

$$\dot{x}_1 = a_1^1 x_1 + a_1^2 x_2 + b_1^{11} x_1^2 + b_1^{12} x_1 x_2 + b_1^{22} x_2^2 \, , \quad (1a)$$

$$\dot{x}_2 = a_2^1 x_2 + a_2^2 x_2 + b_2^{11} x_1^2 + b_2^{12} x_1 x_2 + b_2^{22} x_2^2 \, . \quad (1b)$$

It is easy to construct a linear transformation $x = \Gamma y$ which diagonalizes the linear part of these equations. They then have the form

$$\dot{y}_1 = -\mu_1 y_1 + \nu_1^{11} y_1^2 + \nu_1^{12} y_1 y_2 + \nu_1^{22} y_2^2 \quad , \quad (2a)$$

$$\dot{y}_2 = -\mu_2 y_2 + \nu_2^{11} y_1^2 + \nu_2^{12} y_1 y_2 + \nu_2^{22} y_2^2 \, . \quad (2b)$$

The infinite set of linear equations from which the solution of (2) can be constructed is obtained by defining

$$z_{nm}(t) \equiv y_1^n y_2^m;\ z_{nm}(0) = c_1^n c_2^m \text{ with } y_j(0) \equiv c_j \qquad (3)$$

Then

$$dz_{nm}/dt = ny_1^{n-1} y_2^m \dot{y}_1 + my_1^n y_2^{m-1} \dot{y}_2 ,$$

$$dz_{nm}/dt = -n\mu_1 y_1^n y_2^m + n\nu_1^{11} y_1^{n+1} y_2^m + n\nu_1^{12} y_1^n y_1^{m+1}$$

$$+ n\nu_1^{22} y_1^{n-1} y_2^{m+2} - m\mu_2 y_1^n y_2^m + m\nu_2^{11} y_1^{n+2} y_2^{m-1}$$

$$+ m\nu_2^{12} y_1^{n+1} y_2^m + m\nu_2^{22} y_1^n y_2^{m+1} ,$$

or

$$dz_{nm}/dt = -(m\mu_2 + n\mu_1) z_{nm} + (n\nu_1^{11} + m\nu_2^{12}) z_{n+1,m}$$

$$+ n\nu_1^{22} z_{n-1,m+2} + (n\nu_1^{12} + m\nu_2^{22}) z_{n,m+1} \qquad (4)$$

$$+ m\nu_2^{11} z_{n+2,m-1} .$$

The Laplace transform of these equations are, with $Z_{nm}(\lambda) = L\ z_{nm}(t)$

$$(\lambda + m\mu_2 + n\mu_1) Z_{nm} - (n\nu_1^{11} + m\nu_2^{12}) Z_{n+1,m} - n\nu_1^{22} Z_{n-1,m+2}$$

$$- (n\nu_1^{12} + m\nu_2^{22}) Z_{n,m+1} - m\nu_2^{11} Z_{n+2,m-1} \qquad (5)$$

$$= c_1^n c_2^m .$$

We write this set of equations in matrix form

$$Az = C \text{ so that } z = A^{-1} C . \qquad (6)$$

In the matrix elements we order the indices in the following manner

$$\begin{array}{ccc} 10 & 20 & 30 \\ 01 & 11 & 21 \\ & 02 & 12 \\ & & 03 \end{array} \quad \text{etc.}$$

It is easy to see that the matrix, A, has the following structure, where · implies a non-vanishing matrix element and 0 represents a vanishing one:

	10	01	20	11	02	30	21	12	03	40	31	22	13	04	...
10	.	0	.	.	.	0	0	0	0	0	0	0	0	0	- -
01	0	.	.	.	.	0	0	0	0	0	0	0	0	0	- -
20	0	0	.	0	0	.	.	.	0	0	0	0	0	0	- -
11	0	0	0	.	0	.	.	.	.	0	0	0	0	0	- -
02	0	0	0	0	.	0	.	.	.	0	0	0	0	0	- -
30	0	0	0	0	0	.	0	0	0	.	.	.	0	0	- -
21	0	0	0	0	0	0	.	0	0	.	.	.	.	0	- -
12	0	0	0	0	0	0	0	.	0	0	.	.	.	.	- -
03	0	0	0	0	0	0	0	0	.	0	0	.	.	.	- -

(7)

The most important feature of this matrix is that it is a triangular matrix. Since simple formulae exist for the inverse of a triangular matrix an algorithm can be constructed for the calculation of z_{01} and z_{10}. Note that A has the form

$$A = \begin{pmatrix} A_{11} & A_{12} & 0 & 0 \\ 0 & A_{22} & A_{23} & 0 \\ 0 & 0 & A_{33} & A_{34} \end{pmatrix}, \qquad (8a)$$

where A_{nn} is a diagonal square (n+1) order matrix, $A_{n,n+1}$ is an (n+1) x (n+2) order matrices. The explicit forms for these two classes of submatrices are clear from

$$A_{11} = \begin{pmatrix} (\lambda+\mu_1) & 0 \\ 0 & (\lambda+\mu_2) \end{pmatrix}, \quad A_{22} = \begin{pmatrix} (\lambda+2\mu_1) & 0 & 0 \\ 0 & (\lambda+\mu_1+\mu_2) & 0 \\ 0 & 0 & (\lambda+2\mu_2) \end{pmatrix} \tag{8b}$$

$$A_{33} = \begin{pmatrix} (\lambda+3\mu_1) & 0 & 0 & 0 \\ 0 & (\lambda+2\mu_1+\mu_2) & 0 & 0 \\ 0 & 0 & (\lambda+\mu_1+2\mu_2) & 0 \\ 0 & 0 & 0 & \lambda+3\mu_2 \end{pmatrix}, \text{ etc.} \tag{8c}$$

$$A_{12} = \begin{pmatrix} -\nu_1^{11} & -\nu_1^{12} & -\nu_1^{22} \\ -\nu_2^{11} & -\nu_2^{12} & -\nu_2^{22} \end{pmatrix}, \tag{9a}$$

$$A_{23} = \begin{pmatrix} -2\nu_1^{11} & -2\nu_1^{12} & -2\nu_1^{22} & 0 \\ -\nu_2^{11} & -\nu_1^{11}-\nu_2^{12} & -\nu_1^{12}-\nu_2^{22} & \nu_1^{22} \\ 0 & -2\nu_2^{11} & -2\nu_2^{12} & -2\nu_2^{22} \end{pmatrix}, \tag{9b}$$

$$A_{34} = \begin{pmatrix} -3\nu_1^{11} & -3\nu_1^{12} & -3\nu_1^{22} & 0 & 0 \\ -\nu_2^{11} & -2\nu_1^{11}-\nu_2^{12} & -2\nu_1^{12}-\nu_2^{22} & -2\nu_1^{22} & 0 \\ 0 & -2\nu_2^{11} & -\nu_1^{11}-2\nu_2^{12} & -\nu_1^{12}-2\nu_2^{22} & -\nu_1^{22} \\ 0 & 0 & -3\nu_2^{11} & -3\nu_2^{12} & -3\nu_2^{22} \end{pmatrix}, \text{etc.}$$

III. ON THE INVERSE OF TRIANGULAR MATRICES WHOSE MATRIX ELEMENTS ARE THEMSELVES MATRICES

The inverse of a triangular matrix A, whose elements are also matrices has a form analogous to (I.20) as can be guessed by direct calculation. Let

$$A = \begin{pmatrix} A_{11} & A_{12} & A_{13} & A_{14} & \cdots \\ 0 & A_{22} & A_{23} & A_{24} & \cdots \\ 0 & 0 & A_{33} & A_{34} & \cdots \\ 0 & 0 & 0 & A_{44} & \cdots \end{pmatrix}. \quad (1)$$

By direct multiplication it can be verified that

$$A^{-1} = \begin{pmatrix} A_{11}^{-1} & -A_{11}^{-1} A_{12} A_{22}^{-1} & A^{-1}_{11}(A_{12}A_{22}^{-1}A_{23} - A_{13})A_{33}^{-1} & - A_{11}^{-1}C_{1234}A_{44}^{-1} & \cdots \\ 0 & A_{22}^{-1} & -A_{22}^{-1} A_{23}A^{-1}_{33} & A_{22}^{-1}(A_{23}A_{33}^{-1}A_{34}-A_{24})A_{44}^{-1} & \cdots \\ 0 & 0 & A_{33}^{-1} & -A_{33}^{-1}A_{34}A_{44}^{-1} & \cdots \\ 0 & 0 & 0 & A^{-1}_{44} & \cdots \\ \cdot & \cdot & \cdot & \cdot & \cdot \end{pmatrix} \quad (2)$$

where

$$C_{1234} \equiv A_{12}A_{22}^{-1}A_{23}A_{33}^{-1}A_{34} - A_{12}A_{22}^{-1}A_{24} - A_{13}A_{33}^{-1}A_{34} + A_{14}. \quad (3)$$

Notice that there is a 1-1 correspondence between every term in this expression and (I.20). Let us compare term by term:

$\frac{1}{11} \equiv a_{11}^{-1}$ corresponds to A_{11}^{-1} ; $12/11{:}22 \equiv a_{11}^{-1}\, a_{12}\, a_{22}^{-1}$ corresponds to $A_{11}^{-1}\, A_{12}\, A_{22}^{-1}$. det $(123)/11{:}22{:}33 \equiv a_{11}^{-1}\, a_{12}\, a_{22}^{-1}\, a_{23}\, a_{33}^{-1} - a_{11}^{-1}\, a_{13}\, a_{33}^{-1}$ corresponds to $A_{11}^{-1}(A_{12}A_{22}^{-1}A_{23} - A_{13})A_{33}^{-1}$. The difference in the pattern of terms generated by (I.20) and those generated from (III.2) is that the ordering of the matrices A_{ij} in (2) cannot be changed while that ordering is not important in the expansion of (I.20).

However, we can take advantage of the known expansions of the determinants to give the pattern for the elements of A^{-1}. The way to find non-vanishing matrix elements of (2) is to (a) construct the corresponding term in (I.20) (b) replace the inverse of all diagonal terms a_{jj}^{-1} by A_{jj}^{-1}. Order a given term in the expansion of terms in matrix element $A_{jk}^{(-1)}$ as follows

$$A_{jj}^{-1} A_{j,j_1} A_{j_1,j_1}^{-1} A_{j_1,j_2} A_{j_2,j_2}^{-1} A_{j_2,j_3} \text{ -- } A_{j_n,k} A_{kk}^{-1} \quad ,(4)$$

with

$$j<j_1< j_2< j_3 <\ldots<j_n<k \quad .$$

To see how this applies let us calculate the term $A_{11}^{-1} C_{1234} A_{44}^{-1}$ using the determinant form (I.20).

$$[\det 1234]/11:22:33:44 = [(12)(23)(34) + (22)(33)(14)$$

$$- (12)(24)(33)-(22)(13)(34)]/11:22:33:44$$

$$= a_{11}^{-1} a_{12} a_{22}^{-1} a_{23} a_{33}^{-1} a_{34} a_{44}^{-1} + a_{11}^{-1} a_{14} a_{44}^{-1} \quad (5)$$

$$- a_{11}^{-1} a_{12} a_{22}^{-1} a_{24} a_{44}^{-1} - a_{11}^{-1} a_{13} a_{33}^{-1} a_{34} a_{44}^{-1} .$$

We have put the subscripts in an increasing order and intersperse sequences .. a_{12} a_{24}.. with a_{22}^{-1} to make the pattern ..a_{12} a_{22}^{-1} a_{24} .. etc. If we identify a_{jj}^{-1} with A_{jj}^{-1} and a_{ij} (for $i\neq j$) with A_{ij} we find that (5) becomes $A_{11}^{-1} C_{1234} A_{44}^{-1}$ as defined by (3).

The form of A^{-1} simplifies when A can be written as (II.8a). Then

$$A^{-1}=\begin{pmatrix} A_{11}^{-1} & -A_{11}^{-1}A_{12}A_{22}^{-1} & A_{11}^{-1}A_{12}A_{22}^{-1}A_{23}A_{33}^{-1} & -A_{11}^{-1}A_{12}A_{22}^{-1}A_{23}A_{33}^{-1}A_{34}A_{44}^{-1} & \cdots \\ 0 & A_{22}^{-1} & -A_{22}^{-1}A_{23}A_{33}^{-1} & A_{22}^{-1}A_{23}A_{33}^{-1}A_{34}A_{44}^{-1} & \cdots \\ 0 & 0 & A_{33}^{-1} & -A_{33}^{-1}A_{34}A_{44}^{-1} & \cdots \\ 0 & 0 & 0 & A_{44}^{-1} & \cdots \end{pmatrix} \quad (6)$$

The solution of the equation AZ=C for a set of rate equations in 2 variables $..y_1, y_2$ with quadratic nonlinearities is then

$$\begin{matrix} Y_1 \\ Y_2 \end{matrix} = A_{11}^{-1}\begin{pmatrix} c_1 \\ c_2 \end{pmatrix} - A_{11}^{-1}A_{12}A_{22}^{-1}\begin{pmatrix} c_1^2 \\ c_1c_2 \\ c_2^2 \end{pmatrix} + A_{11}^{-1}A_{12}A_{22}^{-1}A_{23}A_{33}^{-1}\begin{pmatrix} c_1^3 \\ c_1^2c_2 \\ c_2c_1^2 \\ c_2^3 \end{pmatrix} - .. \tag{7}$$

In the case of cubic nonlinearities the matrix A has the form

$$A = \begin{pmatrix} A_{11} & 0 & A_{13} & 0 & 0 & 0 & . \\ 0 & A_{22} & 0 & A_{24} & 0 & 0 & . \\ 0 & 0 & A_{33} & 0 & A_{35} & 0 & . \\ 0 & 0 & 0 & A_{44} & 0 & A_{46} & . \\ & . & . & . & . & A_{55} & . \end{pmatrix}, \tag{8}$$

from which it follows that

$$A^{-1} = \begin{pmatrix} A_{11}^{-1} & 0 & -A_{11}^{-1}A_{13}A_{33}^{-1} & 0 & A_{11}^{-1}A_{13}A_{33}^{-1}A_{35}A_{55}^{-1} & \cdots \\ 0 & A_{22}^{-1} & 0 & -A_{22}^{-1}A_{24}A_{44}^{-1} & 0 & \cdots \\ 0 & 0 & A_{33}^{-1} & 0 & -A_{33}^{-1}A_{35}A_{55}^{-1} & \cdots \\ 0 & 0 & 0 & A_{44}^{-1} & 0 & \cdots \\ 0 & 0 & 0 & 0 & A_{55}^{-1} & \cdots \end{pmatrix} \tag{9}$$

and

$$\begin{pmatrix} Y_1 \\ Y_2 \end{pmatrix} = A_{11}^{-1}\begin{pmatrix} c_1 \\ c_2 \end{pmatrix} - A_{11}^{-1}A_{13}A_{33}^{-1}\begin{pmatrix} c_1^3 \\ c_1^2c_2 \\ c_1c_2^2 \\ c_2^3 \end{pmatrix} + A_{11}^{-1}A_{13}A_{33}^{-1}A_{35}A_{55}^{-1}\begin{pmatrix} c_1^5 \\ c_1^4c_2 \\ c_1^3c^2 \\ c_1^2c_2^3 \\ c_1c_2^4 \\ c_2^5 \end{pmatrix} \cdots\cdots \tag{10}$$

Equations (7) and (10) yield, upon Laplace transform inversion the formal solutions to eq. 1 and to a corresponding equation with cubic nonlinearity only. The solutions generally yield secular terms (products of powers of t and exponentials) since the denominators of the Laplace transforms have multiple roots. While this is not serious when the exponentials decay with time, they lead to a serious problem when the time dependence in the exponent is purely imaginary so that the modulus of the exponential term is equal to one.

In the latter case which is common to dynamical models of single or coupled oscillators we avoid the secular terms by using ideas developed in references 3, 5, and 6. The matrix inverse formula (eq. 6 above) will again be basic to our analysis. The problem we treat here as an example is the dynamics of a quartic anharmonic oscillator. Its second order equation of motion is equivalent to two coupled first order equations. As in the case of eq. 1 these equations have a simple solution (in terms of elliptic functions); however, since they are among the significant examples to demonstrate the method, we examine them here.

IV. THE ANHARMONIC OSCILLATOR WITH QUARTIC ANHARMONICITY

Consider the anharmonic oscillator, of unit mass, characterized by the Hamiltonian

$$H = \frac{1}{2}p^2 + \frac{1}{2}\omega_0^2\phi^2 - \frac{1}{4}R\phi^4. \tag{1}$$

Hamilton's equations of motion are

$$\dot{p} = -\partial H/\partial\phi = -\omega_0^2\phi + R\phi^3 \text{ and } \dot{\phi} = \partial H/\partial p = p\ , \tag{2}$$

which are equivalent to Newton's equation

$$\ddot{\phi} + \omega_0^2\phi = R\phi^3\ . \tag{3}$$

Let the initial displacement and momentum be

$$\phi(0) = \phi_0 \text{ and } p(0) = p_0 \quad . \tag{4}$$

The Hamiltonian is a constant of the motion with the value

$$H = \frac{1}{2} p_0^2 + \frac{1}{2} \omega_0^2\phi_0^2 - \frac{1}{4} R\phi_0^4 , \tag{5}$$

and for a certain regime of initial conditions $\phi(t)$ is a periodic function of the time.

The traditional way of solving Newton's equation (3) by perturbation theory is to expand ϕ in powers of R. Then to eliminate the secular terms that appear naturally in the perturbation theory, one expands the frequency of a periodic motion in a power series in R so then

$$\nu = \omega_0 + \omega_1 R + \omega_2 R^2 + \ldots , \tag{6}$$

as was shown in reference 5. This series converges only when

$$|\theta| = |R\phi_0^2/\omega_0^2| < 1 \cdot \tag{7}$$

An alternative scheme, the "backward scheme" for solving (3) without secular terms was discussed in references 5 and 6. In the case of coupled anharmonic oscillators it was shown in those references that the "backward scheme" provided the possibility of avoiding the small denominator problem which is a serious difficulty with traditional perturbation theory. In the backwards scheme as applied to a single oscillator one assumes that the frequency of the oscillator in periodic motion is known, say to be equal to ν and that the initial displacement is a derived quantity. With coupled oscillators several frequencies are postulated to be known multiples of a basic frequency and the initial displacement of the oscillators are regarded to be derived quantities.

The basic single oscillator equation in the backward scheme, as applied to eq. 3, is chosen to be

$$\ddot{\phi} + \nu^2\phi = \varepsilon\{(\nu^2 - \omega_0^2)\phi + R\phi^2\} , \tag{8}$$

with ε chosen to be the small parameter of a perturbation calculation rather than R. One finally sets ε=1 after a certain "book-keeping" operation is completed. We rewrite (8) as two first order equations and then search for fourier series expansions in powers of exp $i\nu t$.

The Hamiltonian (1) has the alternative form

$$H = \frac{1}{2} p^2 + \frac{1}{2} \nu^2\phi^2 + \varepsilon\{\frac{1}{2}(\omega_0^2-\nu^2)\phi^2 - \frac{1}{4} R\phi^4\} , \tag{9}$$

which yields (8). However if we introduce the variables

$$v \equiv \nu\phi + ip \text{ and } w \equiv \nu\phi - ip = v^* , \tag{10}$$

Hamilton's equations in the new variables are

$$\dot{v} + i\nu v = -i\varepsilon\{s(v+w)-u(v+w)^3\} , \tag{11a}$$

$$\dot{w} - i\nu w = i\varepsilon\{s(v+w)-u(v+w)^3\} , \tag{11b}$$

with

$$S \equiv (\omega_0^2-\nu^2)/2\nu \text{ and } U \equiv R/8\nu^3 . \tag{12}$$

We now embed the coupled nonlinear rate equations (11a) and (11b) into a set of linear equations by defining

$$z_{\lambda\rho} \equiv v^\lambda w^\rho \text{ with } \lambda,\rho = 0,1,2,\ldots . \tag{13}$$

Then from (11a) and 11b)

$$\begin{aligned} \dot{z}_{\lambda\rho} - i\nu(\rho-\lambda)z_{\lambda\rho} = {} & -i\varepsilon S[(\lambda-\rho)z_{\lambda\rho} + \lambda z_{\lambda-1,\rho+1} - \rho z_{\lambda+1,\rho-1}] + \\ & i\varepsilon u[\lambda z_{\lambda-1,\rho+3} + (3\lambda-\rho)z_{\lambda,\rho+2} + (\lambda-3\rho)z_{\lambda+2,\rho} \\ & -\rho z_{\lambda+3,\rho-1} + 3(\lambda-\rho)z_{\lambda+1,\rho+1}] . \end{aligned} \tag{14}$$

Our required function $\phi(t)$ is

$$\phi(t) = \frac{1}{2}\nu^{-1}(w+v) = \frac{1}{2}\nu^{-1}(z_{1,0} + z_{0,1}). \tag{15}$$

By substituting the series expansion for $z_{\lambda,\rho}(t)$,

$$z_{\lambda\rho}(t) = \sum_{j=0}^{\infty} \varepsilon^j \sum_{n=-\infty}^{\infty} A_{\lambda\rho}(j,n)\eta^n \text{ with} \tag{16}$$

$$\eta \equiv \exp i(\nu t+\delta) \tag{17}$$

$$A^*_{\lambda\rho}(j,n) = A_{\rho\lambda}(j,-n), \text{ since } w=v^* \tag{18}$$

into (14) and equating coefficients of like powers $\varepsilon^j\eta^n$ on both sides of the equation we obtain a set of linear equations for the coefficients $A_{\lambda\rho}(j,n)$,

$$(n-\rho+\lambda)\nu A_{\lambda\rho}(j,n) =$$

$$-S[(\lambda-\rho)A_{\lambda\rho}(j-1,n) + \lambda A_{\lambda-1,\rho+1}(j-1,n)-\rho A_{\lambda+1,\rho-1}(j-1,n)]$$

$$+U[\lambda A_{\lambda-1,\rho+3}(j-1,n) + (3\lambda-\rho)A_{\lambda,\rho+2}(j-1,n)+ \tag{19}$$

$$3(\lambda-\rho)A_{\lambda+1,\rho+1}(j-1,n) + (\lambda-3\rho)A_{\lambda+2,\rho}(\lambda-1,n)-\rho A_{\lambda+3,\rho-1}(j-1,n)].$$

From (16), $A_{\lambda\rho}(0,\rho-\lambda)$ is the only nonvanishing zeroth order coefficient.

At j=1 we consider first the "resonance condition"; i.e. $n=\rho-\lambda$ in (19) so that the left hand side of the equation vanishes. Then

$$0 = S-3U[A_{\lambda+1,\rho+1}(0,\rho-\lambda)/A_{\lambda,\rho}(0,\rho-\lambda)] , \tag{20}$$

which must be satisfied for all integral λ and ρ. This equation is satisfied by

$$A_{\lambda\rho}(0,n) = \delta_{n,\rho-\lambda}f^{\lambda+\rho} \text{ with} \tag{21}$$

$$f^2 = (S/3U) = 4\nu^2(\omega_0^2-\nu^2)/3R. \tag{22}$$

Following reference (3) a parameterless recurrence formula is obtained by introducing a new set of coefficients $c_{\lambda\rho}(j,n)$ defined by

$$A_{\lambda\rho}(j,n)e^{in\delta} \equiv c_{\lambda\rho}(j,n)(U/\nu)^j f^{\lambda+\rho+2j} . \tag{23}$$

Then

$$c_{\lambda\rho}(0,n) = \delta_{n,\rho-\lambda}e^{in\delta} . \tag{24}$$

The resulting recursion relation for the c's becomes

$$(n-\rho+\lambda)c_{\lambda\rho}(j,n) =-3[(\lambda-\rho)c_{\lambda\rho}(j-1,n)+\lambda c_{\lambda-1,\rho+1}(j-1,n)-\rho c_{\lambda+1,\rho-1}(j-1,n)]$$

$$+ \lambda c_{\lambda-1,\rho+3}(j-1,n) + (3\lambda-\rho)c_{\lambda,\rho+2}(j-1,n) + 3(\lambda-\rho)c_{\lambda+1,\rho+1}(j-1,n) +$$

$$(\lambda-3\rho)c_{\lambda+2,\rho}(j-1,n)-\rho c_{\lambda+3,\rho-1}(j-1,n) . \tag{25}$$

Since $c_{\lambda\rho}(0,n)$ is known we obtain $c_{\lambda\rho}(1,n)$ by choosing j=1 and substituting (24) into the right hand side of (25). For simplicity we restrict ourselves here to the case of zero phase angle δ so that $\dot{\phi}(0)=0$. Then if $k\neq 0$

$$c_{\lambda\rho}(1,\rho-\lambda+k) = \frac{1}{4}(\lambda\delta_{k,4} + \rho\delta_{k,-4} - 2\rho\delta_{k,2} - 2\lambda\delta_{k,-2}) \cdot \quad (26)$$

The case k=0 is obtained by letting j=2 and $n=\rho-\lambda$ in eq. (25). Then

$$c_{\lambda+1,\rho+1}(1,\rho-\lambda) - c_{\lambda\rho}(1,\rho-\lambda) = \frac{1}{4} \cdot \quad (27)$$

The appropriate solution of this recurrence formula is as discussed in reference 3,

$$c_{\lambda\rho}(1,\rho-\lambda) = (\rho+\lambda)/8. \quad (28)$$

We are now in a position to obtain the general solution of the set of linear equation (25) as required for the determination of the Fourier coefficient in eq. 16 for $z_{\lambda\rho}$.

The recurrence formula (25) can be simplified by introducing the generating function

$$g_{\lambda,\rho}(\mu,n) \equiv g_{\lambda,\rho}(n) \equiv \sum_{j=1}^{\infty} \mu^j \, c_{\lambda,\rho}(j,n) \quad (29)$$

Then

$$\begin{aligned}
&[\mu^{-1}(n-\rho+\lambda)+3(\lambda-\rho)]g_{\lambda\rho}(n)+3[\lambda g_{\lambda-1,\rho+1}(n)-\rho g_{\lambda+1,\rho-1}(n)] \\
&-\lambda g_{\lambda-1,\rho+3}(n)-(3\lambda-\rho)g_{\lambda,\rho+2}(n)-3(\lambda-\rho)g_{\lambda+1,\rho+1}(n) \qquad (30) \\
&-(\lambda-3\rho)g_{\lambda+2,\rho}(n) + \rho g_{\lambda+3,\rho-1}(n) = (n-\rho+\lambda)c_{\lambda,\rho}(1,n)
\end{aligned}$$

where the $c_{\lambda,\rho}(1,n)$ on the right hand side of the equation is the known function (26) and (28).

It will not be necessary to determine the $c_{\lambda,\rho}(j,n)$ for $j>1$, since, as will now be shown, our required function $z_{\lambda\rho}(t)$ can be expressed in terms of the $g_{\lambda,\rho}$. From (16), interchanging the order of summation

$$z_{\lambda\rho}(t) = \sum_{n=-\infty}^{\infty} e^{i\nu t} \sum_{j=0}^{\infty} \varepsilon^j A_{\lambda\rho}(j,n)\, e^{i\nu\delta} \quad (31a)$$

$$= f^{\lambda+\rho} \sum_{n=-\infty}^{\infty} e^{i\nu t} \sum_{j=0}^{\infty} [\varepsilon U f^2/\nu]^j \, c_{\lambda\rho}(j,n) \tag{31b}$$

$$= f^{\lambda+\rho} \sum_{n=-\infty}^{\infty} e^{i\omega n} \{\delta_{n,\rho-\lambda} \, e^{in\delta} + g_{\lambda\rho}(\mu,n)\} \tag{31c}$$

where

$$\mu = \varepsilon f^2 U/\nu. \tag{32}$$

For fixed μ and n the set of linear equations (30) have the structure

$$AG = C \text{ so that } G = A^{-1}C \tag{33}$$

where C is the column vector (written in row form for convenience).

$$C=\{(n+1)c_{1,0}(1,n),(n-1)c_{0,1}(1,n),(n+2)c_{2,0}(1,n),nc_{1,1}(1,n),(n-2)c_{02}(1,n),$$
$$(n+3)c_{3,0}(1,n),(n+1)c_{2,1}(1,n),(n-1)c_{1,2}(1,n),(n-3)c_{0,3}(1,n),....\} \tag{34}$$

and A^{-1} is the inverse of the A triangular matrix

$$A = \begin{pmatrix} A_{11} & 0 & A_{13} & 0 & 0 & 0 \\ 0 & A_{22} & 0 & A_{24} & 0 & 0 \\ 0 & 0 & A_{33} & 0 & A_{35} & 0 \\ . & . & . & . & . & . \end{pmatrix} \tag{35}$$

whose elements are matrices.

The $A_{\alpha\alpha}$ and $A_{\alpha,\alpha+2}$ have the structure:

$$A_{11} = \begin{pmatrix} f_1(0) & 3 \\ -3 & f_1(1) \end{pmatrix} \tag{36}$$

with

$$f_\alpha(m) = (n+\alpha-2m)\mu^{-1} + 3(\alpha-2m) \tag{37}$$

$$A_{22} = \begin{pmatrix} f_2(0) & 6 & 0 \\ -3 & f_2(1) & 3 \\ 0 & -6 & f_2(2) \end{pmatrix} \text{and generally} \tag{38}$$

$$A_{\alpha\alpha} = \begin{pmatrix} f_\alpha(0) & 3\alpha & 0 & 0 & \dots & 0 & 0 \\ -3 & f_\alpha(1) & 3(\alpha-1) & 0 & \dots & 0 & 0 \\ 0 & -6 & f_\alpha(2) & 3(\alpha-2) & \dots & 0 & 0 \\ \cdot & \cdot & \cdot & \cdot & \cdot & \cdot & \cdot \\ \cdot & \cdot & \cdot & \cdot & \cdot & f_\alpha(\alpha-1) & 3\cdot 1 \\ 0 & 0 & \cdot & \cdot & \cdot & -3\alpha & f_\alpha(\alpha) \end{pmatrix} \tag{39}$$

with $A_{\alpha\alpha}$ being an $(\alpha+1) \times (\alpha+1)$ square matrix. The $A_{\alpha,\alpha+2}$ are $(\alpha+1) \times (\alpha+3)$ order matrices:

$$A_{1,3} = \begin{pmatrix} -1 & -3 & -3 & -1 \\ 1 & 3 & 3 & 1 \end{pmatrix} \tag{40a}$$

$$A_{2,4} = \begin{pmatrix} -2 & -6 & -6 & -2 & 0 \\ 1 & 2 & 0 & -2 & -1 \\ 0 & 2 & 6 & 6 & 2 \end{pmatrix} \tag{40b}$$

and in general

$$A_{\alpha,\alpha+2} \equiv \begin{pmatrix} a^\alpha_{-1}(0) & a^\alpha_0(0) & \cdot & \cdot & \cdot & a^\alpha_{\alpha+1}(0) \\ a^\alpha_{-2}(1) & a^\alpha_{-1}(1) & \cdot & \cdot & \cdot & a^\alpha_\alpha(1) \\ a^\alpha_{-3}(2) & a^\alpha_{-2}(2) & \cdot & \cdot & \cdot & a^\alpha_{\alpha-1}(2) \\ \cdot & \cdot & \cdot & \cdot & \cdot & \cdot \\ a^\alpha_{-\alpha-1}(\alpha) & a^\alpha_{-\alpha}(\alpha) & & \cdot & \cdot & a^\alpha_1(\alpha) \end{pmatrix} \tag{41a}$$

where

$$
\begin{aligned}
a_0^\alpha(j) &= -3(\alpha-2j) && j = 0,1,..,\alpha \\
a_1^\alpha(j) &= -3\alpha+4j && j = 0,1,..,\alpha \\
a_{-1}^\alpha(j) &= -\alpha+4j && j = 0,1,..,\alpha \\
a_2^\alpha(j) &= -\alpha+j && j = 0,1,..,\alpha-1 \\
a_k^\alpha(j) &\equiv 0 \text{ if } |k| > 2.
\end{aligned}
\qquad (41b)
$$

All other $A_{\alpha\beta} \equiv 0$ (with $\beta \neq \alpha$ and $\beta \neq \alpha+2$).

The first few required matrix inverses are easily found to be

$$A^{-1}_{11} = (\mu/D_2)\begin{pmatrix} n-1-3\mu & -3\mu \\ 3\mu & n+1+3\mu \end{pmatrix} \qquad (42a)$$

$$D_2 \equiv (n^2-1)-6\mu$$

$$A^{-1}_{22} = (\mu/D_3)\begin{pmatrix} n(n-2-6\mu)+18\mu^2 & -6(n-2)\mu+36\mu^2 & 18\mu^2 \\ 3(n-2)\mu-18\mu^2 & n^2-4-24\mu-36\mu^2 & -3(n+2)\mu-18\mu^2 \\ 18\mu^2 & 6(n+2)\mu+36\mu^2 & n(n+2+6\mu)+18\mu^2 \end{pmatrix} \qquad (43a)$$

$$D_3 = n[(n^2-4)- 24\mu] \qquad (43b)$$

$$A^{-1}_{33} = D_4^{-1}\begin{pmatrix} \gamma_1\gamma_2\gamma_3+27\gamma_1+36\gamma_3 & -9(\gamma_2\gamma_3+27) & 54\gamma_3 & -162 \\ 3\gamma_2\gamma_3+81 & \gamma_0\gamma_2\gamma_3+27\gamma_0 & -6\gamma_0\gamma_3 & 18\gamma_0 \\ 18\gamma_3 & 6\gamma_0\gamma_3 & \gamma_0\gamma_1\gamma_3+27\gamma_3 & -3(\gamma_0\gamma_1+27) \\ 162 & 54\gamma_0 & 9(\gamma_0\gamma_1+27) & \gamma_0\gamma_1\gamma_2+27\gamma_2+36\gamma_0 \end{pmatrix}$$

with

$$D_4 = (n^2-9)(n^2-1)\mu^{-4} - 12(5n^2-9)\mu^{-3} + (18)^2\mu^{-2}$$

$$\gamma_0 = (n+3)\mu^{-1}+9 \; ; \; \gamma_2 = (n-1)\mu^{-1}-3$$

$$\gamma_1 = (n+1)\mu^{-1}+3 \; ; \; \gamma_3 = (n-3)\mu^{-1}-9 \; .$$

Proceeding in a manner analogous to that used in the derivation of eq. (10) of Section III we find that

$$\begin{pmatrix} g_{1,0}(u,n) \\ g_{0,1}(u,n) \end{pmatrix} = A_{11}^{-1} \begin{pmatrix} (n+1)c_{1,0}(1,n) \\ (n-1)c_{0,1}(1,n) \end{pmatrix} - A_{11}^{-1}A_{13}A_{33}^{-1} \begin{pmatrix} (n+3)c_{3,0}(1,n) \\ (n+1)c_{2,1}(1,n) \\ (n-1)c_{1,3}(1,n) \\ (n-3)c_{0,3}(1,n) \end{pmatrix} + \ldots \qquad (44)$$

These equations combined with (31c) and (15) yield the required algorithm for the solution of our problem. With the frequency ν having been chosen originally, the initial condition $\phi(0)$ and $\dot{\phi}(0)$ became derived quantities.

We close our presentation here by exhibiting the general formulæ for the inverses of the $A_{\alpha\alpha}$ for all integral α. We simplify our notation to prevent the required formulæ from occupying too much space. Let

$$A_{n-1} \equiv A_{n,n} \quad , \; a_{j+1} \equiv f_n(j)$$

$$b_j = 3(n-j+1), \; c_j = -3j \qquad (45)$$

so that $A_{n-1,n-1}$ or A_n has the form

$$A_n = \begin{pmatrix} a_1 & b_1 & 0 & 0 & \ldots & 0 \\ c_1 & a_2 & b_2 & 0 & \ldots & 0 \\ 0 & c_2 & a_3 & b_3 & \ldots & 0 \\ . & . & . & . & \ldots & . \\ 0 & . & . & . & \ldots & a_n \end{pmatrix} \equiv (a_{ij}) \quad . \qquad (46)$$

The elements of $a_{ij}^{(-1)}$ are easily verified to be

$$a_{ij}^{(-1)} = \frac{1}{D_1} \begin{cases} (-1)^{i+j} A_{i-1}\, b_i\, b_{i+1} \cdot\cdot\cdot\, b_{j-1}\, D_{j+1} & \text{if } j>i \\ (-1)^{i+j} A_{i-1}\, D_{i+1} & \text{if } j=i \\ (-1)^{i+j} A_{j-1}\, c_j\, c_{j+1} \cdot\cdot\cdot\, c_{i-1}\, D_{i+1} & \text{if } j<i \end{cases} \tag{47}$$

where

$$D_n = a_n \;;\; D_{n-1} = \begin{vmatrix} a_{n-1} & b_{n-1} \\ c_{n-1} & a_n \end{vmatrix} \;;\; D_{n-2} = \begin{vmatrix} a_{n-2} & b_{n-2} & 0 \\ c_{n-2} & a_{n-1} & b_{n-1} \\ 0 & c_{n-1} & a_n \end{vmatrix} \;;\; \text{etc.} \tag{48}$$

$$a_o = D_{n+1} = 1$$

and the A_m and D_m satisfy the recurrence relation

$$A_m = a_m A_{m-1} - c_{m-1} b_{m-1} A_{m-2} \text{ with } A_1 = a_1,\ A_2 = a_1 a_2 - b_1 c_1 \tag{49}$$

$$D_m = a_m D_{m+1} - c_m b_m D_{m+2} \qquad \text{with } D_n = a_n,\ D_{n-1} = a_n a_{n-1} - b_{n-1} a_{n-1} \tag{50}$$

when n = 4

$$A_4^{-1} = \frac{1}{D_1} \begin{bmatrix} A_o D_2 & -A_o b_1 D_3 & A_o b_1 b_2 D_4 & -A_o b_1 b_2 b_3 D_5 \\ -A_o c_1 D_3 & A_1 D_3 & -A_1 b_2 D_4 & A_1 b_2 b_3 D_5 \\ A_o c_1 c_2 D_4 & -A_1 c_2 D_4 & A_2 D_4 & -A_2 b_3 D_5 \\ -A_o c_1 c_2 c_3 D_5 & A_1 c_2 c_3 D_5 & -A_2 c_3 D_5 & A_3 D_5 \end{bmatrix} \tag{51}$$

The details of the solution of our problem as well as that of a pair of coupled oscillators will be given elsewhere.

In conclusion, the author wishes to thank Robert Helleman for many interesting discussions on nonlinear dynamics.

REFERENCES

1. T. Carleman, "Application de la théorie des équations intégrales singulières aux équations differetielles de la dynamique, Arkiv För Mathematik, Astronomi och Fysik, 22B, No. 7 (1931).
2. T. Carleman, Application de la théorie des équations intégrales linéaires aux systèmes d'équations differentielles non linéaires Acta Mathematica 59, 63 (1932).
3. E. W. Montroll and R. H. G. Helleman, On a Nonlinear Perturbation Theory Without Secular Terms II, A.I.P. Conference Proc. No. 27, 75 (Am. Inst. of Phys. 1976).
4. R. Bellman and J. M. Richardson, Quart. Appl. Math. 20, 333 (1963).
5. R. H. G. Helleman and E. W. Montroll, "On a Nonlinear Perturbation Theory Without Secular Terms, I", Physica 74, 22 (1974).
6. C. R. Eminhizer, R. H. G. Helleman, and E. W. Montroll, "On a Convergent Nonlinear Perturbation Theory without Small Denominators or Secular Terms, J. Math. Phys. 17, 121 (1976).

INTEGRATION OF LINEARIZED NON-LINEAR EVOLUTION EQUATIONS

K. M. Case
The Rockefeller University, New York, New York

The problem considered is that of solving the linear equation which arises when a non-linear evolution equation is linearized around some particular solution. It is shown that if the original equation is of completely integrable Hamiltonian form there are an infinity of explicit solutions of the linearized equation. These are almost always linearly independent.

INTRODUCTION

Frequently, one encounters the following problem: We have some non-linear evolution equation and a particular solution of it. It is desired, for example to study stability of quantization, to discuss the equation linearized around the particular solution. Now it is well known that if the original equation is translationally invariant, one can immediately write down a solution of the linearized equation. Thus, if the original equation is invariant under spatial translation and u is a solution, then $\partial u/\partial x$ is a solution of the linearized equation. Similarly, if we have time translation invariance, $\partial u/\partial t$ is a solution. In general, these two solutions are linearly independent and non-trivial. There is one special case: if $u = u(x - ct)$, then these solutions are proportional.

Here we wish to point out that for a very large class of evolution equations many more explicit solutions of the linearized equations are readily obtained. These solutions are related to the densities of conserved functionals. The class of evolution equations involved appears to include all of those which are known to be completely integrable by the Inverse Scattering Transform method and even some for which this is not known. Since in the former case we know that there are an infinity of conserved functionals, we obtain an infinity of explicit solutions of the linearized equations.

THE FUNDAMENTAL THEOREM

In essence the theorem relates solutions of the linearized equation to constants of motion. It probably goes back at least to Poisson[1]. However, until recently it had little practical significance. For most systems the constants of motion were the simple ones obtained through Noether's theorem from elementary symmetry properties.

These were usually momentum, energy, and angular momentum. As indicated in the Introduction, one knows how to use these to construct solutions of the linearized equations. However, the recent discovery of completely integrable systems with an infinite

ISSN: 0094-243X/78/360/$1.50

number of degrees of freedom gives much more significance to the result. We state this as a

THEOREM:

Consider a Hamiltonian system. We have a Hamiltonian H and an appropriately defined Poisson Bracket [,] such that the equations of motion are of the form

$$\frac{dF}{dt} = [F, H]. \tag{1}$$

Let G be a conserved functional, i.e.

$$[G, H] = 0, \tag{2}$$

then a solution of Eq. (1) linearized around a particular solution $F^{(0)}$ is:

$$F^{(1)} = [F^{(0)}, G]. \tag{3}$$

In words: The infinitesimal contact transformation generated by G yields a solution of the linearized equation.

We give a proof in a simple case.

Proof: Let the equations of motion be

$$\frac{dq_i}{dt} = [q_i, H] = \frac{\partial H}{\partial p_i}, \tag{4}$$

$$\frac{dp_i}{dt} = [p_i, H] = \frac{-\partial H}{\partial q_i}, \quad i = 1, 2, \ldots, N.$$

Imagine we have found a solution $(q^{(0)}, p^{(0)})$ and seek solutions near this of the form

$$q = q^{(0)} + \varepsilon q^{(1)}, \ p = p^{(0)} + \varepsilon p^{(1)} . \tag{5}$$

To first order in ε, we get the linear equations

$$\frac{dq_i^{(1)}}{dt} = \sum_j \left\{ q_j^{(1)} \frac{\partial^2 H(q^{(0)}, p^{(0)})}{\partial p_i^{(0)} \partial q_j^{(0)}} + p_j^{(1)} \frac{\partial^2 H}{\partial p_i^{(0)} \partial p_j^{(0)}} \right\}, \tag{6}$$

$$\frac{dp_i^{(1)}}{dt} = -\sum_j \left\{ q_j^{(1)} \frac{\partial^2 H}{\partial q_i^{(0)} \partial q_j^{(0)}} + p_j^{(1)} \frac{\partial^2 H}{\partial q_i^{(0)} \partial p_j^{(0)}} \right\},$$

We maintain that a solution of Eqs. (6) is:

$$q_i^{(1)} = [q_i^{(0)}, G], \; p_i^{(1)} = [p_i^{(0)}, G]. \qquad (7)$$

Now

$$[q_i^{(0)}, G] = \frac{\partial G}{\partial p_i^{(0)}}, \; [p_i^{(0)}, G] = \frac{-\partial G}{\partial q_i^{(0)}}. \qquad (8)$$

Hence, we need to show that

$$[[q_i^{(0)}, G], H] = \sum_j \left\{ \frac{\partial G}{\partial p_j^{(0)}} \frac{\partial^2 H}{\partial p_i^{(0)} \partial q_j^{(0)}} - \frac{\partial G}{\partial q_j^{(0)}} \frac{\partial^2 H}{\partial p_i^{(0)} \partial p_j^{(0)}} \right\}, \qquad (9a)$$

and

$$[[p_i^{(0)}, G], H] = -\sum_j \left\{ \frac{\partial G}{\partial p_j^{(0)}} \frac{\partial^2 H}{\partial q_j^{(0)} \partial q_i^{(0)}} - \frac{\partial G}{\partial q_j^{(0)}} \frac{\partial^2 H}{\partial q_j^{(0)} \partial p_j^{(0)}} \right\}. \qquad (9b)$$

By the Jacobi identity

$$[[q_i^{(0)}, G], H] = - [[G, H], q_i^{(0)}] - [[H, q_i^{(0)}], G].$$

The first term on the right vanishes since it was assumed that G is constant. Using the first of Eqs. (4), we see that

$$[[q_i^{(0)}, G], H] = [\frac{\partial H}{\partial p_i^{(0)}}, G], \tag{10a}$$

which on expanding the bracket on the right yields Eq. (9a). Similarly,

$$[[p_i^{(0)}, G], H] = [[p_i^{(0)}, H], G] = -[\frac{\partial H}{\partial q_i^{(0)}}, G]$$

$$= -\sum_j \left\{ \frac{\partial^2 H}{\partial q_j^{(0)} q_i^{(0)}} \frac{\partial G}{\partial p_j^{(0)}} - \frac{\partial^2 H}{\partial q_i^{(0)} \partial p_j^{(0)}} \frac{\partial G}{\partial q_j^{(0)}} \right\}, \tag{10b}$$

which is indeed Eq. (9b).

THE TODA LATTICE

The equations as originally formulated [2] are in Hamiltonian form. They are:

$$\dot{q}_n = p_n,$$

$$\dot{p}_n = -[\exp\{-(q_{n+1} - q_n)\} - \exp\{-(q_n - q_{n-1})\}]. \tag{11}$$

While our application can be either to the infinite or periodic case, we will here restrict ourselves to the former. Thus, $-\infty < n < \infty$ and we assume $p_n \to 0$, $q_n \to$ constant as $|n| \to \infty$ (sufficiently rapidly). Flashka [3] has shown that Eqs. (11) can be put in the Lax [4] form

$$\frac{\partial L}{\partial t} = [B, L], \tag{12}$$

with the matrices L and B given by

$$L(n, m) = a(m)\delta(n - 1, m) + a(n)\delta(n + 1, m) + b(n)\delta(n, m), \tag{13}$$

and

$$B(n, m) = a(n)\delta(n + 1, m) - a(m)\delta(n - 1, m). \tag{14}$$

Here,

$$a(n) = \frac{\exp\left\{-(q_n - q_{n-1})/2\right\}}{2}, \quad b(n) = \frac{-p(n-1)}{2}. \tag{15}$$

This leads to the solution of the Eqs. (11) via the Inverse Scattering Transform method using as eigenvalue problem

$$L\psi = \lambda\psi, \tag{16}$$

and determining time dependence from

$$\frac{\partial\psi}{\partial t} = B\psi. \tag{17}$$

From Eqs. (12), (16), and (17) it is readily shown that the eigenvalues of L are conserved functionals. However, there are many more (and simpler) ones.

Formally let us consider

$$\mathrm{Tr}\, L^N. \tag{18}$$

Then,

$$\frac{\partial}{\partial t}\mathrm{Tr}\, L^N = N\,\mathrm{Tr}\,\frac{\partial L}{\partial t}L^{N-1}$$

$$= N\,\mathrm{Tr}\,[B, L]\,L^{N-1}$$

$$= N[\mathrm{Tr}\,BL^N - \mathrm{Tr}\,LBL^{N-1}]$$

$$= N[\mathrm{Tr}\,BL^N - \mathrm{Tr}\,BL^N] = 0,$$

and we seem to have constants for arbitrary N. Unfortunately, however, the traces in the expression (18) do not exist. This is readily remedied. Thus, consider the matrix L_0 obtained from L be replacing the elements by their limits as $|n| \to \infty$, i.e.,

$$L_0(n, m) = \frac{1}{2}\left\{\delta(n-1, m) + \delta(n+1, m)\right\}.$$

Then $F_N = \mathrm{Tr}(L^N - L_0^N)$ exists, and by the above argument is a conserved functional - for all N.

Using these, we construct solutions of the Toda equations linearized around a given solution by:

$$q_n^{(1)} = [q_n^{(0)}, F_N(q^{(0)}, p^{(0)})] = \frac{\partial F_N}{\partial q_n^{(0)}}, \tag{19}$$

$$p_n^{(1)} = [p_n^{(0)}, F_N(q^{(0)}, p^{(0)})] = \frac{-\partial F_N}{\partial q_n^{(0)}} .$$

Up to unessential multiplicative constants the first three of our conserved functionals are:[5]

$$F_1 = \sum_{n=-\infty}^{\infty} p(n), \tag{20}$$

$$F_2 = \sum_{n=-\infty}^{\infty} [\exp\{-(q_n - q_{n-1})\} - 1] + \frac{p^2(n)}{2} \tag{21}$$

and

$$F_3 = \sum_{n=-\infty}^{\infty} [\exp\{-(q_n - q_{n-1})\}][p(n-1) + p(n)] + \frac{p^3(n)}{3} . \tag{22}$$

The first two of these are the "classical" constants. Indeed, F_1 is the momentum and F_2 is the energy (Hamiltonian). Corresponding to these, we have as solutions of the linearized equations

$$q_n^{(1)} = 1, \; p_n^{(1)} = 0, \tag{23}$$

and

$$q_n^{(1)} = \dot{q}_n^{(0)}, \; p_n^{(1)} = \dot{p}_n^{(0)} . \tag{24}$$

From the non-classical constant F_3, we obtain

$$q_n^{(1)} = [\exp\{-(q_{n-1}^{(0)} - q_{n-2}^{(0)})\} + \exp\{-(q_n^{(0)} - q_{n-1}^{(0)})\}] + (p_n^{(0)})^2,$$

$$p_n^{(1)} = [p_{n-1}^{(0)} + p_n^{(0)}]\exp\{-(q_n^{(0)} - q_{n-1}^{(0)})\} - [p_n^{(0)} + p_{n+1}^{(0)}] \times \exp\{-(q_{n+1}^{(0)} - qn^{(0)})\}.$$

EQUATIONS OF LAX FORM

Lax[6] has considered equations of the form

$$\frac{\partial u}{\partial t} = \frac{\partial}{\partial x} \frac{\delta H}{\delta u(x)} \quad , \tag{26}$$

where H is some functional of u. Poisson Brackets between two functionals F_1, F_2 are defined by

$$[F_1, F_2] = \int \frac{\delta F_1}{\delta u(x')} \frac{\partial}{\partial x'} \frac{\delta F_2}{\delta u(x')} dx' \; . \tag{27}$$

Then, Eq. (26) is of Hamiltonian form

$$\frac{\partial u}{\partial t} = [u, H] \tag{28}$$

and conserved functionals, i.e., such that

$$\frac{dF}{dt} = 0 \tag{29}$$

satisfy

$$\int \frac{\delta F}{\delta u} \frac{\partial u}{\partial t} = \int \frac{\delta F}{\delta u} \frac{\partial}{\partial x} \frac{\delta H}{\delta u} dx = [F, H] = 0. \tag{30}$$

Eq. (26) linearized around some solution u has the form

$$\frac{\partial v}{\partial t} = \frac{\partial}{\partial x} Nv \tag{31}$$

where the linear operator N is defined by

$$Nv = \frac{d}{d\varepsilon} H(u + \varepsilon v)\Big|_{\varepsilon=0} \quad . \tag{32}$$

Our theorem tells us that if F[u] is a conserved functional, we have as a solution of Eq. (31)

$$\begin{aligned} v(x,t) = [u, F] &= \int \frac{\delta u(x')}{\delta u(x)} \frac{\partial}{\partial x'} \frac{\delta F}{\delta u(x')} dx' \\ &= \int \delta(x - x') \frac{\partial}{\partial x'} \frac{\delta F}{\delta u(x')} dx \\ &= \frac{\partial}{\partial x} \frac{\delta F}{\delta u(x)} \; . \end{aligned} \tag{33}$$

Thus, the derivative of the conserved density satisfies the linearized equation.

The three "classical" conserved functionals corresponding to Eq. (26) are

$$I_1 = \int u\, dx$$

$$I_2 = \int \frac{u^2}{2}\, dx$$

$$I_3 = H$$

with functional derivatives $\phi_1 = 1$, $\phi_2 = u$, $\phi_3 = \delta H/\delta u$. The corresponding solutions of Eq. (31) are $\partial\phi_1/\partial x = 0$, which is trivial and

$$\frac{\partial \phi_2}{\partial x} = \frac{\partial u}{\partial x}\,, \tag{34}$$

and

$$\frac{\partial \phi_3}{\partial x} = \frac{\partial}{\partial x}\frac{\delta H}{\delta u} = \frac{\partial u}{\partial t}\,, \tag{35}$$

which are the solutions mentioned in our introduction.

For special forms of H there may, however, be more conserved functionals. For example, a typical equation of the form of Eq. (26) is

$$\frac{\partial u}{\partial t} = u^P \frac{\partial u}{\partial x} + \frac{\partial}{\partial x}\int_{-\infty}^{\infty} G(x' - x)\,\frac{\partial u}{\partial x'}\, dx', \tag{36}$$

with $G(x) = -G(-x)$. Here,

$$H = \int_{-\infty}^{\infty} \frac{u^{P+2}dx}{(P+1)(P+2)} + \frac{1}{2}\int\int_{-\infty}^{\infty} \frac{\partial u}{\partial x}\,\hat{G}(x' - x)\,\frac{\partial u}{\partial x'}\, dx\, dx' \tag{37}$$

with $\partial\hat{G}/\partial x = G(x)$. Thus,

$$\frac{\delta H}{\delta u} = \frac{u^{P+1}}{P+1} + \int_{-\infty}^{\infty} G(x' - x)\,\frac{\partial u}{\partial x'}\, dx'. \tag{38}$$

The linearized form of Eq. (36) is then

$$\frac{\partial v}{\partial t} = \frac{\partial}{\partial x}\left\{u^P v + \int_{-\infty}^{\infty} G(x' - x)\,\frac{\partial v}{\partial x'}\,dx'\right\}. \qquad (39)$$

Examples:

1) If p = 1 and G(x) = -δ´(x), Eq. (36) becomes the Korteweg-de Vries equation

$$\frac{\partial u}{\partial t} = u\,\frac{\partial u}{\partial x} + \frac{\partial^3 u}{\partial x^3}. \qquad (40)$$

This, it is known, has an infinite number of conserved functionals. Thus, we have an infinite number of solutions of the specialization of Eq. (39)

$$\frac{\partial v}{\partial t} = \frac{\partial u}{\partial x}\,v + \frac{\partial v}{\partial x}\,u + \frac{\partial^3 v}{\partial x^3}\,. \qquad (41)$$

The simplest non-classical of these corresponds to the conserved functional

$$I_4 = \int_{-\infty}^{\infty}\left\{\frac{u^4}{12} - u\left(\frac{\partial u}{\partial x}\right)^2 + \frac{3}{5}\left(\frac{\partial^2 u}{\partial x^2}\right)^2\right\}dx\,. \qquad (42)$$

Thus,

$$\phi_4 \equiv \frac{\delta I_4}{\delta u} = \frac{u^3}{3} + \left(\frac{\partial u}{\partial x}\right)^2 + 2u\,\frac{\partial^2 u}{\partial x^2} + \frac{6}{5}\,\frac{\partial^4 u}{\partial x^4}\,. \qquad (43)$$

In addition to solutions $v = u_t$, $v = u_x$, we then have the solution

$$v = \partial\phi_4/\partial x\,. \qquad (44)$$

2) If p = 1 and G = 1/x,[7] Eq. (26) becomes the Benjamin-Ono equation

$$\frac{\partial u}{\partial t} = u\,\frac{\partial u}{\partial x} + \frac{\partial}{\partial x}\int_{-\infty}^{\infty}\frac{\partial u/\partial x'}{x' - x}\,dx'\,. \qquad (45)$$

This is known[8] to have at least one more conserved functional than the classical ones. It is

$$I_4 = \int_{-\infty}^{\infty}\left\{\frac{u^4}{12} + \frac{2\pi^2}{3}\left(\frac{\partial u}{\partial x}\right)^2\right\}dx + \int\int_{-\infty}^{\infty}\frac{u(x)u(x')\partial u/\partial x'}{x' - x}\,dx\,dx'. \qquad (46)$$

From this we conclude that if u is a solution of Eq. (45) then $v = \partial\phi_4/\partial x$ is a solution of

$$\frac{\partial v}{\partial t} = u_x v + uv_x + \frac{\partial}{\partial x}\int_{-\infty}^{\infty} \frac{\partial v/\partial x'}{x' - x}\, dx' , \qquad (47)$$

where

$$\phi_4 = \frac{u^3}{3} - 3\pi^2 \frac{\partial^2 u}{\partial x^2} + \int_{-\infty}^{\infty} \frac{[u(x)+u(x')]\partial u/\partial x'}{x' - x}\, dx' . \qquad (48)$$

LINEAR INDEPENDENCE

We should consider the question of whether the solutions obtained are linearly independent. The answer seems to be that they are, except for one very special (but unfortunately very interesting) case. This is when $u = u(x - ct)$--the single soliton case. We saw that solutions $v = u_x$ and $v = u_t$ are proportional. The situation for the general conserved polynomial functional is readily discussed in the K-de V case. Here [9] the density of the n'th conserved polynomial can be written in the form

$$\phi_n = M^{n-1}\, 1 , \qquad (49)$$

where the operator M is defined by

$$M\psi = \frac{\partial^2 \psi}{\partial x^2} + \frac{u\psi}{3} + \frac{1}{3}\int_{-\infty}^{x} u(x')\, \frac{\partial \psi}{\partial x'}\, dx' . \qquad (50)$$

Thus, $\phi_1 = 1$, $\phi_2 = u/3$ and

$$\phi_3 = \frac{1}{3}\left[\frac{\partial^2 u}{\partial x^2} + \frac{u^2}{2}\right] . \qquad (51)$$

However, if in Eq. (40) we put $u = u(x - ct)$ and integrate, we obtain

$$-cu = \frac{\partial^2 u}{\partial x^2} + \frac{u^2}{2} . \qquad (52)$$

Thus,

$$\phi_3 = \frac{-cu}{3} , \qquad (53)$$

and in general

$$\phi_n = (-c)^{n-2}\, u/3, \; n \geq 2 \; . \qquad (54)$$

The reason for this degeneracy is, of course, that for solutions of this special form the time derivative is equivalent to the space derivative. Another manifestation of the special character of these solutions is that any integral involving powers and spatial derivatives of u which does not involve x or t explicitly is a constant of motion.

In other cases the different conserved functionals give rise to different solutions of the linearized equations. For example, in the n soliton case we know that, asymptotically, u has the form

$$u(x, t) = \sum_i u_i(x - c_i t), \; c_i \neq c_j, \; i \neq j. \qquad (55)$$

Then, certainly, $u_t \neq u_x$.

ADDITIONAL CONSTANTS

It is obvious that given a set of conserved functionals we can <u>in principle</u> construct more. Thus, suppose we have F_i and F_j such that

$$[F_i, H] = 0 = [F_j, H]. \qquad (56)$$

Then the Jacobi identity

$$[[F_i, F_j], H] + [[F_j, H], F_i] + [[H, F_i], F_j] = 0 \qquad (57)$$

implies

$$[[F_i, F_j], H] = 0, \qquad (58)$$

i.e., $[F_i, F_j]$ is conserved.

Thus, we have a new conserved functional. However, this may or may not be of interest. Three possibilities can arise:

a) The functionals may all be in involution, i.e.,

$$[F_i, F_j] = 0, \text{ all } i, j. \qquad (59)$$

The new functionals are then trivial.

<u>Examples</u>:

1) For the Benjamin-Ono equation we have written down four conserved functionals. The similarity to the K-de V equation might raise hopes that there are many more. It is natural then to apply

the theorem of this section to our four known constants. Unfortunately, it turns out that these are indeed all in involution.

2) For the K-de V equation it has been shown[10,6] that all the known constants are in involution.

b) The new constants $[F_i, F_j]$ might be expressible as a linear combination of the previously known set. While the result would be non-trivial it would be useless from the point of view of the present application. I know of no such case to date.

c) The new constants $[F_i, F_j]$ are indeed linearly independent of the known set. This would then be very interesting. However, again I know of no such occurrence in problems of the type considered here.

FOOTNOTES

1. See, for example, E. T. Whittaker, "A Treatise on the Analytical Dynamics of Particles and Rigid Bodies", Cambridge University Press, Fourth Edition, Sec. 144, (1937).
2. M. Toda, Prog. Theor. Phys. Suppl., No. 45., 174 (1970).
3. H. Flashka, Prog. Theor. Phys., <u>51</u>, No. 3, 703 (1974).
4. P. D. Lax, Comm. Pure Appl. Math., <u>21</u>, 467-490 (1968).
5. See, for example, K. M. Case, J. Math. Phys. <u>16</u>, 1435. Sec. IV (1975).
6. P. D. Lax, Comm. Pure Appl. <u>28</u>, 141-188 (1975).
7. By 1/x we mean the principal value interpretation.
8. K. M. Case, Stanford Research Institute report JSS-77-31. J. D. Meiss and N. R. Pereira, Lawrence Berkeley Laboratory Preprint.
9. H. H. Chen, Y. C. Lee, and C. S. Liu, University of Maryland, Preprint. (Submitted to Phys. Rev. Letters).
10. C. S. Gardner, J. Math. Phys. <u>12</u>, 1548-1551 (1971).

APPENDIX

MULTI-SOLITON SOLUTIONS OF THE BENJAMIN-ONO EQUATION

For two of our three specific examples, the K-de V equation and the Toda lattice, many multi-soliton solutions are known. Hence, the uselessness of our result for the one soliton case is not very important. However, to my knowledge, only single soliton solutions to the Benjamin-Ono equation have been published. Here we would like to point out the N-soliton solutions do indeed exist.

Explicitly, the two soliton solution of

$$\frac{\partial u}{\partial t} = 2u\,\frac{\partial u}{\partial x} + \frac{\partial}{\partial x}\,\frac{P}{\pi}\int_{-\infty}^{\infty} \frac{\partial u/\partial x'}{x' - x}\,dx'$$

is

$$u = i\left(\frac{1}{x-a_+} + \frac{1}{x-a_-}\right) + \text{c.c.}$$

where

$$a_\pm = \alpha t + \beta \pm \eta\ .$$

$$\eta = \frac{1}{\varepsilon}\{1+\varepsilon^2(\gamma+\varepsilon t)^2\}^{1/2}\ ,$$

$$-\beta_i > |\gamma_i|\ 2^{1/2} > 0\ .$$

THE DISK DYNAMO

Edward Bullard
Scripps Institution of Oceanography
La Jolla, California 92093

ABSTRACT

The disk dynamo consists of a rotating disk connected by brushes to a coil coaxial with the disk. If a couple is applied to the disk the dynamo can generate current. If it is provided with a shunt across the coil and with a series impedance it gives a current which reverses in an apparently random manner. A pair of interconnected dynamos, without shunts shows similar reversals. The disk dynamo is a system with 3 degrees of freedom, it is the simplest known example of pseudo-stochastic behavior in a mechanical system. The paper describes the development of this behavior as the couple, the shunt and the series impedance are successively added to the machine. The resemblance of the behavior of a disk dynamo to that of a dynamo in a sphere of conducting fluid is discussed. The reversals are compared to those of the earth's magnetic field.

1. INTRODUCTION

The purpose of this note is to describe a very simple machine, with only three degrees of freedom, which behaves in a way analogous to some of the more complex continuous systems described in this volume. The machine is the disk dynamo. Its simplicity enables attention to be concentrated on the behavior rather than on mathematical techniques. The equations for the machine will be set out but detailed proofs of the properties of their solutions will not be given. Proofs for most of the statements in sections 2 to 5 will be found in Bullard[1], for section 6 in Robbins[2,3] and for section 7 in Cook and Roberts[4].

The study of the machine was initially motivated by a wish to understand dynamos consisting of an electrically conducting fluid sphere, which present great analytical difficulties. The first mention of a disk dynamo in this connection was by Inglis[5] who also introduced a dynamo which he described as 'a fanciful merry-go-round'. His discussion is almost entirely intuitive and qualitative.

It has been found that the disk dynamo behaves in a way that is not only analogous to the behavior of dynamos composed of bodies of homogeneous fluid but also shows reversals analogous to those of the earth's magnetic field. Such reversals can plausibly be supposed to occur for a fluid dynamo, but have not been proved to do so.

2. THE DISK DYNAMO

A disk dynamo is shown in figure 1. It consists of an electrically conducting disk which can be rotated on an axle by an applied couple. If it rotates in an axial magnetic field, a radial

ISSN: 0094-243X/78/373/$1.50

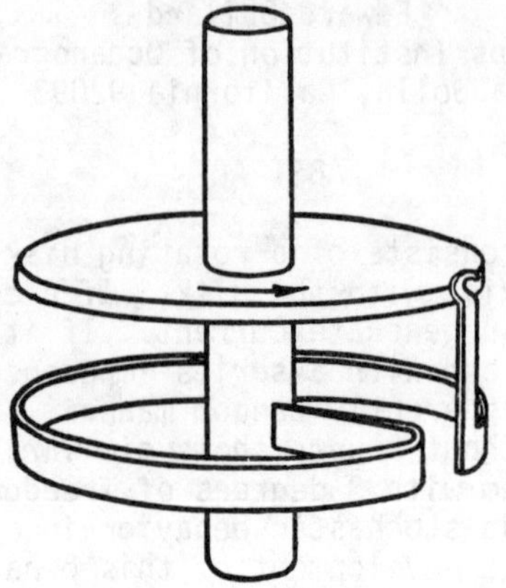

Figure 1. A disk dynamo

e.m.f. will be produced between the axle and the periphery of the disk. If this were all, the e.m.f. would be balanced by an electric charge on the edge of the disk and no current would flow. If one end of a stationary coil coaxial with the disk and axle is joined to the edge of the disk by a sliding contact (a brush) and the other end is joined to another sliding contact on the axle, as in figure 1, a current will flow through the coil and an axial magnetic field will be produced. This is the field already mentioned. At large distances from the machine the field falls to zero as the inverse cube of the distance and almost all field lines return to the machine. No external source of field or current is required and no part of the machine is ferromagnetic. The device is a self-exciting dynamo; it can be regarded as an idealisation of the engineer's 'homopolar dynamo'.

The theory is simplified by a slight further idealisation. The machine of figure 1 has current flowing predominately in the part of the disk between the axle and the brush. It is easier to formulate the equations for the system if the current in the disk is axially symmetric, this can be achieved by substituting a highly conducting ring for the brush. The single brush was shown in figure 1 since it gives a more easily understood diagram. The advantage of the ring is that the current in the disk can be represented by a single variable and is not a function of azimuth. A similar ring can be substituted for the brush on the axle, but here the difference between the point contact and the ring is unimportant.

3. DISK DYNAMO WITH ONE DEGREE OF FREEDOM

Disk held stationary. The results for this are trivial. Let

the current, which is the only variable, be I, the inductance of the coil L and its resistance R. The equation governing the current is

$$L\dot{I} + RI = 0 ,$$

which has the solution

$$I = I_0 \exp(-Rt/L) ,$$

where t is the time and I_0 is an arbitrarily chosen initial value for the current. As would be expected in a stationary system, the current decays exponentially to zero. The only steady state is that with zero current, this state is stable in the sense that if a disturbing field or current is introduced it will decay when its cause is removed.

Disk rotating at constant angular velocity. Let the angular velocity be ω, then it is easily shown that the current, which is still the only variable, is given by

$$L\dot{I} + RI = M\omega I , \qquad (1)$$

where $2\pi M$ is the mutual inductance of coil and disk. The solution for constant ω is

$$I = I_0 \exp[(M\omega - R)t/L] .$$

Thus if $\omega < R/M$ the current decays exponentially to zero and the solution with zero current is stable. If $\omega > R/M$ the solution with zero current is unstable and any current, however small, grows exponentially with time constant $L/(M\omega - R)$. If $\omega = R/M$ any arbitrary initial current will be maintained constant.

Since (1) is linear and homogeneous in I, the dynamo can produce a current, and therefore a magnetic field, in either direction. It cannot however switch from one direction to the other since, if I is zero, so is $\dot{I}$ and all higher derivatives.

Reversal of ω prevents the machine generating current. The direction required for the dynamo to maintain a current depends on the sense in which the coil is wound; the positive direction of ω is taken as that in which the dynamo produces current and is shown by the arrow in figure 1.

To produce a bounded, non-zero current requires the long term average of ω to be exactly R/M, a departure leads either to a zero current or to one that increases without limit. A periodic variation of ω about a mean value of R/M produces bounded, periodic fluctuations in the current[6].

4. DISK DYNAMO WITH TWO DEGREES OF FREEDOM

Constant couple applied to the axle. If the disk is driven by a constant couple, G, the angular velocity, ω, becomes a variable;

it and the current are determined by equation (1) above and by the dynamical equation

$$C\dot{\omega} = G - MI^2 , \tag{2}$$

where C is the moment of inertia of the disk. There are now two non-linear terms, ωI in (1) and I^2 in (2). It is these that produce the behavior that is relevant to the topic of this volume. Since (1) is linear and homogeneous in I and (2) is quadratic, the dynamo can still produce a current in either direction and still cannot switch from one direction to the other. A reversal of the couple prevents dynamo action. Positive G is taken as the direction in which it will cause the dynamo to sustain a current; it is the same as the positive direction of ω .

As in 3.2 there is a state with no current, but now the disk accelerates and, as soon as ω exceeds R/M , this state becomes unstable. At the critical angular velocity

$$\omega_C = R/M$$

a steady current

$$I_C = (G/M)^{\frac{1}{2}}$$

can be maintained. The value of the current is no longer arbitrary but depends on G .

Departure of the initial conditions from (I_C, ω_C) results in periodic variations in I and ω . For small departures from the steady state the variations are simple harmonic with period $2\pi(CL/2GM)^{\frac{1}{2}}$; it is to be noted that this bears no relation to the electrical time constant L/R . Large departures from the steady state give a highly non-linear, but still periodic behavior, as is shown in figures 2 and 3. The behavior is easily intelligible. If the initial current exceeds the equilibrium value the disk is slowed, when ω falls below the critical value, the current cannot be supported and falls to a small value. The disk then accelerates till its angular velocity again exceeds ω_C , the current then builds up and the cycle is repeated. If the initial departure from equilibrium is increased, the peaks of current become narrower and the troughs flatter and lower. The angular velocity can reverse during the cycle but the current cannot.

If a viscous damping term, $k\omega$, proportional to ω is added to (2) the system settles to a steady state with angular velocity ω_C and current $(1 - k\omega_C/G)I_C$. If $k\omega_C > G$ the critical speed cannot be reached and the current sinks to zero. With the damping term the system illustrates bifurcation and exchange of stability. For small G the state with zero current and angular velocity G/k is stable. If G is increased ω will reach its critical value, the state with no current then becomes unstable and two stable equilibrium states, with equal and opposite currents, branch from it.

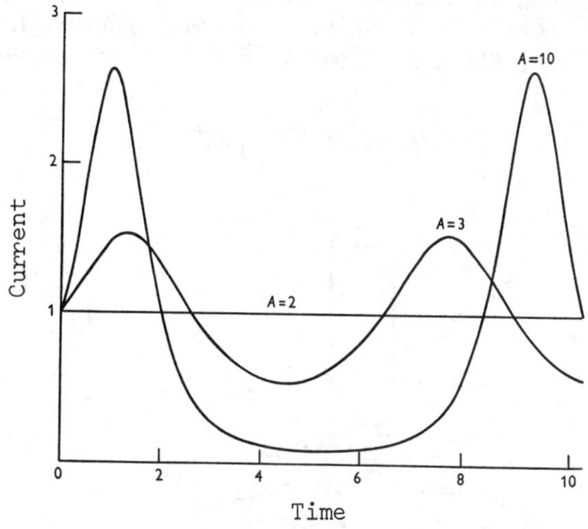

Figure 2. Variation of current, in units of $(G/M)^{\frac{1}{2}}$, for the dynamo of figure 1 driven by a constant couple. The unit of time is $(CL/2GM)^{\frac{1}{2}}$. A is a parameter giving the departure from the equilibrium state, for which A = 2 (from Bullard[1]).

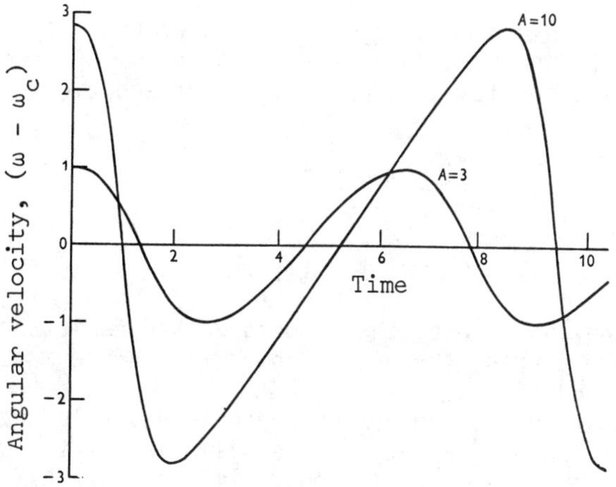

Figure 3. Variation of angular velocity, in units of $(GL/2CM)^{\frac{1}{2}}$, for the conditions of figure 2. The variation is symmetric above and below ω_c .

The behavior shown in figures 2 and 3 is highly non-linear but it is periodic and shows no reversals. Two small changes in the system produce striking changes in the solution. These are described in sections 5 and 6.

5. DISK DYNAMO WITH A SHUNT

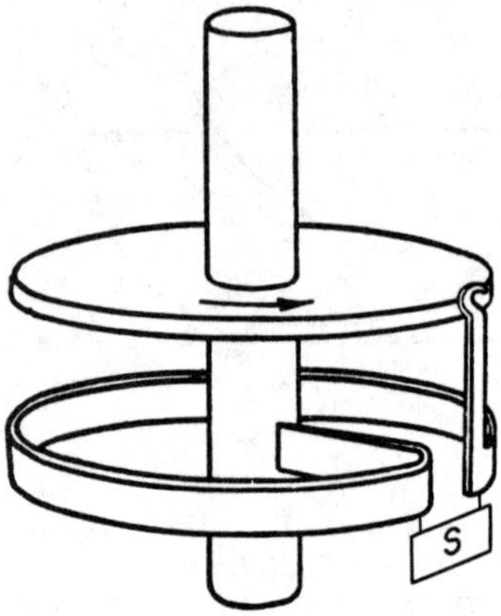

Figure 4. Disk dynamo with a shunt, S.

Figure 4 shows the dynamo of figure 1 with a shunt connected across the coil. If the shunt possesses resistance, R_s , and inductance, L_s , and carries a current I_s , the equations of the system are

$$L\dot{I} + RI = M\omega I \tag{3}$$

$$L_s\dot{I}_s + R_sI_s = M\omega I \tag{4}$$

$$C\dot{\omega} = G - MI(I + I_s) \ . \tag{5}$$

Here L and R are, as before, the inductance and resistance of the coil, I is the current through it and $2\pi M$ is the mutual inductance of coil and disk.

As before the state with no current becomes unstable when ω exceeds R/M . If $L/R > L_s/R_s$ the two equilibrium states with current are stable for small oscillations. If $L/R < L_s/R_s$ the equilibrium states with current are unstable. A numerical solution is shown in figure 5, a small initial disturbance gives oscillations which grow and become highly non-linear. With the passage of time the period increases, the maxima of current (for both I and I_s) become higher and the peaks become narrower; in the troughs between the peaks the currents become smaller and the proportion of the time spent in them becomes larger. While the currents are small the disc accelerates under the applied couple until it is pulled

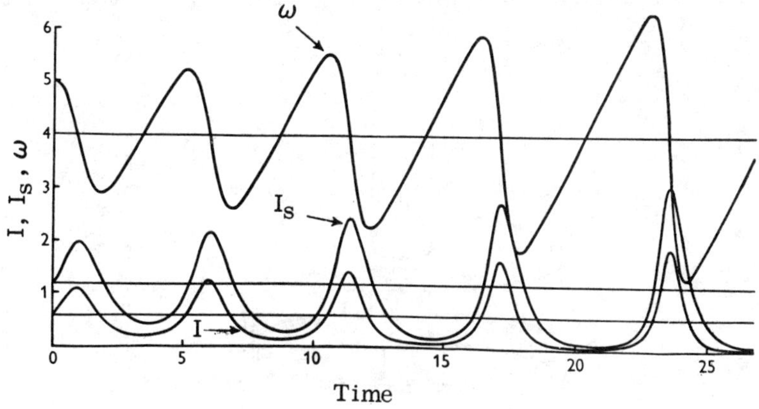

Figure 5. Current and angular velocity for a disk dynamo with a shunt. The units of time and angular velocity are $(CL/GM)^{\frac{1}{2}}$ and $(GL/CM)^{\frac{1}{2}}$ (Bullard[1]).

back by a burst of current in the coil. As time goes on the bursts get rarer, narrower and of greater amplitude.

Since (3) is identical with (1) no reversal in current can take place. If noise is added to the current, reversals will occur no matter how small the noise is; however small it is, the current in the interval between peaks will ultimately become small enough for reversals to occur. If the noise reverses the field, the next burst of current will be in the opposite direction to the previous one (unless the noise has again reversed the small current before the burst comes). Such oscillations (produced by rounding errors in a digital computation) were found in unpublished work by J.H. Wilkinson; they are irrelevant to the present topic.

The generation of ever sharper bursts of current by the shunted dynamo is unexpected; it is however a regularly developing phenomenon without the random reversals which are so characteristic of the earth's magnetic field and of many of the systems described in this book.

6. DISK DYNAMO WITH SHUNT AND A SERIES RESISTANCE

The absence of reversals for the dynamo of the previous section is a consequency of $\dot{I}$ in (3) vanishing when I does. This occurs because every term in (3) contains either I or $\dot{I}$. What is needed is another term, that is another e.m.f., which is not proportional to I or $\dot{I}$. Two ways of providing this are known; one using two coupled dynamos was devised by Rikitake[7] and is described in section 7, the other, due to Malkus[8] requires only the addition of an impedance between the brush and the coil of a single shunted dynamo. The arrangement is shown in figure 6. It is remarkable that those interested in these dynamos in the 1950's and 1960's did not realize the importance of this simple change in the machine of section 5.

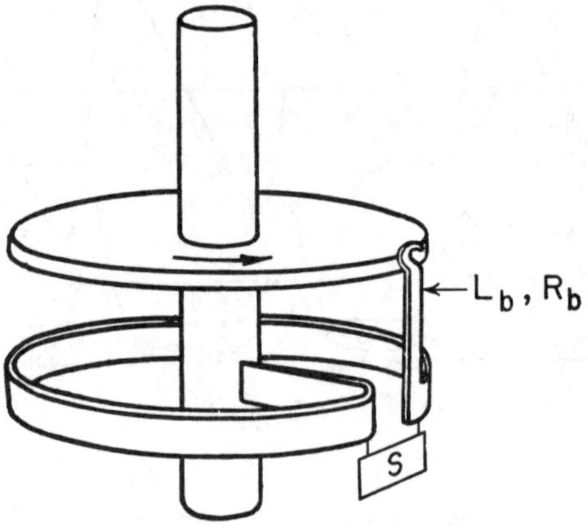

Figure 6. Disk dynamo with shunt and series impedance.

If L_b and R_b are the inductance and resistance of the series impedance (3) becomes

$$(L + L_b)\dot{I} + (R + R_b)I + L_b\dot{I}_s + R_bI_s = M\omega I \ , \qquad (6)$$

where I and I_s are, as before, the currents in the coil and the shunt. Equations (4) and (5) remain as before. Since I_s is not in phase with I (unless $L/R = L_s/R_s$), reversal can be expected for some range of the parameters.

The properties of these equations have been worked out in some detail by Robbins[2,3]. She has also added a viscous damping term, $k\omega$, where k is a positive constant, to (5); this sets a limit to ω and makes the state with zero current and angular velocity G/k an equilibrium state, but does not change the qualitative nature of the solutions. .

Robbins shows that there are four regimes

(1) $L/R > L_s/R_s$ and $L_b/R_b > L_s/R_s$

(2) $L_b/R_b > L_s/R_s > L/R$

(3) $L_s/R_s > L/R$ and $L_s/R_s > L_b/R_b$

(4) $L/R > L_s/R_s > L_b/R_b$.

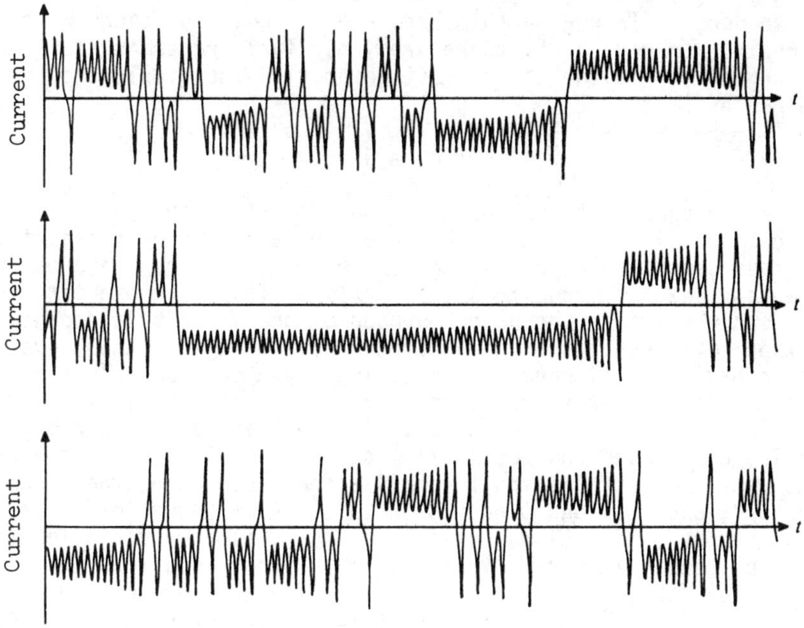

Figure 7. Pseudo-random oscillations of the current in the coil of the dynamo of figure 6 (from Robbins[3]).

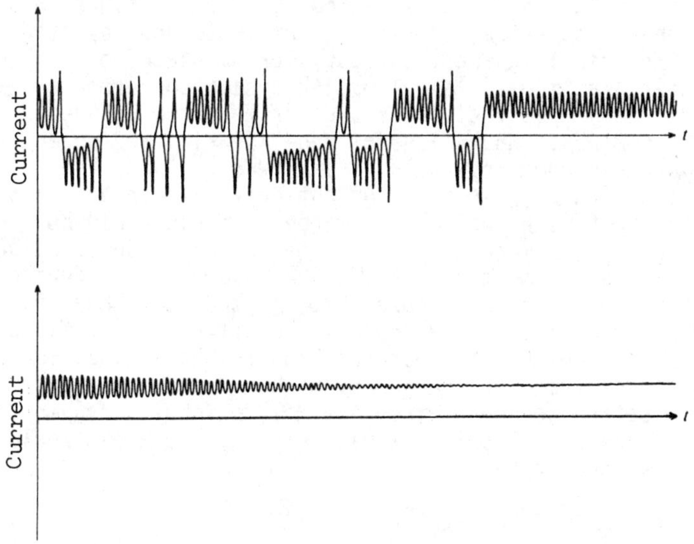

Figure 8. Oscillations settling to a steady state (from Robbins[3]).

In regimes (2) and (4) no reversals occur after transients have had time to decay. In region (3) there are periodic solutions with reversals. In region (1) there are irregularly reversing solutions.

Region (1) is the one of most interest. The simplest case is obtained by setting

$$L_s = 0 .$$

This is the case that Robbins has studied most extensively. The critical question is whether the solutions all stay near one of the two steady state solutions which, as before, have equal and opposite currents. She has shown that there exist solutions whose path in phase space stays in the neighbourhood of one equilibrium point and then switches to the neighbourhood of the other without ever being captured by either. The examples given in Robbins' thesis are not suitable for reproduction but are qualitatively similar to those in figure 9. The currents in the coil and shunt show growing oscillations around one equilibrium state culminating in a reversal of current and a transition to oscillations about the other. An example of oscillations produced in this way is shown in figure 7. The number of oscillations between reversals is very variable, in the sample shown it varies from 1 to 66. Other solutions exist, such as that in figure 8, where the erratic behavior does not persist and where, after 18 reversals, the solution is captured by one of the equilibrium points. The elaborate arguments of Robbins concerning the conditions under which these solutions exist in region 1 will not be reproduced here.

Robbins suggests that pseudo-random behavior is impossible for a system with less than three degrees of freedom since in a two dimensional phase space trajectories cannot cross and repeated transitions from circling about one equilibrium state to circling about the other are impossible. The disk dynamo therefore provides an example of the simplest class of systems that can behave in an apparently random way. What kinds of non-linearity are necessary to produce pseudo-random reversals is unknown.

In all these calculations a constant driving couple is assumed. Clearly the analogy with the earth's magnetic field would be improved if the dynamo were supposed to be driven by an idealised heat engine. This was mentioned by Inglis[5] and has been further developed by Backus[9]. He has shown that, if the ohmic heat is returned to the hot side of the engine, the efficiency, defined as the ohmic heat divided by the heat supplied to the engine, can exceed 100%. The dynamo considered by Backus has no shunt, it would be interesting to see whether the Malkus-Robbins dynamo driven by a Carnot engine with a constant rate of supply of heat would produce pseudo-random reversals.

7. COUPLED DISK DYNAMOS

In 1958, long before Malkus and Robbins had discovered the pseudo-random reversals of the single disk dynamo, Rikitaki[7] had shown that a pair of dynamos with the coil of each connected to the brushes of the other and with no shunts, would produce reversals when the two disks were driven by constant couples. Allan[10] showed that the reversals had an apparently random distribution in time. The double dynamo attracted considerable interest as a model of, or analogy to, the reversals of the earth's magnetic field. Numerical and analytical investigations were given by Allan[10], Mathews and Gardner[11], Sommerville[12], and Cook and Roberts[4].

If the two dynamos are similar and the couples applied to them are equal, the equations are

$$L\dot{I}_1 + RI_1 = M\omega_1 I_2 \tag{7}$$
$$L\dot{I}_2 + RI_2 = M\omega_2 I_1 \tag{8}$$
$$C\dot{\omega}_1 = G - MI_1 I_2 \tag{9}$$
$$C\dot{\omega}_2 = G - MI_1 I_2 \ , \tag{10}$$

where the suffixes 1 and 2 refer to the first and second dynamos. It might be thought that, as this system has four variables, two currents and two angular velocities, it would be more complicated than the dynamo of section 6. Owing to the high symmetry of the system this is not so; in fact (9) and (10) show that

$$\omega_1 - \omega_2 = \text{constant} \ ,$$

which reduces the effective number of variables to three.

There are two equilibrium states with equal and opposite currents. A first order calculation shows that the motions in the neighbourhood of the two equilibrium states with steady currents and angular velocities are of two kinds, one a transient which decays to zero and the other an undamped oscillation. A second order calculation by one of Liapounov's direct methods shows that the oscillatory terms are in fact unstable and that a trajectory of the system in phase space encircles one stable state several times and then switches rapidly to circle round the other, without ever being captured by either. An example from numerical calculations by Cook and Roberts[4] is shown in figure 9.

Viewed as a function of time the oscillations resemble those of the Malkus-Robbins dynamo as is shown in figure 10.

Allan[10] has shown that if the track of a volume element in phase space be followed, its volume will continually decrease. From the numerical calculations it appears that this is achieved by almost all trajectories approaching a limit surface and that, in spite of the contraction in volume, the trajectories of neighbouring particles may become more and more dispersed. Robbins[2] suggests that this dispersal, which implies a great sensitivity to initial conditions, is a necessary condition for apparently stochastic behavior. Cook and Roberts[4] suggest that the limit surface has two sheets. One equilibrium point is associated with each sheet but lies in a hole in the sheet and is inaccessible. A trajectory

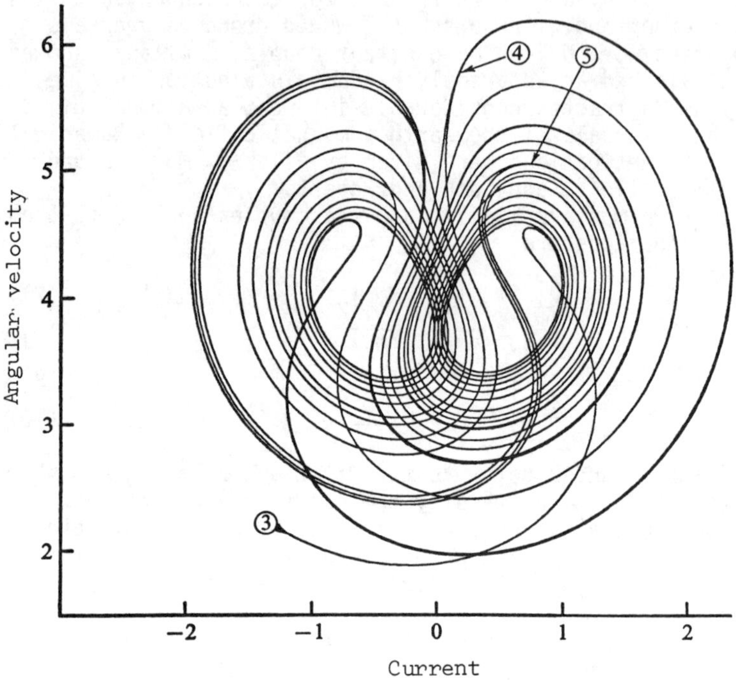

Figure 9. Projection of the path of a double disk dynamo in phase space on the plane of the angular velocity and one current (from Cook and Roberts[4]).

exactly on the limit surface could not switch from one sheet to the other and it is surprising that the numerical solutions do not show a change in character as the limit is approached.

Bullard and Gubbins[6] have obtained the curious result that if the disks of two coupled dynamos are constrained to oscillate with a phase difference of 90°, then one generates a close approximation to direct current whilst the other produces alternating current. Busse[13] and Gubbins[14] have pointed out that such mechanisms are inefficient and that the production of steady currents and fields by periodic motions are unlikely to be important in the earth.

Lebovitz[15] has considered the properties of an interconnected ring of dynamos.

8. COMPARISON OF THE DISK DYNAMO WITH A SPHERICAL DYNAMO

A disk dynamo differs from a homogeneous sphere of fluid in two ways; first it is multiply connected and second it has the symmetry of a watch face, or of a cylinder with a circumferential arrow, rather than that of a sphere. In view of these differences, the ability of a disc dynamo to produce an electric current and a

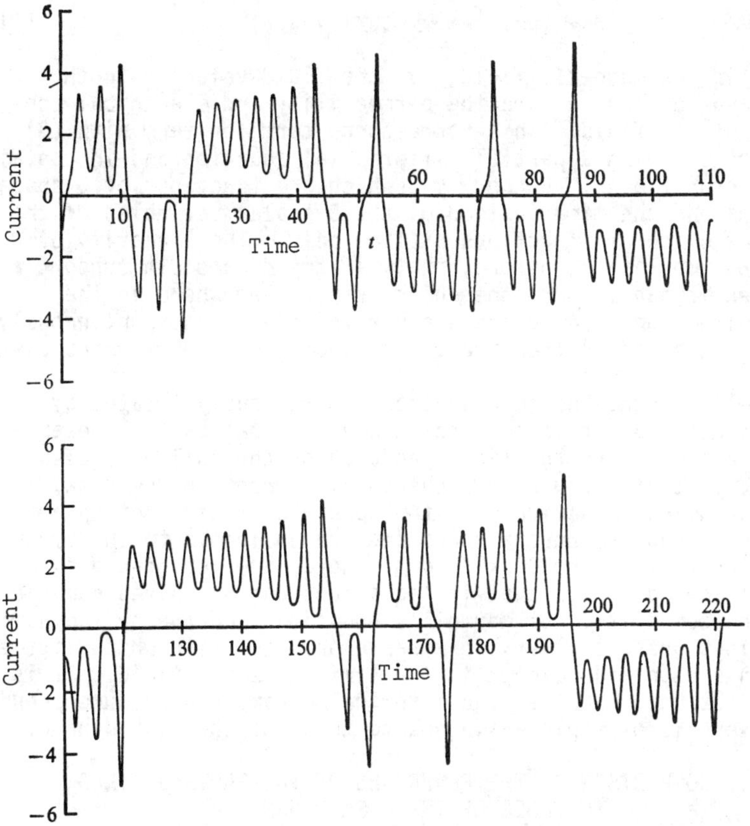

Figure 10. Oscillations of the current from one of a coupled pair of similar disk dynamos (from Cook and Roberts[4]).

magnetic field does not imply that a spherical mass of moving conducting fluid can do so. It is not obvious that the relatively complex structure of the disc dynamo can be replaced by a sphere, even if a complex motion is allowed in place of the very simple motion of the disc. This matter was resolved in different ways by Backus and by Herzenberg who showed that dynamos are possible in spheres subject to relatively simple motions. Even certain axisymmetric motions in a sphere can act as dynamos. The literature is large, but the existence of several reviews (Gubbins[16], Busse[17,18]). makes it unnecessary to go into detail here.

In spite of the differences in structure it turns out that there is a close relation between the disk dynamo and the spherical dynamo. If the form of the motion in the sphere is specified, then for both there is a critical speed below which no current is generated. The form of Maxwell's equations used in dynamo theory is

$$\dot{B} = (\mu\sigma)^{-1}\nabla^2 B + \text{curl}(v \times B) \tag{11}$$

where B is the magnetic field, v the fluid velocity (both vectors) and μ and σ are the permeability and electrical conductivity of the fluid. This corresponds term for term with (3) but differs by being a partial differential equation and in containing two vector fields, B and v, which are functions of 3 space coordinates and the time instead of the 3 scalar variables of the disk dynamo which are functions of time only. The linearity of (11) in B implies, as before, that, if the dynamo can support a field, then it can support one which points everywhere in the opposite direction. To decide whether the field can spontaneously reverse is much harder than the corresponding problem for the disk dynamo.

Maxwell's equations in a sphere are frequently treated by dividing the field into a poloidal and a toroidal part. These parts are analogous to the field produced by the coil of a disk dynamo and to that produced by the radial current in the disk.

The dynamical equations for the sphere of fluid contain a driving force and the Lorenz force and are analogous to the dynamical equation (5) for the disk dynamo. Here, however, the disk dynamo has the advantage that its equations can be solved numerically whereas the combination of Maxwell's equations and the hydrodynamical equations present a very formidable problem, with which little progress has been made except in the limit of small field. It is therefore not known whether the spherical dynamo can produce pseudo-random reversals of field analogous to those of the disk dynamo.

9. COMPARISON OF THE REVERSALS OF THE EARTH'S FIELD TO THOSE OF THE DISK DYNAMO

It is well established that the earth's magnetic field has reversed its direction on numerous occasions in the past (for reviews of reversals and related matters see Bullard[19] and Vacquier[20]). A sequence covering the last 1.9 million years is shown in figure 11 and a longer sequence in figure 12. The main features are that the reversals occur suddenly (in a few thousand years) after a period during which the direction and magnitude of the field has only fluctuated slightly about its mean. The behavior is strikingly similar to that of the disk dynamo in figure 7. Over the last 40 million years the mean period between reversals has been 340,000 years. Over a longer period the series appears not to be a stationary stochastic process. As can be seen from figure 12 the average interval between reversals between 80 and 46 million years before the present was longer than it has been since, this is believed to be a genuine phenomenon and not an artifact due to undetected reversals in the older rocks. Between 111 and 85 years before the present no reversals are known, there are similar long periods without reversals in the Jurassic and the Permian. These

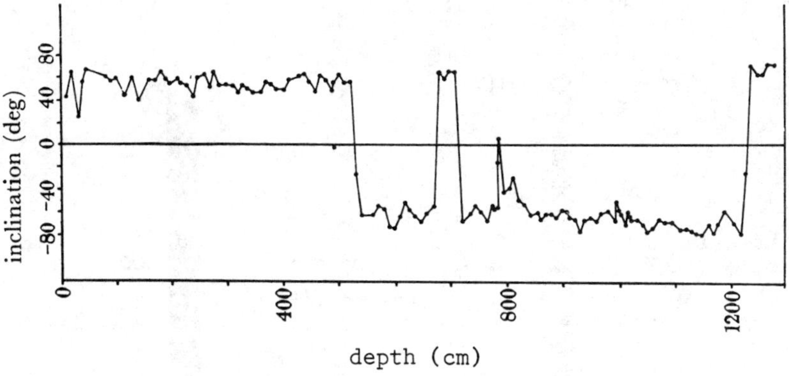

Figure 11. Dip of the magnetic field determined from a core obtained in the north Pacific. The core covers the last 1.9 million years (from Ninkovich *et al.*[21]).

periods are reminiscent of the long section without reversals in figure 7 or, perhaps of the settling into a steady state in figure 8. The main difference between the reversals of the earth's field and those of the disk dynamo is that the earth does not show growing fluctuations preceding a reversal.

It would be of great interest to know whether a spherical dynamo can give similar reversals and to determine their statistical properties.

The sun also shows reversals in its magnetic field, these are more nearly periodic than those of the earth, and have the interesting property that the reversals of the general field (a poloidal field) are out of step with those of the sunspots, which probably represent a toroidal field within the sun. This behavior is in some degree intelligable in terms of the theory of a spherical dynamo in which the poloidal field is built up by the interaction of the toroidal field with motions in the sun. The poloidal field could then itself interact with the motion to destroy the toroidal field and then build it up with the opposite sign. The regularity of the sunspot cycle since early in the 18th century may be misleading; there is rather strong evidence that during the period 1645-1700 there were few sunspots, little auroral activity and, presumably, no reversals of the sun's field[23].

10. CONCLUSION

The disk dynamo has a double interest. It is the simplest system which is known to show pseudo-random behavior and, since it has only 3 degrees of freedom, its properties are more easily understood than are those of continuous systems. Its other interest is

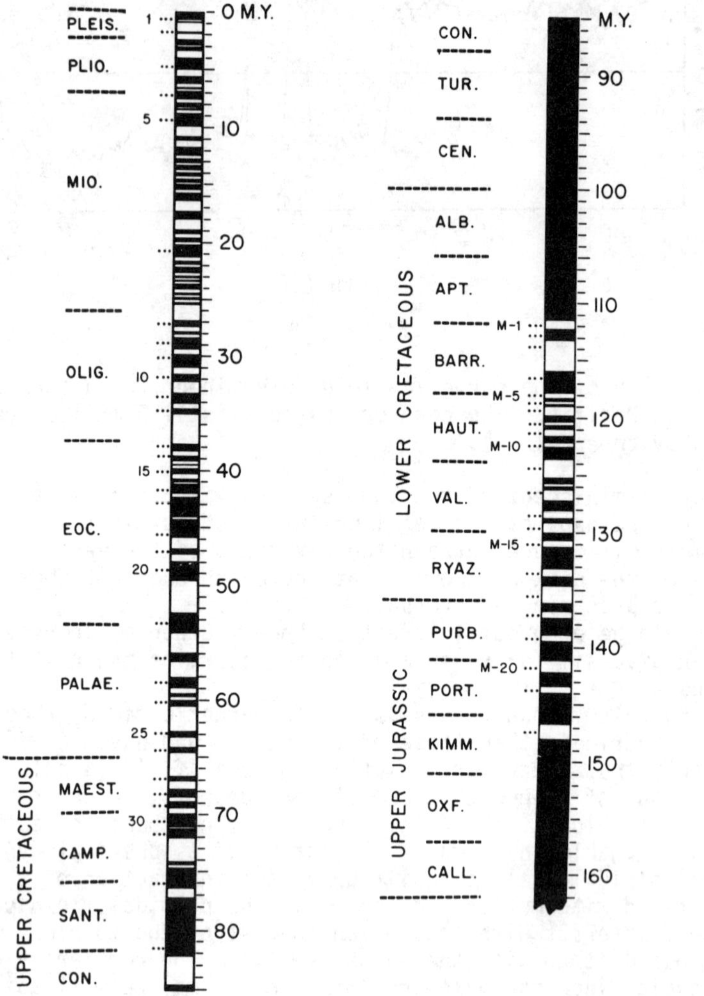

Figure 12. Reversals of the earth's field over the last 160 million years (from Larsen and Pitman[22]).

that it mimics the behavior of the earth's magnetic field and holds out a hope of rational explanations of its eccentricities. In 1692 Edmond Halley[24] wrote of the motions and magnetic effects of the earth's core 'the nice Determination of this, and of several other Particulars in the Magnetic System is reserv'd for remote posterity'. The unexpected analogy between the earth and the disk dynamo encourages the hope that understanding of some of these matters is at last being approached.

ACKNOWLEDGMENTS

Permission to reproduce the figures has been granted by the authors and by the Cambridge University Press (Figs. 1-3, 5, 7-10); the Elsevier Publishing Company (Fig. 11); and the Geological Society of America (Fig. 12).

REFERENCES

1. E. Bullard, Proc. Camb. Phil. Soc. 51, 744-760 (1955).
2. K.A. Robbins, Ph.D. thesis, M.I.T. (1975).
3. K.A. Robbins, Math. Proc. Camb. Phil. Soc. 82, 309-325 (1977).
4. A.E. Cook and P.H. Roberts, Proc. Camb. Phil. Soc. 68, 547-569 (1970).
5. D.R. Inglis, Rev. Mod. Phys. 27, 212-248 (1955).
6. E.C. Bullard and D. Gubbins, Geophys. Monogr. Amer. Geophys. Un. 16, 325-328 (1972).
7. T. Rikitake, Proc. Camb. Phil. Soc. 54, 89-105 (1958).
8. W.V.R. Malkus, Eos, Trans. Amer. Geophys. Un. 53, 617 (1972).
9. G. Backus, Proc. Nat. Acad. Sci. 72, 1555-1558 (1975).
10. D.W. Allan, Proc. Camb. Phil. Soc. 58, 671-693 (1962).
11. J.H. Mathews and W.K. Gardner, Naval Res. Lab. Rep. No. 5886 (1963).
12. R.C.J. Sommerville, Woods Hole Oceanographic Institution Report No. 67-54, vol. 2, 132-139 (1967).
13. F.H. Busse, J. Geophys. Res. 80, 278-280 (1975).
14. D. Gubbins, Geophys. Res. Letters 2, 409-412 (1975).
15. N.R. Lebowitz, Proc. Camb. Phil. Soc. 56, 154-173 (1960).
16. D. Gubbins, Rev. Geophys. Space Phys. 12, 137-154 (1974).
17. F.H. Busse, Applications of bifurcation theory (edit. P.H. Rabinowitz) 175-202. Academic Press (1977).
18. F.H. Busse, Ann. Rev. Fluid. Mech. 10, 435-462 (1978).
19. E.C. Bullard, Phil. Trans. Roy. Soc. A, 263, 481-524 (1968).
20. V. Vacquier, Geomagnetism in marine geology. Elsevier (1972).
21. D. Ninkovich, N. Opdyke, B.C. Heezen and J.H. Foster, Earth Planet. Sci. Lett. 1, 476-492 (1966).
22. R.L. Larson and W.C. Pitman, Bull. Geol. Soc. Amer. 83, 3645-3662 (1972).
23. J.A. Eddy, Science 192, 1189-1202 (1976).
24. E. Halley, Phil. Trans. Roy. Soc. 17, 563-578 (1692).

THE ERRATIC ELECTRON: NONLINEAR EFFECTS IN THE THEORY OF THE ELECTRON

by

A. O. Barut
Department of Physics
The University of Colorado
Boulder, CO 80309

ABSTRACT

Nonlinear effects due to the radiation reaction (or self-energy) and spin effects which occur both in relativistic classical mechanics, and in Dirac electron are discussed. Some properties of the resulting erratic motion of the electron (zitterbewegung) around an averaged motion are used to calculate the numerical value of the Planck's constant $\hbar$. Also a calculation of the mass ratio of muon and electron is indicated on the basis of the nonlinear radiation reaction term.

1. INTRODUCTION

Our picture of the electron as a mass point moving according to Newton's equations under the influence of the Lorentz force is incomplete due to the nonlinear effects of the radiation reaction and due to spin effects, even in the classical domain. Usually a sharp distinction is made between the classical and quantum behavior of the electron, and the difficulties of classical mechanics at short distances are attributed to the fact that below a certain distance the quantum regime will take over. The study of nonlinear effects would be essential, in my opinion, in understanding of both the structure of the electron, and the origin of quantum mechanics. This paper is devoted to the border area between classical mechanics and quantum mechanics.

2. THE POSTULATES OF H. A. LORENTZ

The basic principles of the theory of the electron and the electromagnetic field in interaction goes back to the work of H. A. Lorentz [1]. Although modified in form the essential features are still remarkably the same. Lorentz's three fundamental principles are (in modern notation):

(i) Electron is characterized by two parameters, charge e and mass m. In matter the trajectory of each charge obeys the relativistic equation of motion

$$m \ddot{Z}_\mu(S) = \frac{e}{c} F_{\mu\nu}(Z) \dot{Z}^\nu(S) , \qquad (1)$$

where $Z_\mu(S)$ is the world-line of the electron as a function of the

ISSN: 0094-243X/78/390/$1.50

invariant parameter S, c is the velocity of light which we shall put equal to unity in the following, and

(ii) $F_{\mu\nu}$ is the electromagnetic field produced by all the charged particles, including the self-field of the particle whose motion we are studying with eq. (1), and obeys the Maxwell's equations

$$F_{\mu\nu}{}^{'\nu}(x) = j_\mu = \sum_{(i)} \int e_{(i)} \dot{z}^{(i)} \; \delta(x-z^{(i)}) \; ds, \qquad (2)$$

where the source of the field $F_{\mu\nu}$ is the sum of all currents produced by the motion of all charges;

(iii) Matter consists of a set of electrically bound electrons.

3. QUANTUM VERSIONS OF LORENTZ'S POSTULATES

Lorentz's assumption (iii) has been of course considerably refined in the meantime; we have very sophisticated models of matter on the atomic and molecular level. The assumptions (i) and (ii) are slightly modified in relativistic quantum theory: Eq. (1) is replaced by the Dirac equation in external fields [2]:

$$(\gamma^\mu p_\mu - m) \; \psi(x) = e\gamma^\mu \psi(x) \; A_\mu(x) \; , \qquad (1')$$

and the right-hand side of eq. (2) has the representation

$$F_{\mu\nu}{}^{'\nu}(x) = j_\mu(x) = \sum_{(i)} e_{(i)} \; \overline{\psi}^{(i)} \; \gamma_\mu \; \psi^{(i)} \; . \qquad (2')$$

Finally, in quantum electrodynamics, eq. (1') remains the same with $\psi(x)$ and $A_\mu(x)$ being (second) quantized field operators, and eq. (2') can then be written more simply as

$$F_{\mu\nu}{}^{'\nu}(x) = e \; \overline{\psi}(x) \; \gamma_\mu \; \psi(x) \; . \qquad (2'')$$

In this operator form many particle states are automatically included.

4. SELF-ENERGY AND MASS

Lorentz's theory of the electron has been enormously successful in unifying many phenomena which were thought to be unrelated. There is however, one difficulty: The self-field of each charge which is included in (1) or (1') is infinite at the position of the particle, at x = z. In most cases one completely ignores the self-field, and this doesn't cause as much trouble as it should.

The reason for this is that the major part of the self-energy of each charged particle is actually included in the inertial mass m of the particle. Mass is a phenomenological concept as far as electrodynamics is concerned. By lumping the effect of the self-field of the charge around itself into an inertial term which then occurs on the left hand side of eq. (1) we have taken into account perhaps a large part of the self field on the right hand side.

But not completely, and that is what the radiation effects and renormalization theory of electrodynamics are all about. The development of the theory of the electron has been really concerned with the question; how much of the self energy is in the mass? [3]

5. CLASSICAL MASS RENORMALIZATION

In classical relativistic electrodynamics one obtains by various limiting procedures a separation of the inertial part of the self-energy from a remainder which we call the radiative part, and arrives at the nonlinear Lorentz-Dirac equation which in fact makes eq. (1) more precise [4]:

$$\begin{aligned}(m+\delta m)\, \ddot{Z}_\mu(s) &= e\,\lambda F^{ext}_{\mu\nu}\, \dot{Z}^\nu + \frac{2}{3} e^2(\dddot{Z}_\mu - \dot{Z}_\mu(\ddot{Z})^2) \\ &\quad - [\frac{2}{3} e^2(\dddot{Z}_\mu - \dot{Z}_\mu(\ddot{Z})^2]_{\lambda=0}\,, \end{aligned} \qquad (3)$$

where λF^{ext} is the field produced by all other particles. In deriving eq. (3) an inertial part of the self-energy ($\delta m\, \ddot{Z}_\mu$) has been transferred to the left hand side, so that the second nonlinear term on the right hand side is the finite remainder, the radiative term. Unfortunately the term δm is infinite in the limit of the point electron. This is the only shortcoming of the classical theory of the electron. It seems that only more information about the internal structure of the electron can solve this problem. The third term in the right hand side of (3) is, we believe, new: It is part of the renormalization procedure. We must renormalize the self-energy in such a way that for a free particle, i.e. when the parameter λ of the external field goes to zero, we should have the free particle equation with the experimental mass:

$$\lambda \to 0 \quad : \quad m_{exp}\, \ddot{Z}_\mu = 0 \qquad (4)$$

6. SPIN STRUCTURE OF THE ELECTRON

That the internal structure of the electron is more complicated than just a point particle is indicated first by the spin structure of the electron.

Although the spin of the electron has been discovered in a quantum mechanical context (doubling of levels in atomic spectra) it is a very essential property of the electron, and we cannot expect to have a consistent classical theory of the electron without taking into effect this property. But the electron is not a naive classical spinning object. All such models fail, because one cannot obtain the correct g-factor 2 for the magnetic moment of the electron, and because in all such mechanical models the degrees of freedom are much larger (≥ 3) than the number of electron spin degrees of freedom which we know from experiment is only 2. Thus the internal structure of the electron becomes a subject of real curiosity and a definite physical question. Even Lorentz had anticipated this situation: "In speculating on the structure of these minute particles we must not forget that there may be many possibilities not dreamt of at present; it may very well be that other internal forces serve to ensure the stability of the system, and perhaps, after all, we are wholly on the wrong track when we apply to the parts of an electron our ordinary notion of force."

Surprisingly, there are other structures between a point electron and an extended electron, as we shall see.

The theory of the relativistic spinning electron leads to more complicated nonlinear equations than (3), [5]. Very few solutions of these nonlinear equations are known; in fact, none with the inclusion of both spinning effects <u>and</u> radiation reaction. These effects should not be dismissed by saying that quantum electrodynamics supercedes the classical effects, because, first of all, the correct classical limit of the Dirac equation gives the classical spin equations [6]; secondly, classical radiation reaction is a non perturbative result; in quantum electrodynamics we only know how to deal with radiative reaction in perturbation theory. In fact, the problem of the classical limit of quantum electrodynamics with self energy effects is an outstanding unsolved problem, even in perturbation theory. The known solutions of the classical spinning particles (without self energy) do reproduce the behavior of the free Dirac particle [4]. The most conspicuous one being the erratic behavior of the electron, called <u>"Zitterbewegung"</u>. [8]

7. THE "ZITTERBEWEGUNG"

With a relativistic particle in a state of definite momentum $|\vec{p}>$ we can associate <u>two different</u> position and velocity operators:

(i) A relativistic (center of mass) position operator $\underset{\sim}{X}$, and the corresponding velocity operator

$$i[H,\underset{\sim}{X}] = \frac{\vec{p}}{p_0} , \tag{5}$$

(ii) A position operator $\vec{x}$ occurring in the wave equation which is conjugate to $\vec{p}$, $\dot{x}_k = - i \frac{\partial}{\partial p_k}$, and the velocity operator

$$i[H,\vec{x}] . \tag{6}$$

The relation between the two position operators may be written as

$$\vec{X} = \vec{x} + \frac{\vec{M}}{m^2} p^0 - \frac{\vec{p}(\vec{p}\cdot\vec{M})}{m^2 p^0} + \frac{(\vec{p} \times \vec{J})}{m^2} , \tag{7}$$

where $\vec{J}$ and $\vec{M}$ are the generators of the Lorentz group acting on the representation space of the Poincaré group for the particle in question. The coordinate $\vec{x}$ is the position of the charge because the electromagnetic field $A_\mu(x)$ couples minimally to the particle at the point x.

In particular, in the rest-frame of the particle, $\vec{p} = 0$, $p^0 = m$, we have $\vec{X} = \vec{x} + \frac{\vec{M}}{m}$. For a particle at $\underline{X} = 0$, the coordinate $\vec{x}$ has two eigenvalues at $\pm 1/m$, and the velocity of the charge is

$$i[H,\vec{x}] = - \frac{i}{m} [H,\vec{M}] . \tag{8}$$

This is the characteristic zitterbewegung: the charge has with respect to the center of mass two eigenvalues of velocity $\pm$ c and two relative positions at $\pm 1/m$, and the different components of relative positions and relative velocities do not commute. It is important to emphasize that the origin of spin may be understood as due to the angular momentum of the charge performing the "Zittermotion" around its center of mass.

There is so far very little information on the properties zitterbewegung in the presence of external fields, or with account of self-field, or for highly localized states.

But this is the regime where classical non-linear equations and quantum mechanical equations touch each other so that we may obtain more information on the origin, limitations and extensions of quantum mechanics. This expectation is based on the following results:

(a) Relativistic spin equations classically have the characteristics of zitterbewegung [5], [7].

(b) Schrödinger or Dirac equations may be developed from the assumption of a stochastic behavior around an average motion, just like zitterbewegung. [9].

(c) Nonlinear effects can give erratic or stochastic deviations from an average trajectory [10].

8. RADIATION FROM ZITTERMOTION AND PLANCK'S CONSTANT h

To these we now add a further clue towards the origin of the quantum principle: A heuristic calculation of the magnitude of the Planck's constant h based on the radiation emitted due to zitterbewegung.

The radiation emitted by the charge due to its 'zittermotion' according to the velocity (8) and the corresponding acceleration is qualitatively very different from the standard radiation calculated from an accelerated electron, the latter corresponds to the average velocity $\vec{p}/p_0$ given in eq. (5). In the case of (8) the velocity has always the same maximal amplitude c in every direction. The square of the velocity is

$$\vec{\dot{Z}}^2 = c^2 \vec{\alpha}^2 = 3c^2 . \qquad (9)$$

In order to calculate the radiation emitted we use the invariant formula

$$P = \frac{2}{3} \frac{e^2}{4\pi\varepsilon_0 c^3} 4\pi \overline{(\ddot{Z}_\mu \ddot{Z}^\mu)} \qquad (10)$$

for the power of emitted radiation. The factor 4π has been introduced because zittermotion is equally probable in every direction. For the free electron the average zitteracceleration will be zero, since it does not radiate. For an accelerated electron we consider a Fourier component of the zittermotion with frequency ω, and assume that the velocity has the form

$$|\dot{Z}| = c\sqrt{3}\, e^{i\omega t} , \qquad (11)$$

because the amplitude of the velocity is fixed. Hence

$$|\ddot{Z}| = i\omega c\sqrt{3}\, e^{i\omega t} , \text{ or } \overline{|\ddot{Z}|^2} = 3c^2\omega^2 . \qquad (12)$$

We insert (12) into (10):

$$P = \frac{e^2}{4\pi\varepsilon_0 c} 8\pi^2\omega^2 .$$

Next we evaluate the energy emitted during a period $\tau = 2\pi/\omega$ in the oscillating (zitter)-frame, or during

$$\tau' = \frac{2\pi}{\omega}\sqrt{1-\overline{\beta_z^2}} . \qquad (13)$$

in the center of mass ($\underline{X}$) of the electron:

$$E_{(X)} = \frac{e^2}{4\pi\varepsilon_0 c} 16\pi^2 \sqrt{1-\overline{\beta_z^2}} . \qquad (14)$$

Here $\overline{\beta_z^2}$ is the average velocity of the zittermotion relative to the center of mass of the electron. We assume that, because each component of $|\dot{Z}_\mu|$ changes between 0 and 1, $\overline{\beta^2} = (1/2)^2$, hence $\sqrt{1-\beta^2} = \sqrt{3}/2$. This gives from (14)

$$E_{(X)} = \frac{e^2}{4\pi\varepsilon_0 c} 8\pi^2 \sqrt{3}\, \omega . \tag{15}$$

Comparing (15) with $E = \hbar_0\omega$, we find a numerical value of $\hbar_0$, or a value for the fine structure constant

$$\alpha^{-1} = \left(\frac{e^2}{4\pi\varepsilon_0\hbar_0 c}\right) = 8\pi^2 \sqrt{3} = 136.7572 . \tag{16}$$

Finally we must transform from the electron's center of mass to the laboratory frame by the Doppler formula. The energy emitted during one period in the laboratory frame becomes

$$E = \hbar_0 \frac{\sqrt{1-\beta^2}}{1-\beta\cos\theta}\, \omega \equiv \hbar\omega \quad . \tag{17}$$

For ordinary atomic velocities the correction in (17) is a factor 1.002, which gives $\alpha^{-1} = 8\pi^2 \sqrt{3}\, 1.002 = 137.03$, in agreement with experiment.

Not all the details of zittermotion in an external field have been used in this calculation. These details are needed for a dynamical calculation of the frequency spectrum of the radiation. But the correct result may indicate that our assumptions leading to the value of $\hbar$ are independent of these details of the motion.

This is to our knowledge the first dynamical mechanism to explain the value of the Planck's constant, hence to understand the quantum principle $E = \hbar\omega$, which up to now has been accepted as an empirical and universal law. As expected any explanation of this principle has to go beyond both classical and quantum theories, and takes place in this mysterious borderline between the two domains. The key, we believe, lies really in the internal structure of the electron.

The small variations of $\hbar$ with velocity should provide experimental tests for the correctness of the ideas presented here.

In contrast to the radiation from zittermotion, the radiation of the electron due to its average motion gives a power proportional to ω^4.

9. NONLINEAR QUANTUMELECTRODYNAMICS

The quantum mechanical version of the nonlinear equation (3) with radiation reaction can be obtained from eqs. (1') and (2") by eliminating A_μ. In the gauge A^μ, $\mu = 0$, we obtain using the causal Green's function D,

$$(-i\gamma^\mu\partial_\mu - m)\,\psi(x) = e^2\gamma^\mu\psi(x)\int dy\, D(x-y)\overline{\psi}(y)\gamma_\mu\psi(y)\Big|_{\text{renormalized}} + e\gamma^\mu\psi(x)A_\mu^{ext}. \tag{18}$$

This is a non-linear integrodifferential equation for $\psi(x)$. The first term on the right-hand side is "renormalized", meaning that its value as $A_\mu^{ext} \to 0$ has to be subtracted, because in this limit we should have a "free" electron. For plane-waves the integral diverges, but for localized solutions the theory is finite, only renormalization must be formulated carefully. In particular, we obtain a formulation of many-body problem if we look for a solution of the form

$$\psi = \psi_1(x) + \psi_2(x) + ..$$

where $\psi_1(x)$ are non-overlapping localized (soliton-like) waves. In this case the mutual interaction potential is correctly given by the D-function: $1/r$ + relativistic corrections.

Unfortunately, no closed form of the renormalized integral in eq. (18) exist, as the finite term in eq. (3), after a $[\delta m\psi(x)]$ term has been separated. But we expect that there should be one.

However, we have a renormalization procedure in quantumelectrodynamics when both fields $A_\mu(x)$ and $\psi(x)$ are quantized. This renormalization procedure is not as unambiguous as is generally believed. Because the procedure involves infinite quantities, we must decide a priori to what final result we wish the theory to renormalize, i.e. into what quantities shall we absorb the infinite quantities. In the usual QED, we renormalize the theory to a free electron. In our eq. (18) this means that we shall transform eq. (18) into

$$(-i\gamma^\mu\partial_\mu - m^R)\,\psi^R = e^R\gamma^\mu\psi^R(x)\,A_\mu^{ext}(x)\;, \tag{19}$$

where the superscript R refers to renormalized quantities related to unrenormalized quantities by infinite factors.

10. TWO-MASS WAVE EQUATION. THE μ-MESON

Suppose, however, we renormalize the theory into a different final physical equation, namely to

$$(-i\gamma^\mu\partial_\mu + \lambda\partial_\mu\partial^\mu - M)\psi' = e'\gamma^\mu\psi'(x)\,A_\mu' \tag{20}$$

where a term $\lambda\partial_\mu\partial^\mu$ has been separated from m^R in (19). Now the free part of (20) has solutions with two masses m_1 and m_2 such that m_2/m_1 is determined completely by λ. There is now an interesting connection to the classical nonlinear equation (3).

The nonlinear self-energy term in eq. (3) implies that the charged particle has an anomalous magnetic moment, a g-factor, $g = 4\alpha/3$ (in units of $e\hbar/2mc$). Eq. (20) also implies an anomalous magnetic moment for the particle. If we now determine λ in (20) so that the anomalous magnetic moment of (20) equals $4\alpha/3$, then one finds that [11]

$$\frac{m_2}{m_1} = \frac{3}{2}\alpha^{-1} + 1 \tag{21}$$

which comes extremely close to the experimental mass-ratio between muon and electron. The experimental anomalous magnetic moment of the electron is $g = 2\,\alpha/2\pi$. The difference between $4\alpha/3$ and α/π is due to the contribution of a spin-flip effect to the anomalous magnetic moment, and can be taken into account by a Pauli term $(a\sigma_{\mu\nu}F^{\mu\nu}\psi')$ in eq. (20). This additional term has no effect on the mass ratio (21), because it is separately conserved.

These two dynamical effects, radiation from zitterbewegung giving $E = \hbar\omega$, and the mechanism for the μ-meson, indicate that nonlinear effects in the theory of electron may yet reveal new physics. In addition, we have discussed elsewhere [12] the possibility of new high energy narrow resonances in the system of two charged particles due to self-energy terms in eq. (18).

REFERENCES

1. H. A. Lorentz, The Theory of Electrons and Its Applications to the Phenomena of Light and Radiant Heat, Dover Publ. N. Y. 1952 (first edition 1909).
2. P. A. M. Dirac, Proc. Roy. Soc. A 117, 610-624 and A 118, 351-361 (1928).
3. For the development of the Electron Theory see F. Rohrlich, in The Physicist's Conception of Nature, (D. Reidel, 1973), p. 331-369.
4. For a recent derivation and earlier references see A. O. Barut, Phys. Rev. D 10, 3335 (1974).
5. H. J. Bhabha and H. C. Corben, Proc. Roy. Soc. 178, 273 (1941).
6. S. I. Rubinov and J. B. Keller, Phys. Rev. 131, 2789 (1963).
7. H. C. Corben, Phys. Rev. 121, 1833 (1961) and Nuovo Cim. 20, 529 (1961).
8. E. Schrödinger, Sitzungsberichte der Berliner Akad. d. Wissenschaften, Phys-Math. Klasse, 24, 418 (1930).
9. E. Nelson, Dynamical Theories of Brownian Motion, Princeton Univ. Press 1967.

10. See contributions in these proceedings.
11. A. O. Barut, Physics Letters 73 B, 310 (1978).
12. A. O. Barut and J. Kraus, Phys. Letters 59 B, 175 (1975) Phys. Rev. D 16, 161 (1977).

DYNAMICS REVISITED, A GLOSSARY

Robert H. G. Helleman
School of Physics
Georgia Institute of Technology
Atlanta, Georgia 30332

ABSTRACT

Several articles in this Volume presuppose some knowledge of the recent exciting developments in Hamiltonian dynamics and its jargon. This brief glossary serves as a connection between one's knowledge of standard, graduate, classical mechanics and these newer developments.

1. HAMILTONIAN SYSTEMS

Consider a general Hamiltonian system of N degrees of freedom, i.e. N *coupled non-linear* second order differential equations

$$\ddot{q}_k = - \partial V(\vec{q},t)/\partial q_k \quad , \; k = 1,.., N \, , \tag{1}$$

which may explicitly depend on the time t. We only consider its bounded solutions here. Such a Hamiltonian system is either "Integrable" or "Non-Integrable":

2. INTEGRABLE SYSTEMS

The system (1) is "*Integrable*"[1-4] if N of the constants of the motion are single valued *differentiable* (analytic) *functions* of $q_1,\dots, q_N, p_1,\dots p_N$. These N functions must be independent of each other and exist globally, i.e. for "all" allowed values of $\vec{q}$, $\vec{p}$. Examples of such functions are the Hamiltonian H in a conservative system and the angular momentum L in the Kepler 2-body problem. [From N such constant functions one can in principle solve for the $p_k(\vec{q})$, ∴ $\dot{q}_k(\vec{q})$, and integrate these equations by quadratures[4]; whence the name integrable.] For integrable systems the canonical transformation theory, described in most standard texts on classical mechanics (e.g. Goldstein's "Classical Mechanics"), yields a *canonical perturbation series with a non-zero region of convergence*. The N single-valued constants of the motion restrict the orbit in 2N-dimensional phase space to an N-dim. surface which one can prove to be an N-dim. *torus*.[4,1] Therefore the solution of (1) can be expressed in terms of N cyclic ("angle-") variables. The complete set of "action-angle" variables can be obtained globally in each of those regions of convergence.

The Hamilton-Jacobi Equation can be solved globally.[3] No "stochasticity", as described in Sec. 5, can arise. Examples: the

ISSN: 0094-243X/78/400/$1.50

Kepler problem, N coupled harmonic (linear-) oscillators, the exponential "Toda-Chain",[8,6,5] all linear systems, all analytic one-variable problems [k = 1 in (1)], all systems that can be "separated" into N analytic one-variable problems, the "Korteweg-de-Vries Equation",[8] all textbook exercises.

3. NON-INTEGRABLE SYSTEMS

The system (1) is *"Non-Integrable"* if it does *not* have this complete set of N *differentiable* single-valued constants of the motion. Some constants of the motion are now highly pathological, *discontinuous, multiple-valued,* functions of the $\vec{q}$, $\vec{p}$. Here the standard *canonical perturbation theory diverges,*[1-3,9] a fact not often stated in the familiar graduate texts. Global action-angle variables cannot be obtained. The Hamilton-Jacobi Equation has no global solution in terms of elementary differentiable functions[3]. There exist many, perhaps small, "stochastic" regions (described in Sec. 5) in phase space. The motion can become very "chaotic" in such regions and certain quantities can be proven to be truly *random* variables [ref. 1, sect. 8; ref. 3, chapt. 3]. Example:

$$\ddot{x} = -x - 2\,x\,y\,, \qquad \ddot{y} = -y + y^2 - x^2\,, \tag{2}$$

the Hénon-Heiles system.[1,2,6,5] [very simple examples can be found also in "area-preserving mappings"[1-6]]. There is some hope that this chaotic behavior may help to justify the use of statistical mechanics in large-N systems.[6-8]

4. MOST HAMILTONIAN SYSTEMS ARE NOT-INTEGRABLE

This perhaps surprising result is contained in theorems by Poincaré and by Siegel [ref. 2, section 3; ref. 3, chap. II, section 2.b]. Hence canonical perturbation theory diverges for most systems. Also, most systems have "stochastic regions", discussed below [all this does not imply that most systems are "ergodic"[4,6,7]].

5. STOCHASTIC REGIONS

These are regions in phase space where the, completely deterministic, *orbits display an extremely sensitive dependence on the initial conditions* $\vec{q}(0)$, $\vec{p}(0)$. In such regions it is not at all uncommon to observe effects of order: 1 for causes of order 10^{-13}.[5,6,3] Although this may seem strange for eqs. (1), familiar examples abound. Consider:

a. A dice-throw (coin-toss) depends quite sensitively on its initial conditions. Whence the saying: "As random as a dice-throw (coin-toss)", which is perfectly deterministic, nevertheless.
b. The Galton-board, i.e. a "pin-ball" machine, yields a stochastic distribution of arrival points at the bottom.
c. A "random-number generator" in a digital computer employs a deterministic algorithm.

Numerical integration of the equations of motion, using ordinary step-methods, is most difficult in such stochastic regions. There exist notorious problems where no known numerical method provides reliable results over time-intervals of physical interest. Analytic results are few, but interesting:[1-4]

6. THE K.A.M. THEOREM

The *Kolmogorov-Arnold-Moser Theorem* is the major analytic result that exists, for nearly-integrable systems. Loosely speaking, the theorem proves that a small perturbation of an integrable system *only slightly* modifies most orbits [yet some stochastic regions can appear, cf. Sec. 5]. The historical use of divergent canonical perturbation series might perhaps be justified for such orbits. The magnitude of the non-integrable perturbation permitted in the proof is microscopically small. Numerical calculations often indicate similar orbits even for macroscopic perturbations [the theorem is stated in chapter 2, sections 3d, e and 4a, b of ref. 3 and Theorem 21.11 of ref. 4; the proofs require series with "accelerated" convergence, à la Newton's method].

7. A RIGOROUSLY STOCHASTIC SYSTEM

For a "hard-sphere" Maxwell-Boltzmann gas of N particles, contained in a box, *"all" allowed orbits are stochastic* in the sense of Sec. 5. This is a result, due to Sinai[4,6,7], which holds for $N \geq 2$ and particle motion in 3-dim.-as well as 2-dim. space [a few particular initial conditions excepted]. If the system had smoother (analytic) interparticle-potentials, with a local minimum, the K.A.M. theorem would apply again to certain orbits, at low energy, and *not all* of phase space would be stochastic for such a system.

8. APPROACH TO EQUILIBRIUM AND ERGODICITY

So far we have only mentioned properties of single orbits or compared two orbits which were initially close together, cf. Sec. 5. However statistical mechanics requires the time evolution, in phase space, of a large collection (distribution) of initial conditions. One has to explain the (experimental) *approach to* "thermal equilibrium" of an average isolated mechanical system, as $t \to \infty$, as well as justify the (highly successful) use of "ensemble theory" to calculate the average properties of a system *in* thermal equilibrium[7,6]. For an "*ergodic*" system[7,4,6], whose orbits cover the constant energy surface densely and uniformly, the latter can be done. The approach to equilibrium requires more than ergodicity however. It turns out that a *completely* "stochastic" system, e.g. the hard sphere system of Sec. 7, does approach equilibrium and is ergodic[4,7]. If the K.A.M. theorem applies somewhere a system cannot approach equilibrium from all initial conditions but might conceivably be ergodic [some integrable systems can be ergodic].[6,4,7]

"EPPUR SI MUOVE", *Galileo Galilei*[9]

REFERENCES

1. M. V. Berry, Regular and Irregular Motion, (Chapter II) [*an excellent introduction into this field*], this Volume.
2. J. Moser, Lectures on Hamiltonian Systems, Memoirs Am. Math. Soc. 81, 1-60 (1968) [*one of the "canonical" texts; more introductory than ref. 3*], and his article in this Volume.
3. J. Moser, Stable and Random Motions in Dynamical Systems, (Princeton University Press, 1973) [*one of the "canonical" texts; chapt. 1: introductory; chapter 2: discusses Hamiltonian systems, among others; the remaining 4 chapters go into more mathematical details*].
4. V. I. Arnold and A. Avez Ergodic Problems of Classical Mechanics, (Benjamin, New York, 1978); V. I. Arnold, Les Méthodes Mathematiques de la Mécanique Classique, MIR, Moscow (1976) [*both: "canonical" texts*].
5. L. J. Laslett, Stochasticity, LBL-3016 (1974), and this Volume [*an interesting selection of examples*].
6. J. Ford, "The Statistical Mechanics of Classical Analytic Dynamics", 215-255 in Fundamental Problems in Statistical Mechanics, ed. E. Cohen (North-Holland, Amsterdam, 1975); also see Adv. Chem. Phys. 24, 155-185 (1973) and his article in this Volume [*reviews connecting dynamics with statistical mechanics*].
7. J. L. Lebowitz and O. Penrose, Modern Ergodic Theory, Physics Today 26, (#2), 1-7 (1973) [*a review, may serve as introduction to some of the more mathematical chapters of refs. 3 and 4*].
8. Proceedings of the 1977 "Volta-Memorial" conference on "Stochastic Behavior in Hamiltonian Systems" [Como, Italy], eds. G. Casati and J. Ford, Springer Verlag N.Y. (1978); Proceedings of the 1979 International Conference on "Nonlinear Dynamics" [New York], ed. R. Helleman, Ann. N.Y. Acad. Sci., to appear (1979); [*samplings of research subjects in this field*].
9. "And yet it moves", Galileo Galilei (Speaking of the earth).

LA JOLLA INSTITUTE
P. O. Box 1434
La Jolla, California 92038

A nonprofit research institute for science and technology. La Jolla Institute is a research and consulting corporation consisting of applied scientists with varied backgrounds who maintain a close working relationship with a diverse group of scientists from universities and research centers. Long-range direction and guidance of the Institute is provided by trustees, most of whom are professors at several universities. The La Jolla Institute was organized on July 23, 1976 under California general nonprofit corporation law.

TRUSTEES

Marvin L. Goldberger	President - California Institute of Technology
Adolf R. Hochstim	La Jolla Institute
Elliott W. Montroll	Einstein Professor of Physics, University of Rochester
William A. Nierenberg	Director - Scripps Institution of Oceanography
Irwin Oppenheim	Professor of Chemistry, Massachusetts Institute of Technology
Kenneth M. Watson	Professor of Physics, University of California, Berkeley

INSTITUTE OF THEORETICAL PHYSICS

On the campus of the University of California, San Diego. Supported and administered by La Jolla Institute.

SENIOR ADVISORY BOARD

Luis W. Alvarez	Nobel Laureate, Physics Department, University of California, Berkeley
Sir Samuel Edwards	Physics Department, University of Cambridge
Willis E. Lamb	Nobel Laureate, Physics Department, University of Arizona, Tucson
Nicolaas G. van Kampen	Physics Department, University of Utrecht
Steven Weinberg	Physics Department, Harvard University
David Y. Wong	Physics Department, University of California, San Diego

WORKSHOPS OF INSTITUTE OF THEORETICAL PHYSICS*

31 July-18 August 1978	Elementary Particles (Quantum Chromodynamics)
21-25 August 1978	Energy Workshop
3-5 January 1979	Particle Acceleration Mechanisms in Astrophysics
Spring 1979	Nonlinear Models of Ecological Systems

*To be published by American Institute of Physics.

AIP Conference Proceedings

		L.C. Number	ISBN
No.1	Feedback and Dynamic Control of Plasmas (Princeton) 1970	70-141596	0-88318-100-2
No.2	Particles and Fields - 1971 (Rochester)	71-184662	0-88318-101-0
No.3	Thermal Expansion - 1971 (Corning)	72-76970	0-88318-102-9
No.4	Superconductivity in d- and f-Band Metals (Rochester, 1971)	74-18879	0-88318-103-7
No.5	Magnetism and Magnetic Materials - 1971 (2 parts) (Chicago)	59-2468	0-88318-104-5
No.6	Particle Physics (Irvine, 1971)	72-81239	0-88318-105-3
No.7	Exploring the History of Nuclear Physics (Brookline, 1967, 1969)	72-81883	0-88318-106-1
No.8	Experimental Meson Spectroscopy - 1972 (Philadelphia)	72-88226	0-88318-107-X
No.9	Cyclotrons - 1972 (Vancouver)	72-92798	0-88318-108-8
No.10	Magnetism and Magnetic Materials - 1972 (2 parts) (Denver)	72-623469	0-88318-109-6
No.11	Transport Phenomena - 1973 (Brown University Conference)	73-80682	0-88318-110-X
No.12	Experiments on High Energy Particle Collisions - 1973 (Vanderbilt Conference)	73-81705	0-88318-111-8
No.13	π-π Scattering - 1973 (Tallahassee Conference)	73-81704	0-88318-112-6
No.14	Particles and Fields - 1973 (APS/DPF Berkeley)	73-91923	0-88318-113-4
No.15	High Energy Collisions - 1973 (Stony Brook)	73-92324	0-88318-114-2
No.16	Causality and Physical Theories (Wayne State University, 1973)	73-93420	0-88318-115-0
No.17	Thermal Expansion - 1973 (Lake of the Ozarks)	73-94415	0-88318-116-9
No.18	Magnetism and Magnetic Materials - 1973 (2 parts) (Boston)	59-2468	0-88318-117-7
No.19	Physics and the Energy Problem - 1974 (APS Chicago)	73-94416	0-88318-118-5
No.20	Tetrahedrally Bonded Amorphous Semiconductors (Yorktown Heights, 1974)	74-80145	0-88318-119-3
No.21	Experimental Meson Spectroscopy - 1974 (Boston)	74-82628	0-88318-120-7
No.22	Neutrinos - 1974 (Philadelphia)	74-82413	0-88318-121-5
No.23	Particles and Fields - 1974 (APS/DPF Williamsburg)	74-27575	0-88318-122-3
No.24	Magnetism and Magnetic Materials - 1974 (20th Annual Conference, San Francisco)	75-2647	0-88318-123-1

		L.C. Number	ISBN
No.25	Efficient Use of Energy (The APS Studies on the Technical Aspects of the More Efficient Use of Energy)	75-18227	0-88318-124-X
No.26	High-Energy Physics and Nuclear Structure - 1975 (Santa Fe and Los Alamos)	75-26411	0-88318-125-8
No.27	Topics in Statistical Mechanics and Biophysics: A Memorial to Julius L. Jackson (Wayne State University, 1975)	75-36309	0-88318-126-6
No.28	Physics and Our World: A Symposium in Honor of Victor F. Weisskopf (M.I.T., 1974)	76-7207	0-88318-127-4
No.29	Magnetism and Magnetic Materials - 1975 (21st Annual Conference, Philadelphia)	76-10931	0-88318-128-2
No.30	Particle Searches and Discoveries - 1976 (Vanderbilt Conference)	76-19949	0-88318-129-0
No.31	Structure and Excitations of Amorphous Solids (Williamsburg, Va., 1976)	76-22279	0-88318-130-4
No.32	Materials Technology - 1975 (APS New York Meeting)	76-27967	0-88318-131-2
No.33	Meson-Nuclear Physics - 1976 (Carnegie-Mellon Conference)	76-26811	0-88318-132-0
No.34	Magnetism and Magnetic Materials - 1976 (Joint MMM-Intermag Conference, Pittsburgh)	76-47106	0-88318-133-9
No.35	High Energy Physics with Polarized Beams and Targets (Argonne, 1976)	76-50181	0-88318-134-7
No.36	Momentum Wave Functions - 1976 (Indiana University)	77-82145	0-88318-135-5
No.37	Weak Interaction Physics - 1977 (Indiana University)	77-83344	0-88318-136-3
No.38	Workshop on New Directions in Mössbauer Spectroscopy (Argonne, 1977)	77-90635	0-88318-137-1
No.39	Physics Careers, Employment and Education (Penn State, 1977)	77-94053	0-88318-138-X
No.40	Electrical Transport and Optical Properties of Inhomogeneous Media (Ohio State University, 1977)	78-54319	0-88318-139-8
No.41	Nucleon-Nucleon Interactions - 1977 (Vancouver)	78-54249	0-88318-140-1
No.42	Higher Energy Polarized Proton Beams (Ann Arbor, 1977)	78-55682	0-88318-141-X
No.43	Particles and Fields - 1977 (APS/DPF, Argonne)	78-55683	0-88318-142-8
No.44	Future Trends in Superconductive Electronics (Charlottesville, 1978)	77-9240	0-88318-143-6
No.45	New Results in High Energy Physics - 1978 (Vanderbilt Conference)	78-67196	0-88318-144-4
No.46	Topics in Nonlinear Dynamics (La Jolla Institute)	78-057870	0-88318-145-2